71

Fortschritte der Chemie organischer Naturstoffe

Progress in the Chemistry of Organic Natural Products

Founded by
L. Zechmeister

Edited by
W. Herz, G. W. Kirby,
R. E. Moore, W. Steglich,
and Ch. Tamm

Authors:
A. Andersen, S. B. Christensen,
D. Deepak, G. Gäde, A. Khare,
U. W. Smitt, S. Srivastav

Springer-Verlag Wien GmbH

Originally published by Springer-Verlag Wien New York in 1997

Library of Congress Catalog Card Number AC 39-1015

Typesetting Thomson Press (India) Ltd , New Delhi

Graphic design Ecke Bonk
Printed on acid-free and chlorine-free bleached paper

With 11 partly coloured Figures

ISSN 0071-7886
ISBN 978-3-7091-7343-5 ISBN 978-3-7091-6529-4 (eBook)
DOI 10.1007/978-3-7091-6529-4

Contents

List of Contributors

ANDERSEN, A , Department of Medicinal Chemistry, Royal Danish School of Pharmacy, Universitetsparken 2, DK-2100 Copenhagen Ø, Denmark

CHRISTENSEN, Assoc Prof S B , Department of Medicinal Chemistry, Royal Danish School of Pharmacy, Universitetsparken 2, DK-2100 Copenhagen Ø, Denmark

DEEPAK, Prof D , Department of Chemistry, University of Lucknow, Lucknow 226007, India

GADE, Prof G , Zoology Department, University of Cape Town, Rondebosch 7701, Republic of South Africa

KHARE, Prof A , Department of Chemistry, University of Lucknow, Lucknow 226007, India

SMITT, Assoc Prof U W , Department of Medicinal Chemistry, Royal Danish School of Pharmacy, Universitetsparken 2, DK-2100 Copenhagen Ø, Denmark

SRIVASTAV, Dr S , Department of Chemistry, University of Lucknow, Lucknow 226007, India

The Explosion of Structural Information on Insect Neuropeptides

G GADE
Zoology Department, University of Cape Town,
Rondebosch 7701, Republic of South Africa

Contents

1. Introduction

Insects form the largest class of the phylum Arthropoda. There are at least one million known species, so more than 50% of all existing organisms on earth are insects. It is even thought that at least another million insect species have not yet been discovered. Insect-like forms inhabited the terrestrial and freshwater ecosystems about 300 million years ago and their basic features have been so successful that they were able to exploit almost every available habitat except the true marine environment, which is occupied by their arthropod "cousins", the Crustacea.

Metazoan animals like insects had to develop systems for communication between cells, tissues and organs in order to coordinate their responses to internal and external stimuli and to regulate biochemical and physiological processes. Both the nervous and the endocrine systems are well-known cellular components for communication, recruiting chemical messengers for their tasks. In general, the nervous system is used for rapid communication, whereas the endocrine system is involved in the regulation of longer lasting responses. Both systems, however, quite often do not work in isolation from each other, but form a functional, integrated system. This is best seen in the action of the so-called neurosecretory cells which synthesize and release specific chemical messengers, the neuropeptides (there are also aminergic neurosecretory cells, but these will not be dealt with here).

As early as 1922 the Polish scientist Kopeč (*239*) proposed that substances in the brain (in specific neurons though) control the processes necessary for moulting and metamorphosis, thus acting in distant parts of the body. He had extirpated brains from the gypsy moth, *Lymantria dispar*, and shown that the debrained larvae never pupate. This "brain hormone" is now known under the name prothoracicotropic hormone (PTTH), but its sequence in the gypsy moth is still not known.

Historically, the Scharrers coined the term neurosecretion to characterize the activities of those neurons which contained electron-dense granules of about 400 nm in diameter. Ernst Scharrer was studying vertebrate animals and discovered nerve cells with secretory activity in the fish, *Phoxinus laevis* (*401*), whereas his wife Berta Scharrer was studying invertebrate animals, including insects, in which she reported the presence of neurosecretory cells including those in the corpora cardiaca of the cockroach, *Leucophaea maderae* (*398, 399*). The Scharrers were the first to characterize the structural and functional similarities between the vertebrate hypothalamo(nervous)-hypophyseal system and the insect brain-corpora cardiaca-corpora allata complex (*400*).

Today we know that all nerve cells are secretory and that the distinction between "ordinary" neurons containing small synaptic vesicles and the neurosecretory neurons with large-cored vesicles is fluid. Between these two extremes – the ordinary neurons forming synapses and releasing their chemical messengers, the neurotransmitters, into the synaptic cleft, and the neurosecretory cells releasing relatively large quantities of their chemical mediators, the neuropeptides or neurohormones, into the general circulation – all kinds of graded intermediate cells can occur (*331, 455*). Some of these cells directly innervate endocrine or nonendocrine tissues and their function as modulators of nerve or muscle activity is discussed; their messengers may be called neuromodulators.

Although neurosecretory cells were co-discovered in insects, much more attention has been paid to the vertebrates, especially the mammalian system. Consequently, a wide variety of neuropeptides has been shown to be present in vertebrates and has been chemically characterized. For quite a few, even the precursor molecules are known and the gene structures have been elucidated. From these mammalian studies it soon became clear that peptides represent the largest single class of neuroregulatory substances (*195, 433*). After the first discovery, studies to identify (chemically) neuropeptides in insects lagged behind, but this has changed dramatically in the last ten years or so.

Before we outline the progress made in elucidating the primary structures of insect neuropeptides, we first have to discuss briefly the classical, epithelial endocrine glands of insects in the context of development and growth and, subsquently, the main localizations of neurosecretory cells and their release sites.

The life cycle of insects from the fertilized egg to the adult, reproductively-active imago is characterized by growth. Since the insect body is encased in an external skeleton which would prevent growth its volume and surface area must increase from time to time. The growth of the integument is achieved by moulting. A new, larger cuticle is made and the old, confining cuticle is cast away. The latter process is called ecdysis or eclosion (when the resulting insect is an adult one). The whole period between two moults is called a moulting cycle. Changes in morphology, function and life strategy of an insect during its ontogenesis are named metamorphosis.

The morphological changes occurring during metamorphosis can vary quite drastically and three major evolutionary lineages can be distinguished:

1. Ametabolic insects like springtails (Collembola) and silverfishes/firebrats (Zygentoma). Body forms of larvae and adults are identical except for the external genitalia and internal reproductive organs of the adults; adults have no wings and this group is called Apterygota.
2. Hemimetabolic insects like dragonflies (Odonata), cockroaches (Blattaria), grasshoppers (Caelifera) and bugs (Hemiptera). These insects undergo an incomplete metamorphosis. The larvae look very similar to adults, but the latter differ from the larvae in having functional wings. This group is known as Exopterygota.
3. Holmetabolic insects like beetles (Coleoptera), butterflies and moths (Lepidoptera), flies (Diptera) and bees and wasps (Hymenoptera). These insects undergo complete metamorphosis. The larvae look entirely different from the adults and prior to the adult stage a pupal stage is formed. This group is called Endopterygota in reference to the internal development of their wing imaginal disks.

Whichever lineage the insect belongs to, the general hormonal events during moulting are identical. Two non-peptide hormones, the ecdysteroids and the juvenile hormones produced in the two major classical, epithelial endocrine glands are responsible for moulting.

The first glands are the paired corpora allata which are located retrocerebrally and are connected to the brain *via* nerve fibers (Fig. 1). The corpora allata produce and release species-specific juvenile hormones (JH 0-III; JH B3), which chemically are acyclic sesquiterpenoid epoxides

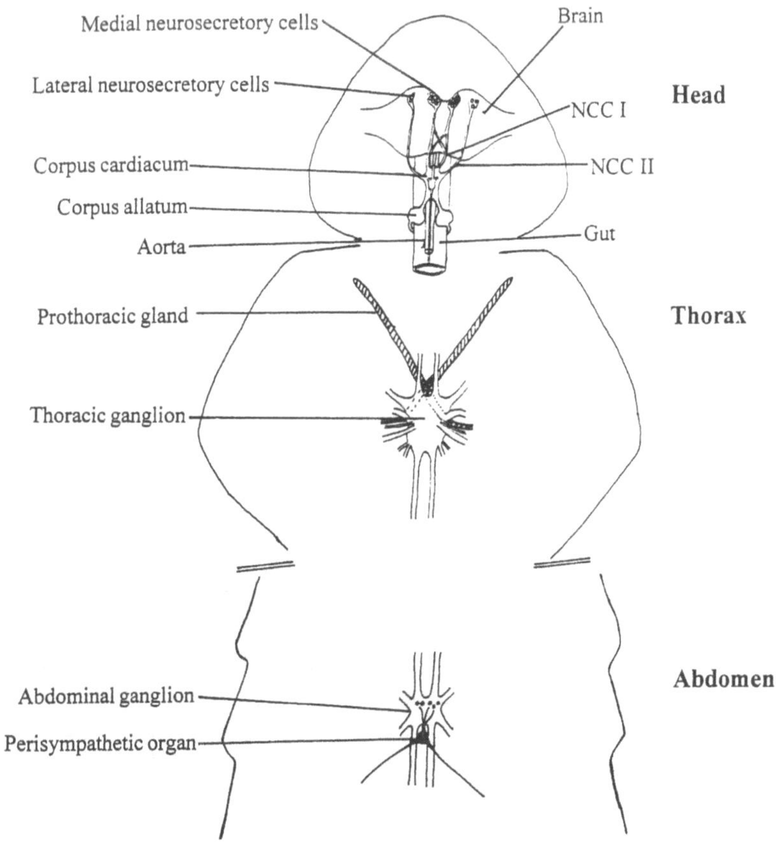

Fig. 1. Schematic diagram of the endocrine system in insects. The epithelial glands (corpus allatum, prothoracic gland) as well as the neurosecretory cells and their release sites (corpus cardiacum, perisympathetic organ) are shown

(Fig. 2). The juvenile hormones are vitally involved in the regulation and control of certain steps of insect development like larval moulting, and also in adult sexual maturation and reproduction (*88*, *89*, *445*).

The second classical, epithelial endocrine glands are the paired prothoracic glands located mainly in the thorax of the larval and pupal insect (Fig. 1). They mainly synthesize and release the steroid ecdysone which is subsequently converted into its active form (20-hydroxyecdysone) by the fat body and by epidermal cells (Fig. 2). The titre of 20-hydroxyecdysone is increased before each moult, but the titre of juvenile hormone determines

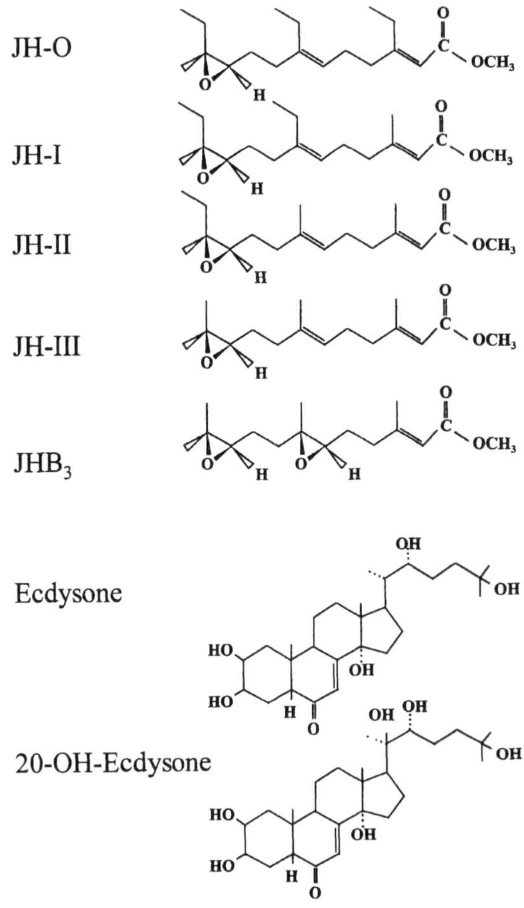

Fig. 2. Structures of the juvenile hormones and main ecdysteroids

the character of the moult. The classical scheme that a high juvenile hormone titre leads to a larval/larval moult, a low titre to a larval/pupal moult and that without juvenile hormone a moult to the adult occurs, is today revised to a somewhat more complicated scheme which is explained in detail elsewhere (*327*). Activity of both gland pairs, however, is controlled and fine-tuned by neuropeptides which are produced in neurosecretory cells of the brain (see Sects. 3.2.2 and 3.2.3).

Whereas prothoracic glands are suggested to be the ecdysteroid source in immature stages, *i.e.* when ecdysteroids are involved in the control of moulting, the gonads and the epidermis represent important sources during late pupal and adult stages, *i.e.* when control of reproduction is the main task (*66*). These alternative sources of ecdysteroids are likely to be regulated by neuropeptides as well.

In general most of the endocrine processes in insects are controlled by neuropeptides. The main centers for neurosecretory cells are in the pars intercerebralis and the median and lateral parts of the protocerebrum, which send axons to the corpora cardiaca (Fig. 1). These retrocerebral structures store and release the neuropeptides produced in the brain's neurosecretory cells and are therefore called neurohaemal organs (located in close proximity to the aorta, thus, ideal for release of neuropeptides into the circulation). In addition, the corpora cardiaca produce their own neuropeptides in their intrinsic neurosecretory cells. In addition to these more classical neurosecretory areas, neurosecretory cells are found throughout the central nervous system, the sympathetic nervous system (including the neurohaemal perisympathetic organs) and also within the peripheral nervous system (*331*, *366*).

A great variety of processes in insects is known to be influenced or regulated by neuropeptides. These processes may be metabolic, behavioral, developmental or reproductive in character. The following list shows some major neuropeptide groups and their actions:

1. Myotropins, which modify spontaneous muscle contractions;
2. diuretic and antidiuretic peptides, which are involved in ion- and water balance;
3. adipokinetic and hypertrehalosaemic peptides, which control fat, carbohydrate and protein metabolism;
4. eclosion hormone which initiates behavioral patterns associated with ecdysis and its timing;
5. allatotropins/allatostatins, which stimulate/inhibit the synthesis of juvenile hormones by the corpora allata;
6. prothoracicotropic hormones, which stimulate moulting by initiating ecdysone biosynthesis and release by the prothoracic gland;
7. diapause hormone, which arrests development in eggs of certain moth species;
8. oostatic hormone, which inhibits maturation of the ovaries;
9. neuropeptides which activate the synthesis of sex pheromones.

Before details on the individual categories of neuropeptides are given, methods important in the research on neuropeptides are discussed very briefly and appropriate examples of the applications of these methods are described.

2. General Methods Used for Isolation, Identification and Characterization of Insect Neuropeptides

2.1. Biological Assays

The existence and detection during isolation of the majority of insect neuropeptides was initially monitored by bioassays. There is a whole range of bioassays available now, including those measuring physiological actions (like energy mobilization and diuresis) as well as behavioral events (like those for the eclosion hormone). As examples of bioassays, the very popular tests for adipokinetic and for myotropic substances are given here in some detail.

2.1.1. Adipokinetic Bioassay

In 1969 two research groups (*15*, *279*) observed that injection of extracts from the corpora cardiaca of locusts increased the amounts of lipids (specifically: diacylglycerols) in the haemolymph. As a result, a bioassay was developed in which the concentration of total lipids was routinely measured in the haemolymph with a very reliable and simple method. In our laboratory, for example, we take a 1 µl haemolymph sample from the migratory locust at time zero, then inject the insect with 10 µl of the solution to be analyzed (either a corpus cardiacum extract from a locust or other insect or HPLC fractions after isolation procedures), and a second 1 µl sample of haemolymph is taken 90 min later from the same insect. For analysis of the lipids the sulpho-phosphovanillin method (*493*; modified by *179*) is used; the developed pink colour is easily read in a simple filter photometer at about 450 nm and the lipid concentration quantified by the use of a standard curve. An increase of the concentration of lipids in the post-injection sample compared to the pre-injection value is indicative of a positive response, *e.g.* the presence of an adipokinetic substance.

For further readings on this and related metabolic bioassays see (*107*, *449*).

2.1.2. Myotropic Bioassay

In 1962, Davey (*63*) demonstrated that homogenates from corpora cardiaca of *Periplaneta americana* had an effect on the spontaneous contractile activity of the isolated hindgut by increasing the tonus, frequency and amplitude of contraction. Later, a preparation of the hindgut from the cockroach *Leucophaea maderae* was used for the successful purification of a great number of myotropic peptides from *L. maderae*, the cricket, *Acheta domesticus*, and from *Locusta migratoria* (*169*, *176*). For this, the digestive tract was carefully removed from the cockroach, all adhering tissues such as fat body, trachea and Malpighian tubules pulled away or trimmed off, the hindgut tied at the junction to the midgut and the latter plus foregut cut off. The posterior end of the rectum was tied with thread as well and then the whole preparation suspended in a muscle chamber (5 ml plastic disposable syringe barrel) filled with an aerated saline solution. The preparation was attached to a muscle transducer, which displayed the signal onto an oscillograph. Such a preparation needs about one hour for equilibration; thereafter, the pattern of spontaneous contractions is relatively constant and the preparation can be used for a whole day. Thus, up to 80 samples can be tested per day by monitoring the alteration of the pattern of spontaneous contractile activity (either stimulatory or inhibitory).

2.2. Liquid Chromatography

The introduction of high performance liquid chromatography (HPLC), using micron-sized particles of high mechanical strength as supports for column packing materials, therefore allowing a fast flow of liquid at high pressure, has provided a very versatile tool for purifying proteins and peptides. This is generally achieved at some stage during isolation by reversed-phase HPLC (RP-HPLC), a partition chromatography where the starting mobile phase is more polar than the stationary phase.

The support material is silica whose silanol groups are chemically derivatized with organosilanes such as octadecyl (C-18), for example. RP-HPLC using various ion-pairing reagents such as trifluoroacetic acid (TFA) or heptafluorobutyric acid (HBFA) has been used widely for purifying neuropeptides because of its excellent resolution. For details of this and other LC methods readers are referred to appropriate reviews (*90*, *418*, *427*). Of course, for the isolation of insect neuropeptides it is important to know at the start roughly how much material is expected to be present and whether the peptide-producing tissue can be easily dissected or whether whole heads/animals have to be used for extraction. This will be briefly illustrated by three examples of isolation procedures.

Adipokinetic/hypertrehalosaemic peptides: Corpora cardiaca sometimes store these peptides in impressive quantities of 200 to more than 3000 pmol per gland. Therefore this tissue is dissected and then extracted with 80% methanol. Such methanolic extracts are applied to C-8 or C-18 RP-HPLC columns which are developed in a gradient mode with acetonitrile/water/0.1% TFA. With a single column step these peptides are sufficiently pure for structural work (*107*, *119*). Almost all of the adipokinetic/hypertrehalosaemic peptides, which often differ only by a single amino acid residue, can be separated in a single run due to the spectacular resolving power of RP-HPLC (*114*).

Myotropic neuropeptides: Due to the low concentration of these peptides (maximally about 1 pmol per head) whole heads of cockroaches (*Leucophaea maderae*) were extracted in a mixture of methanol/water/acetic acid (90:9:1; v/v) and subsequently extracted sequentially with ethyl acetate and hexane to remove lipids (for details see *177*). The aqueous solution was lyophilized, dissolved in 0.1% TFA and prepurified on C-18 Sep-Pak cartridges. This extract was subsequently fractionated on a series of 4 HPLC columns with different separation characteristics. The first step was performed on a μBondapak phenyl column, developed with an acetonitrile/water/TFA gradient. Individual active fractions were processed on a C-1 column using the same solvents and thereafter on a C-18 column, again using the same solvents. The final purification step was HPLC in a normal phase mode (I-125 Protein Pak column); the gradient run from 95% to 75% acetonitrile containing 0.01% TFA. After the final step fractions were pure enough for sequencing.

Allatotropin: Schooley's group isolated eclosion hormone, diuretic hormone and allatotropin from whole heads of *Manduca sexta* in a very similar fashion (*212*, *213*, *214*). As an example, the purification of allatotropin is given here (*418*). Due to the minute amounts of peptides expected, 10000 trimmed heads (eyes, proboscis and other chitinous parts were cut off, leaving brains, corpora cardiaca and corpora allata) of pharate adult moths were first defatted by homogenization in acetone. The extract was filtered, the acetone discarded and the residue re-extracted with a strongly acidic buffer (1 M acetic acid containing 20 mM HCl) containing protease inhibitors. After centrifugation the supernatant was chromatographed on a cation exchanger (sulphopropyl Sephadex C-25) which was eluted with 1 M acetic acid, 50 mM ammonium acetate (pH 4), and then with increasing concentrations (from 50 to 800 mM) of ammonium acetate (pH 7). Eclosion hormone was eluted in the 50 mM fraction, allatotropin in the 100 to 200 mM one and diuretic hormone between 400 and 800 mM NH_4OAc. Concentration and desalting of the sample occurred on a large cartridge column containing Vydac C-4 material. The allatotropin was eluted with 60% acetonitrile containing 0.1% TFA. The next step was a semipreparative Vydac C-4 column which was eluted with a 0–60% acetonitrile/water/TFA gradient. Allatotropin eluted between 17–19% acetonitrile and this material was separated again on a semipreparative Vydac C-4 column, but with a gradient of 10–30% acetonitrile and 0.1%

HBFA as the ion-pairing reagent. An analytical cation exchange LC column (TSK SP-5PW), which was equilibrated with 20 mM sodium phosphate buffer (pH 6.25) and developed with a gradient (0–0.5 M) of sodium chloride, was used next. The last step employed a Vydac C-18 analytical column which was eluted with a gradient (10–40%) of acetonitrile/water/TFA and resulted in a sufficiently pure peak for sequence analysis.

2.3. Edman Degradation Sequencing, Mass Spectrometry and Peptide Synthesis

Edman degradation cleaves the N-terminal amino acid from a peptide or protein backbone and prepares the derivatized residue (the PTH amino acid) for identification. Automated sequencers became available in 1970. Since then continued improvements in peptide isolation techniques and sequencer technology have increased the speed of analysis and vastly reduced the amounts of peptides required in the sequencer reaction chamber. Today on-line microbore RP-HPLC separation and optimized identification of PTH amino acids enable the new generation of gas phase or pulsed liquid phase sequencers to operate in the range of about 10 pmol (*262*).

Many proteins and peptides contain post-translationally modified amino acids. A majority of insect neuropeptides, for example, are blocked at the N-terminus by a pyroglutamate residue. Since Edman degradation sequencing needs a free N-terminal amino acid, the pyroglutamate residue has to be cleaved enzymatically by pyroglutamate aminopeptidase. After separating the deblocked from the parent peptide via RP-HPLC the new des-pyroglutamate peptide can be automatically sequenced.

Other post-translational modifications such as phosphorylation, methylation, acetylation, sulfation or glycosylation can also be detected by specific preparations before Edman degradation or with mass spectrometry (see below) or a combination of both techniques (*281*).

Even with the newest generation of sequencers the "repetitive yield", *i.e.* the overall yield of one step in Edman degradation, is about 95%, which means that these machines only give sequencing results to a maximum length of about 30–40 residues. Thus, longer peptides or proteins first have to be chemically or enzymatically fragmented, the fragments isolated by RP-HPLC, and then analyzed in the sequencer. Fragmentation is facilitated by denaturing the protein/peptide under investigation. Guanidine hydrochloride is the denaturing detergent of choice. Since disulfide bonds may hinder digestion, disulfide bridges are cleaved by reduction yielding two cysteines; subsequently the thiol groups are stabilized by alkylation with, for example, iodoacetic acid yielding S-carboxymethyl cysteine.

Various enzymes are commercially available for enzymatic fragmentation. These are characterized as endopeptidases such as trypsin (specifically cleaving Lys and Arg residues) and endoproteinases Asp-N, Arg-C, Glu-C and Lys-C or as exopeptidases such as carboxypeptidases A, B, P and Y and pyroglutamate aminopeptidase (see above). For further details the reader is referred to the special literature (*163*, *226*, *448*). Complementary to enzymatic digestions are chemical fragmentation methods. The most widely used cleavage chemical is cyanogen bromide which specifically cleaves Met-Xaa bonds thereby converting methionine into a C-terminal homoserine residue and creating a new amino terminus NH_2–Xaa. For further reading see KELLNER (*226*).

Mass spectrometric methods are nowadays continuously used solely or in combination with Edman degradation for elucidation of the primary structures of proteins and peptides. Mainly, mass spectrometry is used to measure the mass of the peptide/protein accurately, thereby confirming sequencing results achieved by other methods. A second goal of modern mass spectrometry is to give sequence assignments of smaller peptides or peptide fragments (for production of those see above), especially when post-translational modifications occur.

However, mass spectrometry is not infallible. For example, the amino acid residues Leu, Ile and hydroxypro have the same mass of 113 Da, thus mass spectrometry cannot differentiate between the three compounds. In such a case mass spectrometry has to be used in combination with Edman degradation sequencing. Thus both methods are complementary. A brief outline will illustrate the power of mass spectrometry. For further information the reader is referred to the following references (*12*, *280*, *387*, *428* and *475*).

During the last two decades tremendous improvements have been made with respect to mass spectrometry. Whereas formerly it was not possible to ionize larger proteins and analyze compounds with a mass greater than 1–2 kDa, the introduction of fast atom bombardment (FAB) mass spectrometry made it possible to ionize peptides and small polar proteins up to 15 kDa. In the FAB mode the peptide/protein is taken up in a glycerol matrix which is then bombarded with a beam of argon or xenon atoms resulting in protonated $[M + H]^+$ or deprotonated $[M - H]^-$ ion signals of the peptide depending on whether positive or negative mass spectra were generated. Because FAB is a relatively soft ionization procedure, the molecular ion is rather stable and is scarcely degraded to fragment ions. Thus, only a limited amount of structural information can be obtained directly. However, for sequence analysis tandem mass spectrometry in the FAB mode can be used and has been the method of choice to sequence, for example, some members of the adipokinetic hormone family (*133*, *491*), even to detect post-translational modifications like unusual glycosylation sites in such a peptide (*128*). In this method four sector mass spectrometers are used consisting of two double-focusing mass spectrometers with the geometry of two electric fields (E) and two magnetic fields (B) in either the BEEB or BEBE configuration. In the first double-focusing mass spectrometer (BE or EB) the peptide is ionized and the parent ion filtered to reach eventually the second instrument. In the free-field region between the two instruments the ion is fragmented by collision with helium or argon atoms (collision-induced decomposition = CID; or collisionally activated dissociation = CAD) producing the daughter or product ions which are detected and analyzed in the second double-focusing mass spectrometer (EB).

In the last 10 years new mass spectrometric techniques have been developed which are especially useful for molecular weight measurements, but may be employed for sequencing as well when modifications are used. The method of matrix-assisted laser desorption/ionization (MALDI) mass spectrometry, for example, has outstanding sensitivity (1 pmol or less) and large biopolymers up to about 300 kDa can be determined when a time of flight mass analyzer is used. Characteristic for MALDI mass spectrometry is that short pulses of lasers emitting in the ultraviolet or infrared are focused on a suitable matrix (for example, sinapinic acid is quite often used for peptides), in which the peptide/protein is embedded. The laser energy is absorbed by the matrix molecules and transferred to the sample molecular layers. Thereafter ionization and desorption takes place. The ions are emitted and separated while they fly to the detector. Generally, the most intense signal is the singly charged molecular ion, but doubly and triply charged molecular ions appear as well.

In electrospray mass spectrometry the peptide/protein sample is dissolved in, for example, a mixture of methanol or acetonitrile and water, infused very slowly into a glass capillary at a constant flow rate and introduced into the electrospray source. At this source a spray of fine, highly charged droplets is created at atmospheric pressure in the presence of a strong electric field. The droplets are made to shrink until ions evaporate and enter the mass analyzer, which, most commonly with this technique, is a triple quadrupole. During the electrospray ionization process multiprotonated molecules $(M + nH)^{n+}$ are formed which give rise to a series of consecutive peaks at $(M + n)/n$ along the mass to charge scale of the ion spectra. The occurrence of multiply charged ions allows the determination of proteins up to more than 100 kDa; the sensitivity for the molecular mass of peptides has been shown to be in the picomole or even femtomole range.

Once the structural data are collected, peptides up to 30 to 50 amino acid residues can be synthesized by solid phase techniques using the modern generation of automated peptide synthesizers. The synthetic peptide, in turn, is carefully compared with the natural peptide with regard to chromatographic retention time in different solvent and support systems, mass spectrometric data and biological activity in the appropriate bioassay. Only when these parameters of the natural and synthetic peptide match can one be sure that the correct sequence was determined or assigned.

2.4. Immunological Techniques (RIA, ELISA, Immunocytochemistry)

Analytical immunochemical methods have been used widely to identify and quantify peptidergic substances in insects. The most important techniques in the context of this review are immunocytochemical methods, which detect qualitatively an insect peptide antigen in tissues and cells, as well as the quantitative radioimmunoassays (RIAs) and enzyme-linked immunosorbent assays (ELISAs), which selectively measure minute amounts of peptide antigens among a mixture of potentially interfering material found in complex biological samples such as haemolymph. Only when the concentration of a neuropeptide is increased in the haemolymph after some specific physiological challenge can a true neurohormonal role be established. Thus, the neuropeptide is then released from its production/storage sites into the general circulation to act on peripheral tissues.

Most immunochemical work in insects is carried out with polyclonal antibodies raised in rabbits, but some monoclonal antisera have now been prepared for insect work (*265*). Most important for the success of any immunochemical method is the availability of a high-titred antiserum. In peptide work a synthetic product is the best antigen; however, problems may be encountered with small peptides, because they are not immunogenic. In such a case they have to be conjugated covalently, using carbodiimide or glutaraldehyde, to a larger carrier molecule which is usually a protein such as thyroglobulin, bovine serum albumin or keyhole limpet haemocyanin. Further problems may occur when the peptide does not contain a reactive group. This is, for example, the case with most peptides of the AKH/RPCH-family. One possible solution is to synthesize chemical analogues: either a des-pGlu-analogue was used for conjugation (*49*, *84*), or the pGlu residue was replaced by Tyr (*422*) or Glu (*290*), or the

N-terminal tetrapeptide (pGlu-Leu-Asn-Phe...) was conjugated via a diaminohexane spacer to thyroglobulin (*421*).

Another problem that may occur when using RIA is the preparation of a tracer, a radiolabelled antigen, with high specific activity. Conveniently, radioiodine is utilized to produce a tracer with high specific radioactivity; it is a gamma emitter which can easily be measured in inexpensive gamma counters. However, for example, most AKH/RPCH-family peptides do not contain an iodine-reactive molecule such as Tyr or His residues. Therefore, MOSHITZKY *et al.* (*290*) prepared a derivative, 4-hydroxyphenylpropionyl-[Glu^1]-Lom-AKH-I, which was subsequently iodinated with sodium ^{125}I. Although such a molecule mimics the structure of the antigen and can be used in a RIA, it is not possible to use it for receptor binding studies. Structure-activity experiments (see Sect. 3.1.1) have shown that the N-terminal pGlu is quite essential for exerting biological activity. Recently, a different radiolabelled AKH peptide, a derivative of a moth AKH (Mas-AKH), was made (*492*). First a peptide analogue with p-iodo-Phe at position 4 was synthesized, which was subsequently treated with tritium gas to produce a peptide analogue with tritium at the para position of Phe. The peptide had high specific activity and showed no difference in biological activity to the native non-tritiated peptide.

For further readings about applications of immunochemical methods in insect research and of problems and challenges of RIA, ELISA and immunocytochemistry the interested reader is referred to excellent articles in the book of GILBERT and MILLER (*139*). The contribution of SCHOONEVELD and VEENSTRA (*423*) in this book, for example, clearly indicates the possible limitations of immunocytochemical work and the caveats needed in interpretation of this histochemical technique. Therefore, positively-reacting cells in immunocytochemistry are generally called immunoreactive-"like"; which means that the specific antibody used has recognized a substance immunologically indistinguishable from the antigen. The true chemical identity has to await classical peptide purification and characterization or identification by molecular biological methods.

2.5. Molecular Biological Techniques

Advances in this particular field are extremely rapid and it is beyond the scope of this article to cover the different techniques. Some information in this respect with regard to insect neuropeptides can be found in several overviews (*140*, *395*, *396*).

It is clear, however, that the entire amino acid sequence of a large peptide or protein can nowadays be obtained more easily by deduction from its DNA sequence than determination of the amino acid sequence using protein chemical techniques. However, there are prerequisites and drawbacks as well: first, a partial amino acid sequence has to be known to construct oligonucleotide probes for screening a recombinant DNA (cDNA) or genomic DNA library to be sequenced for positive DNA clones. Sequencing of those DNA clones, in turn, will then give information on the identity of the amino acid sequence of the encoded peptide/protein. Second, DNA sequencing, as Edman degradation sequencing, is limited to the extent that post-translational modifications cannot be detected and identified.

One of the most successful applications of recombinant DNA techniques in insect research has been the provision of information on the amino acid sequences of neuropeptide precursor proteins. In some cases, as with many vertebrate neuropeptide precursors, other new peptide sequences were identified which occurred in the same precursor. In a recent short review, GIRARDIE (*140*) states that the respective genes for insect neuropeptide hormones can be classified as three types:

1. The preprohormone consists of a signal peptide and the neuropeptide. Examples are the eclosion hormone precursor (*183*; Sect. 3.2.4) and the neuroparsin precursor (*245*; Sect. 3.2.5.1). This type of organization has not yet been demonstrated in vertebrates.
2. The preprohormone consists of a signal peptide, the neuropeptide and other structurally unrelated peptides. Examples are the bombyxin and another insulin-related peptide precursor (*197*, *246*; Sect. 3.2.3.) and the precursors for the adipokinetic hormones of locusts (*329*, *424*; Sect. 3.1.1).
3. The preprohormone contains a signal peptide and multiple copies of the same and/or very similar neuropeptides (isoforms). Examples are the FMRFamide-related peptide precursor of the fruitfly *Drosophila melanogaster* (*320*, *402*; Sect. 3.3.8) and the precursor for the allatostatins of the cockroach *Diploptera punctata* (*71*; Sect. 3.2.2.2).

Since neuropeptide precursors are metabolic intermediates and are present in even smaller amounts than their products, recombinant DNA techniques for elucidating their structures are almost a necessity. This is also true for the receptor proteins of insect neuropeptides which are obviously scarce and therefore extremely difficult to identify structurally by protein chemical methods. Up to now, only the receptor for the diuretic hormone from the Malpighian tubules of the moth, *Manduca sexta*, has been cloned and sequenced (*384*; Sect. 3.1.2),

References, pp. 97–128

but future molecular biological work will undoubtedly reveal more receptor structures.

A third area in which molecular biological techniques are very helpful is the production of large peptides/proteins which are impossible or very difficult to synthesize chemically. For this, cDNA is expressed in cells which are infected with recombinant vectors like baculoviruses. Recently, the cDNA encoding human growth hormone was expressed in larvae of *Bombyx mori* employing *B. mori* nuclear polyhedrosis virus (a baculovirus) as an expression vector (*206*). The hormone was synthesized in the larvae and secreted into the haemolymph. It was confirmed that the recombinant growth hormone had the same molecular weight and amino acid sequence at its N-terminal region as the natural growth hormone. Moreover, the biological activity was comparable to that of natural growth hormone suggesting that the active structure of the recombinant growth hormone is identical with that of the natural one. Thus, this insect's larvae and baculovirus system has the potential as an efficient gene expression system for the industrial production of biologically active peptides/proteins including hormones, important for medical and pharmaceutical purposes.

Expression of insect neuropeptides in insects or cell cultures making use of recombinant baculoviruses has been achieved for eclosion hormone (*86, 156*) as well as for the pheromone biosynthesis activating neuropeptide (PBAN; *463*). For further reading on this subject an article by MAEDA (*267*) is recommended.

3. The Insect Neuropeptides

In the next sections, the various neuropeptides of insects will be discussed. Attention is mainly focused on those whose primary structures are known. Since there has been an explosion of characterized neuropeptides during the last few years and since almost every month new information is published, it is entirely possible that the literature and structures dealt with in this review are not complete. This is not because of deliberate omission, but simply because the author has failed to spot those publications.

The various neuropeptides are categorized by their actions. However, quite a few of those peptides elicit more than one biological response, thus have pleiotropic actions. In general, such peptides are discussed with respect to their main action or to the action they are best known for. This also has a bearing on their nomenclature. Although no single nomenclature is perfect, the one proposed by RAINA and GÄDE (*368*) is used here, but in some instances alternative names are included as well.

3.1. Peptides Involved in Homeostasis and Metabolism

3.1.1. *Adipokinetic and Hypertrehalosaemic Peptides*

Insulin and glucagon are well-known metabolic hormones of vertebrates which are involved in homeostasis of carbohydrate and lipid metabolism. The limited structural knowledge about insulin-like peptides in insects is discussed in Sect. 3.2.3. The first report on the existence of a glucagon-like factor in insects came from STEELE (*446*). Extracts of corpora cardiaca elevated the concentration of the haemolymph sugar trehalose (hypertrehalosaemic effect). The active principle was shown to be peptidic and, because of limited sequence identity of mammalian glucagon and some of these metabolic peptides in insects (see later) and similarities in action, the term "trehalogon" was coined (*447*). In a recent review (*148*), however, it is argued that there is "no justification in claiming any homology or evolutionary relationship" between the insect peptides and vertebrate glucagons.

In 1969 a different effect of extracts of corpora cardiaca was reported in the locusts *Schistocerca gregaria* (*279*) and *Locusta migratoria* (*15*). Here the concentration of haemolymph lipids was elevated (adipokinetic effect). In 1976 the decapeptide adipokinetic hormone, now called Lom-AKH-I, was isolated from 3000 corpora cardiaca by size exclusion chromatography on controlled-pore glass and thin layer chromatography on silica gel (*450*). Structure elucidation was achieved by a combination of enzymatic cleavage and mass spectrometry. The structure (see Table 1) was clearly related to that of the previously described red pigment-concentrating hormone from the shrimp *Pandalus borealis* (Pab-RPCH) (*92*). This structural similarity was the reason for naming this group of peptides the AKH/RPCH-family of peptides. During recent years new members of this family have been described from many insect orders. Isolation was achieved mainly by single-step RP-HPLC (see Sect. 2.2) and structure elucidation was carried out by Edman degradation after deblocking the N-terminal pyroglutamate residue or by various mass spectrometric techniques, mainly FAB-MS. Due to the relatively high concentration of AKH-type peptides per corpus cardiacum, the entire primary structure was resolved using, for example, only 4 glands from the grasshopper *Phymateus leprosus* (*127*) which compares quite favorably with the high amount of material necessary during the first AKH structural study (*450*). About 30 different peptides are known at present (Table 1) and that makes this family one of the largest. Such peptides have been identified from representative species of most insect orders (*106*) and attempts have been made to use the sequence information to construct phylogenetic trees

Table 1 *Primary structures of peptides of the adipokinetic hormone/red pigment-concentrating hormone (AKH/RPCH) family*

Code name	Species	Sequence	Reference(s)
Lom-AKH-I	*Locusta migratoria*	pQLNFTPNWGTamide	*429, 450*
	Schistocerca gregaria		*450*
Phm-AKH	*Phymateus morbillosus*	pQLNFTPNWGSamide	Gäde, Kellner, and Rinehart, unpublished
Del-CC	*Decapotoma lunata*	pQLNFSPNWGNamide	*118*
Cam-HrTH-I	*Carausius morosus*	pQLTFTPNW*GTamide	*128*
Cam-HrTH-II	*C morosus*	pQLTFTPNWGTamide	*133*
	Sipyloidea sipylus		*105*
	Extatosoma tiaratum		*134*
Phl-CC	*Phymateus leprosus*	pQLTFTPNWGSamide	*127*
Taa-HoTH	*Tabanus atratus*	pQLTFTPGWGYamide	*200*
Hez-HrTH	*Heliothis zea*	pQLTFSSGWGNamide	*198*
Rom-CC	*Romalea microptera*	pQVNFTPNWGTamide	*122*
Bld-HrTH	*Blaberus discoidalis*	pQVNFSPGWGTamide	*160*
	Nauphoeta cinerea		*131*
	Leucophaea maderae		*134*
	Gromphadorhina portentosa		*134*
	Blattella germanica		*134, 468*
Plc-HrTH-I**, II	*Platypleura capensis*	pQVNFSPSWGNamide	*123*
	Munza trimeni		*123*
	Cacama valavata		*471*
	Diceroprocta semicincta		*471*
	Magicicada sp		*376*
Mas-AKH	*Manduca sexta*	pQLTFTSSWGamide	*491*
	H zea		*199*
	Bombyx mori		*188*

Table 1 (*continued*)

Code name	Species	Sequence	Reference(s)
Psi-AKH	*Pseudagrion inconspicuum*	pQVNFTPGWamide	*202*
	Ischnura senegalensis		*202*
Lia-AKH	*Libellula auripennis*	pQVNFTPSWamide	*108*
	Ceratogomphus pictus		Janssens, Kellner, and Gade, unpublished
	Pantala flavescens		Janssens, Kellner, and Gade, unpublished
Emp-AKH	*Empusa pennata*	pQVNFTPNWamide	*110*
	Sphodromantis sp		*110*
Ani-AKH	*Anax imperator*	pQVNFSPSWamide	*124*
	Aeshna subpupillata		Janssens, Kellner, and Gade, unpublished
	Anotogaster sieboldii		Janssens, Kellner, and Gade, unpublished
Pea-CAH-I	*Periplaneta americana*	pQVNFSPNWamide	*14, 394, 430, 478*
	Blatta orientalis		*134*
	Leptinotarsa decemlineata		*125*
	Trinervitermes trinervoides		*257*
	Mastotermes darwiniensis		*257*
Grb-AKH	*Gryllus bimaculatus*	pQVNFSTGWamide	*132*
	Acheta domesticus		*61, 481*
	Gryllodes sigillatus		*113*
	R microptera		*122*
Tem-HrTH	*Tenebrio molitor*	pQLNFSPNWamide	*135*
	Zophobas rugipes		*135*
	Onymacris plana		*116*
	O rugatipennis		*116*
	Physadesmia globosa		*116*
	Polyphaga aegyptiaca		*126*

Pab-RPCH	*Pandalus borealis*	pQLNFSPGWamide	*92*
	Cancer magister		*137*
	Carcinus maenas		*137*
	Orconectes limosus		*137*
Lom-AKH-II	*L migratoria*	pQLNFSAGWamide	*120, 429*
Scg-AKH-II	*S gregaria*	pQLNFSTGWamide	*120, 429*
	S nitans		*120*
	Heterodes namaqua		*113*
	Acanthoproctus cervinus		*113*
	Libanasidus vittatus		*113*
	Anabrus simplex		REYNOLDS and SCHOOLEY, unpublished
Mem-CC	*Melolontha melolontha*	pQLNYSPDWamide	*111*
	Geotrupes stercorosus		*111*
	Pachnoda marginata		*129*
	P sinuata		*129*
Lom-AKH-III	*L migratoria*	pQLNFTPWWamide	*339*
Miv-CC	*Microhodotermes viator*	pQINFTPNWamide	*257*
Poa-HrTH	*P aegyptiaca*	pQITFTPNWamide	*126*
Pea-CAH-II	*P americana*	pQLTFTPNWamide	*394, 430, 478*
	Blatta orientalis		*134*
	Leptinotarsa decemlineata		*125*
Taa-AKH	*T atratus*	pQLTFTPGWamide	*200*
Pht-HrTH	*Phormia terraenovae*	pQLTFSPDWamide	*136*
	Drosophila melanogaster		*397*

* There is a hexose substituted on the Trp

** In all species of cicadas two peptides are isolated by HPLC, Edman degradation sequencing yielded the same sequence, at the moment the modification on peptide I is not known

(*117, 130*). It appears that Pab-RPCH is conserved in crustaceans; insect species, however, show a high degree of structural variability. All members are from 8 to 10 amino acids long, are N-terminally blocked by a pyroglutamate residue and C-terminally blocked by an amide. At position 4 (Phe or Tyr) and 8 (Trp) aromatic residues are present; most variations in constituent amino acids are conservative (Table 2). The majority of peptides is not charged under physiological conditions, but certain dipteran species and members of scarabaeid beetles contain peptides with a negatively charged Asp residue at position 7 (see Table 1). The family shows even more post-translational modification than only the blocked termini. For example, the stick insect *Carausius morosus* contains two decapeptides (see Table 1) one of which is glycosylated as shown by mass spectrometry (*128*). The glycosylation site is not the usual Ser/Thr (O-glycosylation) or Asn (N-glycosylation), but Trp is involved. Recently, it was reported that human RNase also uses Trp as a glycosylation site and, by ^{13}C and ^{1}H nuclear magnetic resonance spectroscopy, it was shown that the substituent was an aldohexopyranosyl residue which was C-glycosidically linked to the C2 atom of the indole ring of the tryptophan (*168*).

Moreover, in various cicada species from Africa and America two decapeptides have been found which are identical in structures judged by all methods used, including differences between D- and L-isomers. However, they can be separated on RP-HPLC (*123, 376, 471*), thus have to be different. As yet it is not known which modification does occur.

Besides the hyperlipaemic and hypertrehalosaemic effects mentioned above, other activities of peptides of the AKH/RPCH family are known. The major ones are the following:

1. Stimulation of the frequency of the heart beat in *Periplaneta americana* (*462*) which led to the use of this action as a bioassay for the isolation of the peptides Pea-CAH-I and II (*14, 394*) and also to some structure-activity studies (*13*).
2. Increase in muscle tone and frequency of contraction of the spontaneous activity of the isolated leg of a locust; this bioassay was also successfully used to isolate Pea-CAH-I and II (*336, 478*).
3. Inhibition of protein synthesis in *L. migratoria* (*42*), which was also shown to occur in the cricket *Acheta domesticus* (*61*). In the cockroach *B. discoidalis*, however, the endogenous peptide Bld-HrTH stimulates the rate of protein biosynthesis by interacting cooperatively with juvenile hormone (*223*).
4. Inhibition of fatty acid synthesis in *S. gregaria* (*145*). A simpler, more convenient and rapid method measuring the inhibition in fat body of

Table 2. *Common structural features of the AKH/RPCH-family peptides and variations. The frequency of occurrence of residues (in brackets) at each position is given. Analysis is based on the structure of the 30 family members given in Table 1*

Position	1	2	3	4	5	6	7	8	9	10	
	pQ(30)	L(18)	N(20)	F(29)	T(16)	P(25)	N(13)	W(30)	G(13)	T(5)	amide
		V(10)	T(10)	Y(1)	S(14)	S(2)	G(9)			N(4)	
		I(2)				T(2)	S(5)			S(2)	
						A(1)	D(2)			Y(1)	
							W(1)				

L. migratoria of the synthesis of lipid from [1–^{14}C] acetate was developed recently (*253*, *254*).

5. Inhibition of RNA synthesis in fat body of *L. migratoria* (*232*).

6. In *B. discoidalis* the peptide Bld-HrTH is also thought to regulate the synthesis of haemes for mitochondrial cytochromes, although not directly; furthermore, Bld-HrTH appears to be responsible for the induction of gene expression for a cytochrome P450 enzyme (*221*).

In conclusion, the AKH/RPCH peptides exert multiple physiological effects in various insect model systems. Mainly, they act on the metabolic status of the fat body. Most physiological research is done on the functions of adipokinetic hormones in locusts during flight (*112*, *146*, *147*). The hormones have direct effects on the mobilization of carbohydrates and lipids and/or the utilization of such substrates by the flight muscles, but have additional indirect effects on the transport of lipids as lipoproteins to the flight muscles and on the enzyme system of lipoprotein lipase in the flight muscles. This enzyme is responsible for "unloading" of the diacylglycerol from lipoproteins and making it finally available for oxidation to power the contraction of the flight muscles.

There are numerous reports, for locusts as well as other insects, on the involvement of AKH/RPCH peptides in activation of phosphorylase, of lipase, in the production of cyclic AMP, the usage of calcium for signal transduction *etc.*, but this will not be discussed here. Rather short accounts on structure-activity relationships, biosynthesis, localization by immunocytochemical techniques, release and breakdown are given.

Studies on how the biological information is encoded within the structure of various members of the AKH/RPCH family and some synthetic analogues have been conducted employing bioassays. Such studies on structure-activity relationships have been done on the lipid-mobilizing activity in locusts (*109*, *115*, *150*, *151*, *451*) and in *M. sexta* (*101*), on the carbohydrate-mobilizing activity in *P. americana* (*104*, *109*, *114*, *121*) and *B. discoidalis* (*99*, *159*) and on the phosphorylase-stimulating activity in *M. sexta* (*489*). Major differences apparently exist between those insects containing one endogenous peptide, *M. sexta* (Mas-AKH) and *B. discoidalis* (Bld-HrTH), and those containing two or three endogenous peptides, *L. migratoria* (Lom-AKH-I, II, III) and *P. americana* (Pea-CAH-I, II).

The receptors in *M. sexta* and *B. discoidalis* are apparently more selective, since quite a few of the tested, naturally-occurring analogues (= bioanalogues) were poorly active in those systems. On the other hand, for most bioanalogues up to a 50-fold higher dose was needed to achieve a half-maximal response (ED_{50} value) than for the endogenous peptides in

L. migratoria and *P. americana*. This may be indicative of the presence of more than one receptor type and, therefore, a broader spectrum of binding. Support for a multiple receptor hypothesis comes from various other experiments. For example, the three peptides from *L migratoria* have different potencies in different biological assays. Lom-AKH-III is more potent as an inhibitor of fatty acid uptake and RNA synthesis than Lom-AKH-I, but it is less potent in lipid-mobilization and activation of fat body phosphorylase (*253*, *338*). Since optimal responses for the acetate uptake assay are obtained with locust fat bodies of young insects (< 8-day adults), but for hyperlipaemia in older than 15-day adults, it is assumed that receptor populations may change during adult development (*253*). Moreover, certain single amino acid replacement analogues (at positions 1 and 2) for the endogenous peptide Pea-CAH-I in *P. americana* showed biphasic dose response curves characteristic of two receptors with differing affinities for the analogues (*121*).

Such single replacement studies also revealed that in the cockroaches *P. americana* (*121*) and *B. discoidalis* (*99*), the aromatic amino acid side chains at positions 4 and 8 are absolutely essential and that the amidated C-terminus and the pGlu at the N-terminus are very important as well. Since these are general structural features of the family it is very likely that all receptors are similar in that respect. Another result of these single replacement studies was that replacement at positions 6 and 7 in Pea-CAH-I had very little effect on the activity. These results are consistent with the prediction that a β-turn is formed around residues 5 to 8 (*149*, *477*). The corner residues 6 and 7 would not directly interact with the receptor; however the turn would be present primarily to orient the N-terminal pentapeptide residues and the C-terminal Trp-amide for interaction with the receptor (*121*). Studies on the conformation of some peptides of the AKH/RPCH family appear to confirm these predictions. Although in water such small peptides show a random coil conformation, increasing concentrations of SDS progressively stabilized the emergence of a single structure, as evidenced by circular dichroism spectroscopy, which would be described as a type of β-turn (*477*, O. CUSINATO, A.F. DRAKE, G. GÄDE and G. J. GOLDSWORTHY, unpublished results). Nuclear magnetic resonance studies on the octapeptide Emp-AKH dissolved in dimethylsulfoxide indicated a β-turn encompassing residues 5 to 8, with evidence of a β-sheet conformation for residues 1 to 5 (*494*).

The biosynthesis of adipokinetic hormones, including the genes and precursors, is best understood in the desert locust, *Schistocerca gregaria* (*335*). Direct protein isolation and sequencing methodology was used as well as molecular cloning. It is now believed that each adipokinetic hormone (even when three exist in one species, as in *L. migratoria*) is encoded on a separate gene. Small mRNA's, each of about 500 nucleotide in length have been found for the decapeptide Lom-AKH-I and the octapeptide Scg-AKH-II; they encode the two precursor proteins, prepro-AKH-I of 63 amino acids and prepro-AKH-II of 61 amino acids. The organization of the two preprohormones is very similar: there is a 22-mer signal peptide, followed by the sequence for either Lom-AKH-I (10 amino acids) or Scg-AKH-II (8 amino acids), followed by a Gly residue used for amidation and a Lys-Arg processing site and a 28-mer peptide called the α-chain in prepro-AKH-I and called the β-chain in prepro-AKH-II. After cleavage of the signal peptide the linear prohormones form dimeric precursors by oxidation. There are three dimeric precursors P_1, P_2 and P_3: two homodimers (2 pro-AKH-I and 2 pro-AKH-II) and a heterodimer (1 pro-AKH-I plus 1 pro-AKH-II). The processing of these dimeric precursors yields as products monomeric AKHs and dimeric AKH precursor-related peptides (APRPs), of which there are three different ones: $APRP_1$, consisting of two α-chains, $APRP_2$ consisting of two β-chains and the heterodimer $APRP_3$ consisting of an α-chain and a β-chain. The steps necessary for the prohormone processing have recently been elucidated in an *in vitro* system (*383*). It has been shown that the corpora cardiaca contain an endoproteolytic activity which cleaves at the C-terminal side of the Arg residue at the processing site in each chain of the dimer. The product, the C-terminal extended AKH (AKH-Gly-Lys-Arg), is subsequently digested by a carboxypeptided H-like enzyme removing Arg and then Lys. The next step is catalyzed by a peptidyl-glycine-α-amidating monooxygenase producing the amidated AKH from the glycine-extended peptide. It is also suggested that a structural motif, a so-called Ω loop, located 7 amino acids prior to the cleavage site, is necessary for action of the endopeptidase (*382*). When the structure of the precursor P_1 was analyzed in solution by circular dichroism and nuclear magnetic resonance, no evidence for an Ω loop in the N-terminal region could be found (*182*). However, the authors found an α-helical structure at the C-terminal end where another putative processing site (Arg-Lys) is located. This site is not used in prohormone processing and the study thus supports the idea that cleavage sites do not lie in helical regions, but near flexible structures (*182*).

In another *Schistocerca* species, *S. nitans*, sequence analysis of cloned cDNAs derived from 550 nucleotide long mRNAs that code for the prepro-AKHs led to a very similar organisation as for *S. gregaria* (*329*).

The sequences of the three prepro-AKHs of *L. migratoria* have been deduced from three distinct cDNAs. Whereas the precursors for Lom-AKH-I and II are highly homologous to the precursors of their counterparts in the two *Schistocerca* species, the precursor for Lom-AKH-III is different with respect to its "tail" region (the α- or β-chain) and resembles more, at least in length, the situation in non-locusts and crustaceans (see below) (*23*). *In situ* hybridization data revealed that mRNAs for the three AKHs of *L. migratoria* are co-localized in cell bodies of the glandular part of the corpus cardiacum. Remarkably, when the effect of flight activity on AKH gene expression was studied in *L. migratoria*, it became evident that the level of the Lom-AKH-III transcript was increased about 4 times and those for Lom-AKH-I and II 2 times (*23*). These differences of gene expression during flight constitute another example for the conclusion that the different AKHs of one species may be used for different functions.

The prepro-AKH sequence for *M. sexta* was deduced from the nucleotide sequence by using a genomic library for isolating the AKH gene (*32*). A 19-mer signal peptide is followed by the sequence for the nonapeptide Mas-AKH and subsequent to that by a Gly residue (for amidation) and a classical Lys-Arg cleavage site which is followed by a C-terminal peptide of 34 amino acids. This C-terminal "tail" peptide may be the equivalent to the α- or β-chain in the locusts, but the sequences are unrelated. However, the "tail" contains a Cys residue 4 residues from the C-terminus, which may be used for oxidation to form a dimeric structure like the APRPs, but this has not yet been detected.

The fruitfly, *Drosophila melanogaster*, which contains a single octapeptide identical in sequence with the hypertrehalosaemic peptide of *Phormia terraenovae* (Pht-HrTH), contains the same overall architecture of its Pht-HrTH precursor as shown for the species above (*328*). The length of the C-terminal peptide, however, is 46 amino acids; this is even longer than those of the Lom-AKH-III and Mas-AKH precursor, but shorter than the ones for the Pab-RPCH precursor (see below).

The precursor for the only crustacean member of this family of peptides, the red pigment-concentrating hormone (Pab-RPCH), has the same general organization as the precursors from insects. The sequences for prepro-RPCH from the shore crab *Carcinus maenas* (*258*) and the blue crab *Callinectes sapidus* (*231*) have been deduced from nucleotide sequences using cDNA libraries from the neurosecretory X-organs of *C. maenas* or from eyestalk ganglia of *C. sapidus*. The signal peptide contains 25 amino acids in both species, followed by the 8-mer RPCH sequence with Gly and a dibasic (Lys-Arg) processing site and a 74-(*C. maenas*) or 73-mer (*C. sapidus*) "tail" peptide. This so-called RPCH-precursor related peptide (RPRP in analogy to the insect APRPs) is much longer than the APRPs. It also contains cysteine residues and thus could form dimers, but it is not known if dimers exist.

That adipokinetic hormones are located in and synthesized by intrinsic neurosecretory cells of the corpus cardiacum in insects has also been shown by immunocytochemical methods (*49, 84, 420, 422*). In locusts region-specific antibodies with high specificity for either Lom-AKH-I or Lom-AKH-II/Scg-AKH-II revealed that both peptides are co-localized in the same glandular cells of the corpus cardiacum and even in the same secretory granules (*68, 162*). The release of both Lom-AKH-I and II into the haemolymph during flight has been reported and it was suggested that the release is controlled by octopamine and cyclic AMP (*332, 343*). However, other groups could not find octopamine immunoreactive fibers in the locust corpus cardiacum (*233*) and were unable to show AKH release by octopamine (*344*). It was, however, demonstrated that locustatachykinin I (Lom-TK-I) immunoreactive axon terminals were situated in close contact with the glandular corpus cardiacum cells (*309*). Moreover, Lom-TK-I induced the release of Lom-AKH-I when monitored in an *in vitro* system.

In *M. sexta* the endogenous AKH (Mas-AKH) mobilizes lipids for flight in adults and activates phosphorylase in moulting and wandering larvae during starvation; thus in this species this neurohormone is also involved in energy metabolism and acts on fat body cells (*490*). By synthesizing a radiolabelled (tritiated) Mas-AKH analogue (see Sect. 2.4) it was shown that membrane fractions prepared from fat body cells of *M. sexta* specifically bind this analogue (*492*). No receptor binding, however, was found with membranes prepared from brains, heart or flight muscle tissue. Membrane fractions prepared from the pterothoracic ganglion resulted in, albeit low, specific binding. This result is in full agreement with a recent study in which the injection of Mas-AKH into the mesothoracic neuropile area increased the motor activity of those muscles which are innervated by motorneuron dendrites from this area (*282*).

Inactivation and metabolism of AKH-peptides, thus termination of the hormonal signal, in different insect species have been investigated to some extent. In the central nervous system of *S. gregaria*, for example, Lom-AKH-I can be inactivated by a membrane-bound endopeptidase which cleaves the Asn^3/Phe^4 bond (*187*). According to *in vitro* and *in vivo* studies of RAYNE and O'SHEA (*381*), such an endopeptidase is also present on the external surface of the desert locust's fat body cells. Both endogenous AKHs, Lom-AKH-I and Scg-AKH-II, are cleaved at the Asn^3/Phe^4 bond. The fragments, both of which are biologically inactive, are now susceptible to degradation by exopeptidases. Indeed, for the C-terminal fragments of Lom-AKH-I and Scg-AKH-II, breakdown by aminopeptidase activity, which apparently resides in the haemolymph, could be demonstrated, whereas the N-terminal fragments (pGlu-Leu-Asn) were long-lived. Short characterization of the endopeptidase suggests a great deal of similarity to mammallian endopeptidase 24.11.

References, pp. 97–128

Exchanging Phe^4 with Tyr^4 in an analogue of Lom-AKH-I did not affect the activity of the endogenous endopeptidase (*381*). Since all members of the AKH/RPCH family contain either Phe^4 or Tyr^4 (see Table 1), it is safe to speculate that probably all peptides of this family are degraded by the same mechanism.

Another degradation process may take place in *S. gregaria* as well. Homogenates of the Malpighian tubules of this species or incubation of isolated Malpighian tubules take up and/or break down Lom-AKH-I (*431*). It is thought that the first step in the proteolytic degradation is catalyzed by a post-proline cleaving enzyme. Scg-AKH-II, however, containing no Pro^6 residue, is broken down by another endopeptidase which cleaved between Phe^4 and Ser^5. This action is similar to that of chymotrypsin (*432*). Once the endopeptidases have been active, the now unblocked new N- and C-terminus of the fragments can be attacked by exopeptidases of the leucine aminopeptidase and carboxypeptidase A or B-type. Such enzymes have been demonstrated in homogenates of Malpighian tubules (*432*). From these experiments it is assumed that AKHs can enter the Malpighian tubule cells and can be degraded there. Whether this breakdown by internalization is the major route of inactivation of AKHs is questionable. At least it is clear from the other set of experiments described above (*381*) that breakdown of AKHs by a cell-surface located endopeptidase is also occurring.

3.1.2. Diuretic and Antidiuretic Peptides

The osmotic composition of the haemolymph of insects is tightly regulated. The major organs responsible for fluid and ion secretion are the Malpighian tubules, but the hindgut (ileum and rectum) are important as well (Fig. 3). The insect's excretory system can be viewed in general to consist of two parts: the Malpighian tubules form and secrete the primary urine and the hindgut, specifically the rectum, determines, by reabsorption, the quality of the final excreted waste product. Thus, the primary urine from the Malpighian tubules enters the gut at the junction between the midgut and hindgut, where some may move forward for reabsorption in the midgut (*72*). The remaining major part mixes with the gut contents and moves in a posterior direction through the hindgut to the rectum, where most of the selective resorption and absorption of essential metabolites, including ions, and water occurs (*266*, *346*, *347*).

The primary urine produced by the tubules is isosmotic to the haemolymph. The driving force for fluid secretion is by active transport of cations achieved by a proton pump (an H^+–ATPase) and associated Na^+/H^+ and K^+/H^+ antiporters as well as Cl^- channels, all situated in

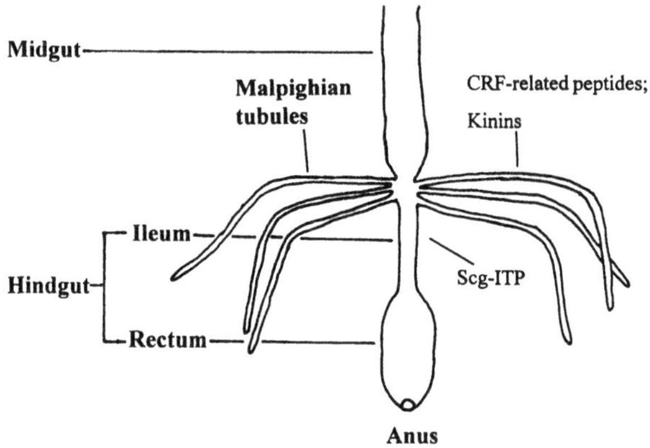

Fig. 3. Schematic diagram of the insect's excretory system indicating upon which part the different neuropeptides are acting

the apical or luminal membranes as reviewed in (*326*, *18*). By this action, either potassium chloride (in nonblood-feeding insects like locusts, beetles, and ants) or sodium chloride (in bloodsucking insects like the bug, *Rhodnius prolixus*, the yellow fever mosquito, *Aedes aegypti*, and the tsetse fly, *Glossina morsitans*) are the major salts which occur in the tubule fluid in sometimes quite high concentrations. The possibly deleterious effect of this high ionic composition is counteracted by the hindgut, where a well-controlled ion reabsorption takes place. Specifically the rectum is capable of producing a excretory product that is hyper- or hypoosmotic to the haemolymph, because the relative rates of water and ion absorption can be varied.

Neuropeptides have been reported to control tubular excretion rates (diuretic effects) as well as to regulate rectal reabsorption (antidiuretic effects). For example, feeding in haematophagous (blood-feeding) insects apparently stimulates release of diuretic peptides resulting in increased secretion rates of the tubules and an overall water loss during this so-called post-prandial diuresis (*266*). In xeric species, however, although diuretic peptides are released, an increased overall water loss may not be noticed; here the accelerated rate of tubular secretion is "masked" by the equally stimulated (by antidiuretic factors) uptake of fluid in the hindgut. The latter scenario results, because of the higher rates of recycled fluid, in a better clearance of toxic wastes and metabolic products and it was on this account that NICOLSON (*325*) proposed the term "clearance hormones" as opposed to diuretic hormones, especially for insects like the Namib Desert beetle which have to conserve water. Thus, as discussed by

Spring (*435*), the definition of "diuretic hormone" is quite ambiguous and has led to substantial confusion. This mainly stems from the different methods used to determine the action in biological assays, *i.e.* water loss from the whole insect, fluid secretion of Malpighian tubules *in situ* or by isolated tubules *in vitro*, measurement of the transepithelial potential in isolated perfused tubules or fluid reabsorption of the rectum *in vitro* (for details, see *326*, *435*, *476*).

Since the concentration of intracellular cyclic AMP (in some cases cAMP is even released into the incubation medium) in the Malpighian tubules is increased by the action of certain diuretic peptides, measurement of cAMP by RIA or competitive protein-binding assays is also frequently used to detect diuretic actions in intact tubules *in vitro*.

In what follows, studies will be reviewed which have dealt with isolation and successful sequence determination of diuretic peptides, but numerous articles on not fully-characterized diuretic peptides will not be discussed.

Using a vertebrate immunochemical approach (antibodies raised against the antidiuretic hormones of many higher vertebrates, *e.g.* arginine vasopressin), immunoreactivity was shown to occur mainly in the suboesophageal and thoracic ganglia of the migratory locust (*359*, *385*). The material was also biologically active in one of the many diuretic assays: it affected the rate of amaranth excretion in the locust. For purification, 51 000 ganglia of *L. migratoria* were homogenised, extracted and isolated on a RP-HPLC column eluted with a acetonitrile/TFA gradient resulting in two zones, F1 and F2, which were immunoreactive, but only F2 material increased dye excretion (*419*). A further 3 to 4 RP-HPLC steps, using different solvents and organic modifiers, purified both immunoreactive compounds sufficiently for peptide analyses. Surprisingly, both factors had identical amino acid composition and identical sequences, although retention times during the different purification steps were always different (*358*). Size-exclusion chromatography, however, revealed a relative molecular mass of about 700 for F1 and 1470 for F2 suggesting that the latter might be a dimer. Finally, it was shown that F2 is the antiparallel dimer of F1, *i.e.* Cys^1 of each chain in the dimer forms a disulfide bridge with Cys^6 of the opposite chain (see Table 3). Comparison with vertebrate arginine vasotocin and arginine vasopressin showed 78 and 67% sequence homology (Table 3). Both native and synthetic F2 had biological activity *in vitro* on Malpighian tubules attached to the midgut, maintaining the urine production which in non-stimulated controls decreases gradually. Concentrations of about 10^{-9} M were effective. Moreover, cyclic AMP production was stimulated by F2 (*357*). Because levels of AVP-like immunoreactivity in the haemolymph altered with

Table 3 *Sequences of arginine vasopressin-like locust diuretic peptide, of corticotropin releasing factor-related insect diuretic peptides and comparison with select vertebrate corticotropin releasing factor (CRF)-related peptides*

Code Name	Species	Sequence	Reference(s)
Mud-DP	*Musca domestica, Stomoxys calcitrans*	NKPSLSIVNPLDVLRQRLLLEIARRQMKENTRQVELNRAILKNVamide	*50*
Pea-DP	*P americana*	TGSGPSLSIVNPLDVLRQRLLLEIARRRMRQSQDQIQANREILQTIamide	*219*
Lom-DP	*L migratoria*	MGMGPSLSIVNPMDVLRQRLLLEIARRRLRDAEEQIKANKDFLQQIamide	*220, 256*
Acd-DP	*A domesticus*	TGAQSLSIVAPLDVLRQRLMNELNRRRMRELQGSRIQQNRQLLTSIamide	*218*
Mas-DP-I	*M sexta*	RMPSLSIDLPMSVLRQKLSLEKERKVHALRAAANRNFLNDIamide	*214*
Mas-DP-II	*M sexta*	SFSVNPAVDILQHRYMEKVAQNNRNFLNRVamide	*19*
Urotensin-I	*suckerfish*	NDDPPISIDLTFHLLRNMIEMARIENEREQAGLNRKYLDEVamide	*251*
Sauvagine	*frog*	QGPPISIDLSLELLRKMIEIEKQEKEKQQAANNRLLLDTIamide	*288*
Corticotropin releasing factor	*rat*	EEPPISLDLTFHLLREVLEMARAEQLAQQAHSNRKLMEIIamide	*388*
Lom-AVP-like DP	*L migratoria*	CLITNCPRGamide CLITNCPRGamide	*358*
Arginine vasopressin	*vertebrates*	CYFQNCPRGamide	
Arginine vasotocin	*vertebrates*	CYIQNCPRGamide	

relative humidity (*359*) and one of the three peaks in diuresis (measured as dye excretion) over a 24h period was correlated with a higher titre of AVP-like peptide in the haemolymph, this peptide was named arginine vasopressin-like insect diuretic hormone. It was thought to be one, of possibly several, of the true diuretic hormones of *L. migratoria*. Unfortunately, neither stimulation of fluid secretion or production of cyclic AMP in isolated Malpighian tubules of *L. migratoria* could be demonstrated in doses of up to 10^{-6} M/10^{-7} M by the synthetic antiparallel dimer F2, which was checked by chromatographic and mass spectrometric methods to be the authentic compound (*56*). Another synthetic locust diuretic peptide, however, which was previously isolated and characterized from whole heads or brains and corpora cardiaca of *L. migratoria* (*220, 256*) stimulated urine production in locust tubules 5-fold and dramatically increased tubule cyclic AMP levels at 5×10^{-8} M (*56*). This *L. migratoria* diuretic peptide is one of a series of peptides which are all related to the mammalian corticotropin releasing factor (CRF) and which are therefore called CRF-related insect diuretic peptides. The first one of this series was isolated in parallel with eclosion hormone (see Sect. 3.2.4) from 10000 trimmed heads of pharate adults of *Manduca sexta* (*214*). Separation on SP-Sephadex was followed by cartridge and semi-preparative RP-HPLC on C-4 material with acetonitrile/TFA and 1-propanol/TFA, followed by ion exchange and subsequent purification on analytical and microbore C-4 with acetonitrile/HFBA and acetonitrile/TFA respectively. As a bioassay throughout purification, "post-eclosion diuresis" (voiding of urine in many lepidopteran species immediately after adult eclosion) in the butterfly, *Pieris rapae*, was used. Newly emerged adult butterflies were ligated behind the neck and beheaded; these insects were then injected with the material to be tested and the activity was scored when clear urine was excreted. The purified material, about 5 nmol from 10000 heads, was sequenced intact and also the tryptic fragments. This yielded a 41-mer peptide in its C-terminal amidated form, called here Mas-DP-I (Table 3). Of the two synthesized forms (amidated or acidic at the C-terminal), the amidated one had the same retention time on RP-HPLC and was about 1000-fold more active than the acidic form in the *Pieris* assay. Furthermore, it promoted a pronounced loss of water through the gut and epidermis in pre-wandering, post-feeding *M. sexta* larvae, but had no direct effect on isolated tubules of these larvae (*214*). Later it was demonstrated that synthetic Mas-DP-I stimulated fluid secretion and production of cAMP in isolated Malpighian tubules of *Acheta domesticus* (*53*). Stimulation of fluid secretion and cAMP production *in vitro* by Malpighian tubules of the butterfly, *P. rapae*, taken from adults within 1–12h of eclosion, was shown by these authors as well. Data of

TROETSCHLER and KRAMER (*458*) revealed a decrease in fluid absorption from the rectum and an increase of intracellular levels of cyclic AMP in the rectum and Malpighian tubules of larval *M. sexta in vivo* by Mas-DP-I. Recently, the direct stimulating effect on Malpighian tubule secretion of adult *M. sexta* by synthetic Mas-DP-I *in vitro* was presented, and it was demonstrated that Mas-DP-I acts as an antidiuretic peptide on the cryptonephric complex of *M. sexta* larvae (*8*). Both effects appear to be mediated via cAMP.

Antisera raised against the N-terminal (Mas-DP-I_{1-21}) and C-terminal (Mas-DP-I_{22-41}) parts of the *Manduca* diuretic peptide both recognised the same two median neurosecretory cells on each side of the protocerebral groove of *M. sexta* larvae and a group of about 80 median neurosecretory cells in the adult (*470*). These data and the positive immunoreactivity of axons leading to the corpora cardiaca and axon terminals in these neurohaemal organs suggest that Mas-DP-I may be released into the haemolymph from these sites and act as a true neurohormone.

Three members of the CRF-related diuretic peptides, one each from *Acheta domesticus*, *Locusta migratoria* and *Periplaneta americana*, were isolated by KAY *et al.* (*218, 219, 220*), using as their primary bioassay the production of cAMP by isolated Malpighian tubules in the species under investigation (or in the locust), but also checking the purified native peptide for stimulation of fluid secretion in its respective Malpighian tubules *in vitro*. Starting materials for the purification were whole heads (1000 from *A. domesticus*, 2000 from *L. migratoria*, and 800 from *P. americana*) which were frozen in liquid nitrogen and powdered. The powder was subsequently extracted with acidified methanol (87% methanol, 5% glacial acetic acid, 8% water) and the fluid concentrated by precipitation with 70% acetone. The resulting pellet was dissolved in 5 mM TFA and then fractionated on a Sep-Pak C-18 cartridge with successive steps of increasing acetonitrile concentration. Diuretic activity of the 40–45% acetonitrile fraction was further purified by HPLC employing a combination of three column chemistries: the first two steps involved RP-HPLC on a semipreparative C-8 and a diphenyl column using acetonitrile/TFA gradient, the next step was a normal size-exclusion Protein-Pak 125 column operated in normal phase, *i.e.* the peptides loaded in a non-polar solvent are forced into polar interactions with the packing material and are eluted with increasing polarity. To confirm purity and to concentrate the purified peptide from the previous step, the last step employed the diphenyl column again. This purification scheme was successful for all three species and in each case resulted in one pure diuretic peptide with an amidated C-terminus as established by automated Edman sequencing combined with either FAB or electrospray mass spectrometry. These peptides, called Acd-DP, Lom-DP and Pea-DP here, are 46-mers and show striking sequence homology with Mas-DP-I (see Table 3). An identical diuretic peptide for *L. migratoria* was purified and sequenced (*256*), using 4600 dissected brains (without optic lobes) plus corpora cardiaca, and testing the fractions during isolation by an ELISA test developed for Mas-DP-I (*103*). Isolation was achieved by a modified protocol of the one employed to isolate Mas-DP-I (*214*); thus 7 chromatography steps were involved despite the relative purity of the starting material.

Essentially only one step of C-8 RP-HPLC purification was used to purify a second diuretic peptide from *M. sexta* when either complexes of corpora cardiaca/corpora allata or dissected clusters of neurosecretory cells from the medial protocerebrum were taken as starting materials (*19*). Edman sequencing, tryptic or endoproteinase Lys-C digests, in association with quadrupole Fourier transform mass spectrometry, identified the primary structure as an amidated 30-mer peptide (Mas-DP-II; Table 3). Biological activity was measured by determining weight loss *in vivo* of adult female *M. sexta*, which were decapitated 24 h after emergence, the wound sealed, and insects assayed the next day; such a weight-loss assay does not discriminate between various pathways for water loss and, thus, it was not known whether Malpighian tubules and/or the rectum were involved (*19*). This was clarified later in two separate studies (*9*, *20*). Mas-DP-II elevates fluid secretion by isolated Malpighian tubules from adult moth at concentrations as low as 4 nM (*20*) or 0.05 nM (*9*). Cyclic AMP production in larval proximal and adult tubules was stimulated as well by Mas-DP-II (*9*), but, in contrast to the effect of Mas-DP-I (*8*), Mas-DP-II was not able to stimulate fluid uptake across the larval cryptonephric complex; thus no *anti*-diuretic effect was measured (*9*). These results are difficult to interpret since Reagan (*384*) had shown that Mas-DP-II binds to and activates Mas-DP-I receptors expressed in COS-7 cells (see below). The phenomenon may be explained by postulating the existence of different receptor subtypes for the distal (cryptonephric) and proximal larval tubules.

The last CRF-related peptide sequenced to date was purified from whole-body extracts of the blowfly, *Musca domestica*, (444 500 individuals) and, separately, the stable fly, *Stomoxys calcitrans*, (*50*). The biological activity was monitored by measuring the ability of fractions to stimulate cAMP production in isolated Malpighian tubules of adult *M. sexta*. Isolation was achieved by seven different column systems and the purified peptide was analyzed by automated Edman degradation and laser desorption and/or electrospray mass spectrometry. The sequence of the 44-mer shown in Table 3 resulted for the material from both insect species. Interestingly, in *M. domestica* the peptide was completely oxidized (Met residue) during isolation, whereas two peaks were isolated and sequenced from *S. calcitrans*, identified as the Met-oxidized and non-oxidized form. In a homologous bioassay, stimulating the rate of fluid secretion of *M. domestica* Malpighian tubules, the synthetic (Met-oxidized) Mud-DP was active at 1 nM concentration. No elevated secretion by another target tissue, the salivary glands of the house fly, was observed.

To date, six insect diuretic neuropeptides are fully characterized which are related to the vertebrate corticotropin releasing factor/urotensin I/sauvagine family (see Table 3). These latter three peptides have at least 45% sequence identity with each other (*252*) and, with the exception of the much shorter Mas-DP-II, the insect CRF-like peptides have at least 40% sequence identity with each other and are about 20–30% identical with the vertebrate counterparts (*52*). When the precursor for Mas-DP-I was characterized (*69*), it became clear the prepro-Mas-DP-I and ovine prepro-CRF only show a low degree of homology (between 28–33%) and a large gap is needed to align the mature and the preceding regions of both precursors (*69*). Moreover, the Mas-DP-I receptor was isolated by expression cloning in COS-7 cells; it possesses seven putative transmembrane domains common to other G-protein coupled receptors and, thus, is coupled to a cAMP second messenger system (*384*). There is a 31% sequence identity between the cloned *Manduca* receptor and the cloned human CRF receptor (*44*). Effects of vertebrate peptides (urotensin I, sauvagine and bovine CRF) on stimulation of fluid in *A. domesticus* tubules were significant (at 10^{-5} M) but small (20%) compared with the maximal possible stimulation in this tissue (*53*). These peptides also elicited small increases in cAMP production in cricket tubules (*in vitro*) (*53*). Similarly, sauvagine, human- and bovine CRF stimulated cAMP production in *Manduca* tubules at 10^{-5} M, but this effect was only 7% of the maximum (*9*). Thus, the limited sequence identity between insect and vertebrate peptides is also mirrored in their action.

Another group of insect neuropeptides, the myokinins (see Sect. 3.3.4), also have diuretic activity. For example, fluid secretion in isolated Malpighian tubules of *A. domesticus* is stimulated by achetakinins, but cAMP does not seem to be involved (*54*). The leucokinins of *Leucophaea maderae* depolarize the transepithelial voltage in isolated Malpighian tubules of *Aedes aegypti* (*161*). The latter bioassay served also as a tool to isolate similar peptides, culekinin depolarizing peptides, from the mosquito, *Culex salinarius* (*158*). Peptides belonging structurally to the kinin family (see Sect. 3.3.4) but are potent stimulators of secretion by

Malpighian tubules of *M. sexta* have also been isolated and sequenced from the abdominal ventral nerve cord of the adult lepidopteran insect, *Heliothis zea* (*22*). It is speculated that these myokinins are probably involved in post-feeding diuresis, to get rid of the excess water derived from the diet, whereas the CRF-related diuretic peptides are more likely to act as clearance peptides, removing metabolic waste products from the haemolymph by creating a high rate of fluid secretion (*55*). With respect to these different putative functions it has been proposed (see *50*) that (1) there is no great evolutionary pressure on structure change for the CRF-related peptides, because metabolic waste management can be viewed as a basic function for all insect species. Thus, these peptides are relatively highly conserved; that (2) the source and physiological state of the diet is different for various species and, therefore, peptides involved in post-feeding diuresis may be more variable and may even be species-specific as seems to be the case for the myokinins.

Most of the primary urine formed in the Malpighian tubules is passed posteriorly into the hindgut which consists of the ileum and the rectum. Functionally, the ileum has the same task as the proximal tubules of the vertebrate kidney, removing large quantities of fluid without affecting the osmolarity of the urine. The rectum has the same function as the distal tubules, loop of Henle and collecting ducts of the vertebrate kidney, selectively reabsorbing water, ions and metabolites and, thereby determining the final composition of the excreta which can be hyper- or hypoosmotic (see *348*, *349*).

Much less is known about the regulation of ion and fluid reabsorption in the hindgut by neuropeptides than regulation of tubule fluid secretion. Except for neuroparsins, which may exert an antidiuretic action (see Sect. 3.2.5.1), no structural data on complete primary sequences have been published (see *11*). A peptide was isolated from the corpora cardiaca of the desert locust, *Schistocerca gregaria*, by a four step separation technique on C-4, C-8 and phenyl-columns using acetonitrile/TFA gradients and partially sequenced (*10*). As a bioassay chloride transport was measured, since an apical electrogenic Cl^- pump is the major rectal ion transport process. Experimentally, ilea were mounted as flat sheets in Ussing-type chambers, voltage-clamped at zero and the short-circuit current (Isc) measured. The isolated peptide was called Scg-ITP (see Table 4), *Schistocerca gregaria* ion transport peptide. It has a molecular mass of 8652 (*11*) and its N-terminal 34 residues show sequence homology with the hyperglycaemic hormones of crustaceans (see Table 4). Interestingly, an immunocytochemical study of stick insect (*Carausius morosus*) brain and retrocerebral complex using an antiserum against *Carcinus maenas* hyperglycaemic hormone had revealed quite a few immunopositive cells

Table 4 *Partial sequence of desert locust ion transport peptide (Scg-ITP) in comparison with part of the sequence from select crustacean hyperglycaemic hormones (CHH)*

Code Name	Species	Sequence	Reference(s)
Scg-ITP	*S gregaria*	SFFDIQ?KGVYDKSIFARLDRI?ED?YNLFREPQ	*10*
Cama*-CHH	*C maenas*	pQIYDTSCKGVYDRALFNDLEHVCDDCYNLYRTSY	*225*
Orl-CHH	*O limosus*	pQVFDQACKGIYDRAIFKKLDRVCEDCYNLYRKPY	*224*

* A four letter code is used for this crustacean peptide to distinguish between Cam = *Carausius morosus* (a stick insect) and Cama = *Carcinus maenas* (a crab)

(*203*). In light of the above results these previous data suggest that the stick insect also contains a neuropeptide which is related to the crustacean CHH-family and may be involved in ion transport in the insect.

3.2. Peptides Regulating Reproduction, Growth and Development

3.2.1. Pheromone Biosynthesis Activating Neuropeptides

Chemicals that are secreted by one individual and affect the physiology or behavior of another member of the same species are termed pheromones (*208*). Sex pheromones are produced by females of many species of Lepidoptera to attract conspecific males. A vast body of information has been accumulated on these sex pheromones, partly because they are vital to assure successful mating and therefore reproduction, partly because of their use in insect control. In 1959 the pheromone produced by the female silkworm moth, *Bombyx mori*, to attract males from a great distance was the first to be purified and identified chemically; it is (10*E*, 12*Z*)-hexadecadien-1-ol, with the trivial name bombykol (*36*).

Since it was observed that (a) sexual activity in both male and female Lepidoptera occurs at defined times of the day (mostly in the scotophase) and that (b) production and release of sex pheromones follows a diel periodicity (*350*), it was apparent that pheromone production was under hormonal control. This was shown to be true for the corn earworm moth, *Helicoverpa* (*Heliothis*) *zea*, by a factor from the brain (*373*). The factor appeared to be a peptide produced in the suboesophageal ganglion of the moth and released, at the onset of the scotophase, into the haemolymph via the corpora cardiaca to travel to the pheromone-producing cells in the ovipositors. There it stimulates production of 11*Z*-hexadecenal, the main pheromone component (*370*). The peptide was isolated and its structure determined from a total of about 20000 brain-suboesophageal ganglia-corpora cardiaca complexes from adult male and female *H. zea* using either a sequence of four RP-HPLC steps (1. C-18; acetonitrile/TFA gradient; 2. C-8; acetonitrile/triethylammoniumphosphate gradient; 3. C-8; acetonitrile/TFA gradient; 4. C-18; acetonitrile/TFA gradient) or three HPLC steps (1. as above; 2. high performance size-exclusion chromatography on a series of 4 Protein-Pak I-125 columns isocratically developed with 40% acetonitrile and 0.1% TFA 3. as 4 above) (*201, 369*).

The pheromonotropic activity was tested during isolation by a rather simple and very sensitive bioassay (*373*): female moths were ligated between head and thorax at least 3 h prior to the test, injected intra-abdominally with the desired material during the scotophase; 3 h later the pheromone gland was extracted and the pheromone quantified by gas chromatography (*374*). After isolation, the major component was sequenced by automated Edman degradation using a pulse-liquid sequencer (also involving carboxypeptidase P to determine the carboxyterminus) and the structure confirmed by plasma desorption mass spectrometry (*201*, *369*). The pheromonotropic neuropeptide, called pheromone biosynthesis-activating neuropeptide (Hez-PBAN) consists of 33 amino acids (Table 5), has a molecular weight of 3900 and only the C-terminal amidated form is biologically very active (2–4 pmol/female needed compared with at least 1000 pmol, when the C-terminus is a free acid) (*375*). The molecule has two methionine residues (Met^5 and Met^{14}), which in the isolated native peptide were both oxidized to methionine sulfoxides; other peaks during the purification step apparently represented the mono- or disulfoxide forms of PBAN (*201*).

Hez-PBAN represents the first member of a new family of insect neuropeptides. The family now includes the pheromonotropic peptides from the silkworm *Bombyx mori* , Bom-PBAN-I and II, which were purified from 6×10^5 (= 4.48 kg fresh weight) heads of adult male silkworms using an 11-step purification procedure (*229*, *230*, *316*) and from the gypsy moth, *Lymantria dispar*, Lyd-PBAN, which was isolated from abut 2000 brain-suboesophageal ganglion complexes in a 5-step HPLC purification protocol using the heterologous bioassay in *H. zea* (*272*). Whereas Bom-PBAN-I and Lyd-PBAN are also 33-mers, as is Hez-PBAN, and have about 82% homology in their primary sequence (see Table 5), Bom-PBAN-II consists of 34 amino acids; it has an additional Arg at the N-terminus compared with Bom-PBAN-I (see Table 5).

A much shorter peptide with pheromonotropic activity has been isolated from 32000 heads of the penultimate instar larvae of the army worm, *Pseudaletia separata*, by a 7-step purification procedure using the heterologous bioassay in *B. mori* (*275*). This 18-mer pheromonotropin called Pss-PT has an identical C-terminal pentapeptide with the other PBANs (with the exception of Thr instead of Ser; see Table 5).

Interestingly, the same pentapeptide sequence (FXPRL-amide, where X is either T, S, G or V) has been found in certain myotropic peptides of the cockroach, *Leucophaea maderae* (*294*), and the locust, *Locusta migratoria*, which stimulate contraction of hind- or foregut and/or oviduct (*406*, *409*, *410*; see Sect. 3.3) and in the diapause hormones of *B. mori* (*185*, *392*; see Sect. 3.2.6). Furthermore, a peptide with the same sequence as Bom-PBAN-I has been isolated as the melanization and reddish colouration hormone (Bom-MRCH) of *B. mori* using an armyworm cuticle melanization test as a bioassay (*277*).

Table 5 *Amino acid sequences of various pheromone biosynthesis-activating neuropeptides (PBANs) and comparison with other selected insect neuropeptides containing the pentapeptide FXPRLamide C-terminus*

Code Name	Species	Sequence	Reference(s)
Hez-PBAN	*H zea*	LSDDMPATPADQEMYRQDPEQIDSRTKYFSPRLamide	*369*
Bom-PBAN-I or (-II)	*B mori*	(R)LSEDMPATPADQEMYQPDPEEMESRTRYFSPRLamide	*229, 230*
Lyd-PBAN	*Lymantria dispar*	LADDMPATMADQEVYRPEPEQIDSRNKYFSPRLamide	*272*
Pss-PT	*Pseudaletia separata*	KLSYDDKVFENVEFTPRLamide	*275*
Bom-DH	*B mori*	TDMKDESDRGAHSERGALCFGPRLamide	*185*
Lem-PK	*L maderae*	pQTSFTPRLamide	*172*
Lom-PK	*L migratoria*	pQDSGDGWPQQPFVPRLamide	*407*

It was shown that the *P. separata* pheromonotropin induces cuticular melanization and also embryonic diapause (*278*). Further results support these data. Quantitative analyses of endogenous PBAN (or MRCH) levels by an enzyme linked immunosorbent assay (ELISA; *138*), in head extracts and haemolymph of larvae of the noctuid moth, *Spodoptera littoralis*, which exhibits morphological color variations when reared under crowded (dark coloration) and isolated conditions (light coloration), suggest that the peptide is involved in color polymorphism (*6*). Thus, a group of peptides showing the FXPRL-amide at their C-terminus and therefore forming a peptide family are widely distributed among various insect groups and are responsible for regulating a number of functions in diverse physiological processes.

Structure-activity studies on both Bom- and Hez-PBAN and their fragments and analogues have revealed some interesting information on how these molecules will interact with their postulated receptor (*243, 317, 371, 372*). The Arg residue at the N-terminus of Bom-PBAN-II is not important for activity; in fact, the whole N-terminal region of Hez-PBAN (amino acid 1–18) was not active in *H. zea*. However, C-terminal fragments (15–33, 19–33, 23–33, 28–33 and 29–33 for Hez-PBAN and 24–33, 25–33, 26–33, 27–33, 28–33, 29–33 for Bom-PBAN-I) display biological activity, indicating that the C-terminus is indispensable for activity. The C-terminal pentapeptide represents the smallest unit required for activity. The C-teminus has to be amidated; the free acid form was at least 1/100-fold less active. When the entire native PBANs which have their two (Hez-PBAN) or three (Bom-PBAN) methionine residues in the sulfoxide forms are assayed, they are more active than the non-sulfoxidized analogues. The increased activity of the sulfoxide forms is suggested to be due to stabilization of PBAN against enzymic deactivation. Interestingly, an internal pentapeptide fragment of Hez-PBAN, which was amidated at its C-terminus (Y-R-Q-D-P-amide) showed very high activity at the low dose of 1 pmol, but was inactive at 100 and 1000 pmol; these results were ascribed to the possible presence of two different types of receptors which could trigger the pheromonotropic response.

Since the C-terminal pentapeptide was very active (Bom-PBAN-I 28–33-amide) each residue was substituted by other amino acid residues. It was shown that Pro^{31}, Arg^{32} and Leu^{33} were essential, suggesting that this part is probably the binding site for a putative receptor. Designing cyclic peptides, containing Lys residues with the carboxyl portion of Bom-PBAN-I in order to get conformationally more rigid peptides, failed to produce very active analogues. However, the cyclo (-N-T-S-F-T-P-R-L) analogue which was used in myotropic studies and shown to have a β-turn in the region of T-P-R-L (*301*) was as active as the C-terminal

References, pp. 97–128

28–33 amide fragment of Bom-PBAN-I. This again demonstrates clearly the close relationship of myotropic and PBAN peptides. During cross-reactivity studies (*244*) it became apparent that the carboxyl-terminal hexapeptide of Bom-PBAN-I elicited myotropic activity comparable to the effect achieved by myotropic peptides, while intact Bom-PBAN-I exhibited much lower activity. All myotropic peptides assayed, however, had high pheromonotropic activity.

The Hez-PBAN gene has been elucidated (*64*). The genome clone of Hez-PBAN was isolated from a genomic library using two mixed probes which represented two overlapping amino acid regions of PBAN. The organization of the Hez-PBAN gene is very interesting since it suggests sequences for two additional, previously unknown insect neuropeptides with pheromonotropic and/or myotropic activity, and, therefore, the gene may represent a prohormone. The proposed open reading frame starts with M-E-F-T-P-R-L (thus including the well-known pentapeptide characteristic for this family) followed by a G (providing the amino group for amidation) and a distant cleavage site (R-R). Thereafter follows the sequence of residues 1 to 14 of PBAN interrupted from the remaining residues 15 to 33 by a 0.63 kilobase intron; the PBAN sequence is followed by G-R, a widely used prohormone processing site in which the G provides the amino group for amidation. Subsequently the sequence is T-M-N-F-S-P-R-L (thus again the characteristic pentapeptide) and is again followed by a putative processing site G-R. One may speculate that besides PBAN the two peptides with the C-terminal pentapeptide sequences F-T/S-P-R-L-amide (thus a hepta- and octapeptide) are released separately and may have specific functions, either in concert or independently of PBAN, for regulating pheromone production and/or ovipositor movement in *H. zea* females (*64*).

The search for the genes for PBAN and for the diapause hormone of *B. mori* (see Sect. 3.2.6), which contains the characteristic C-terminal pentapeptide, has resulted in finding a cDNA encoding a polyprotein precursor which can be processed not only into the diapause hormone, but also into PBAN and 3 other, functionally unknown, neuropeptides (termed: α, β, γ-suboesophageal ganglion neuropeptide) sharing the common C-terminal sequence F-X-P-R/K-L amide (where X is G, T, I or S) (*217, 393*). A schematic representation of the precursor peptide (*217*) showing a 23-mer signal peptide, the sequence of diapause hormone, the 3 putative peptides and of PBAN is shown in Fig. 4. Met^{1} to Cys^{23} is the signal peptide, amino acids 24–27 represent the Bom-DH, followed by Gly for amidation and a processing site, Bom-PBAN-I is localized from residue 126 to 158 (and Bom-PBAN-II from 125 to 158) and the peptides with the conserved pentapeptide sequence were found at residues 118 to

Bombyx mori

Fig. 4. Schematic diagram of the precursor peptide of Bom-PBAN (pheromone biosynthesis-activating neuropeptide). α, β, γ-SGNP = α, β, γ-suboesophageal neuropeptide). Modified after (*217*)

122, 164–168 and 99–103. The last three peptides were synthesized and tested for diapause inducing activity, but were almost inactive (*393*); however, the authors report that one of the components (β-SGNP = SVAKPQTHESLEFIPRL) has higher pheromonotropic activity than Bom-PBAN-I, but the other two peptides were far less active. Interestingly, when these authors re-interpret the gene sequence data of Hez-PBAN (*64*) by assigning the GTG codon not to Met^1 for translation initiation, but to the usual Val residue, they find a sequence of an 18-mer peptide very similar to the Bom-β-SGNP and, surprisingly, to the pheromonotropin of *P. separata* (see Table 5). The Hez-PBAN gene did not code for a diapause hormone, which, of course, has never been found to exist in *H. zea*.

Which steps in the biosynthetic pathways of pheromones may be under control of PBAN is still under debate (see *375*). In general, pheromone production in *H. zea* (major pheromone: 11*Z*-hexadecenal) and *B. mori* (pheromone: bombykol) commences with the production of palmitic acid followed by species-specific steps of desaturation/dehydrogenation, reduction of the acid to alcohol, and, if necessary, oxidation. It is suggested that in *H. zea* PBAN regulates the fatty acid biosynthesis or a step prior to it (*205*), whereas in *B. mori* PBAN promotes the reduction step of the acyl moieties to their corresponding alcohols (*7*).

The availability of synthetic PBAN made it possible to prepare antisera. A highly specific (directed to the N-terminal region of Hez-PBAN 1–33) antiserum was produced and used in an enzyme linked immunosorbent assay (ELISA) (*138*). It was demonstrated that 3- and 7-day old *H. peltigera* moths of both sexes had roughly the same content of PBAN (± 5 pmol/head); PBAN-like immunoreactivity was not present in the first three larval instars, but increased steadily as a function of development from the 4th instar larvae onwards.

Antisera raised against colloidally adsorbed synthetic Hez-PBAN and used in immunocytochemical studies showed three clusters of cells in the mandibular (4 cells), maxillary (12–14 cells) and labial neuromers (*227*). Axons from cells from the labial cluster project to the corpora cardiaca, a possible release site, and to the aorta. Thus there are some indications that PBAN is a true neurohormone, although immunoreactivity has not yet been detected consistently in the haemolymph.

Developing a specific radioimmunoassay (RIA) for PBAN using $[^3H]$-Hez-PBAN and a specific PBAN antiserum was a prerequisite for showing PBAN-like immunoreactivity in various neuronal tissues from females of *H. armigera* (during scoto- and photophase) (*367*). Levels of immunoreactive-PBAN in corpora cardiaca, prothoracic and abdominal (excluding the terminal one) ganglia were higher during the peak hour of pheromone production, thus during the 4–5th hour of scotophase, than the levels in ganglia from insects in the 6-11th hour of photophase. This was interpreted as an increased passage of PBAN from the suboesophageal ganglion to the corpora cardiaca for possible release. In contrast, immunoreactive PBAN levels were higher in the terminal abdominal ganglion during the photophase which may reflect an accumulation before the onset of pheromone production. Future studies with detailed emphasis on the temporal distribution of PBAN have to be undertaken to provide a clear description of storage, passage and release of PBAN from the different neuronal tissues.

3.2.2. Allatotropins and Allatostatins

The corpora allata synthesize and release species-specific juvenile hormones. The activity of the corpora allata, in turn, is regulated by neurosecretory material from the brain (*93*). These are factors which stimulate or inhibit the biosynthesis of juvenile hormone, thus they are either allatotropins or allatostatins.

3.2.2.1. Allatotropins

For detection of active fractions during purification of the allatotropin from the lepidopteran moth, *Manduca sexta*, the following *in vitro* radiochemical bioassay was used (*94, 457*): Corpora allata of female moths, 0 to 4 h after eclosion, were analyzed for incorporation of the labelled methyl moiety from L-[methyl-^{14}C] methionine into juvenile hormone; the labelled hormone is secreted into the medium, then extracted and quantified (*212*). Using a variety of separation steps (see Sect. 2.2) finally 1.5 nmol of pure peptide was obtained from 10 000 trimmed heads of pharatc adult *M. sexta* (*212*). Automated sequence analysis revealed the presence of a 13-residue peptide which was shown to be amidated at the C-terminus (see Table 6). The biological activity of the synthetic peptide was not significantly different from the native peptide. Studies on N-terminal

Table 6 *Amino acid sequences of allatotropin (AT) and allatostatins (AST) determined by isolation* or deduced from cDNA*

Code Name (Alternative Designations)	Species	Sequence	Reference(s)
Mas-AT	*M sexta*	GFKNVEMMTARGFamide	**212*
Mas-AST	*M sexta*	pQVRFRQCYFNPISCF	**240*
Dip-AST-1 (dipstatin 1)	*Diploptera punctata*	LYDFGLamide	*71*
(Pea-AST-1)	*P americana*		Ding, Donly, Tobe, Bendena, unpublished
(BLAST-1)	*B germanica*		**17*
Dip-AST-2 (V, ASB2, dipstatin 2)	*D punctata*	AYSYVSEYKRLPVYNFGLamide	**352, 71*
(Pea-AST-2)	*P americana*		Ding, Donly, Tobe, Bendena, unpublished
Dip-AST-3 (dipstatin 3)	*D punctata*	SKMYGFGLamide	*71*
(Pea-AST-3)	*P americana*		Ding, Donly, Tobe, Bendena, unpublished
Dip-AST-4 (VII, dipstatin 4)	*D punctata*	DGRMYSFGLamide	**479, 71*
Dip-AST-5 (IV, dipstatin 5)	*D punctata*	DRLYSFGLamide	**480, 71*
(BLAST-2)	*B germanica*		**17*
Dip-AST-6 (dipstatin 6)	*D punctata*	ARPYSFGLamide	*71*
(Pea-AST-6)	*P americana*		Ding, Donly, Tobe, Bendena, unpublished
Dip-AST-7(I, dipstatin 7)	*D punctata*	APSGAQRLYGFGLamide	**480, 353, 71*
Dip-AST-8 (III, dipstatin 8)	*D punctata*	GGSLYSFGLamide	**480, 71*
Dip-AST-9 (II, dipstatin 9)	*D punctata*	GDGRLYAFGLamide	**480, 71*
Dip-AST-10 (dipstatin 10)	*D punctata*	PVNSGRSSGSRFNFGLamide	*71*
Dip-AST-11 (VI, dipstatin 11)	*D punctata*	YPQEHRFSFGLamide	**479, 71*
Dip-AST-12 (dipstatin 12)	*D punctata*	PFNFGLamide	*71*
Dip-AST-13 (dipstatin 13)	*D punctata*	IPMYDFGIamide	*71*
Pea-AST-4	*P americana*	SGNDGRLYSFGLamide	Ding, Donly, Tobe, Bendena, unpublished
Pea-AST-5	*P americana*	DRMYSFGLamide	Ding, Donly, Tobe, Bendena, unpublished

Pea-AST-7 (Pea-AST I)	*P americana*	SPSGMQRLYGFGLamide	*474, DING, DONLY, TOBE, BENDENA, unpubl
Pea-AST-3	*P americana*	GGSMYSFGLamide	DING, DONLY, TOBE, BENDENA, unpublished
Pea-AST-9 (Pea-AST II)	*P americana*	ADGRLYAFGLamide	*474, DING, DONLY, TOBE, BENDENA, unpubl
Pea-AST-10	*P americana*	PVSSARQTGSRFNFGLamide	DING, DONLY, TOBE, BENDENA, unpublished
Pea-AST-11	*P americana*	SPQGHRFSFGLamide	DING, DONLY, TOBE, BENDENA, unpublished
Pea-AST-12	*P americana*	SLHYAFGLamide	DING, DONLY, TOBE, BENDENA, unpublished
Pea-AST-13	*P americana*	PYNFGLamide	DING, DONLY, TOBE, BENDENA, unpublished
Blg-AST-3 (BLAST-3)	*B germanica*	AGSDGRLYSFGLamide	*17
Blg-AST-4 (BLAST-4)	*B germanica*	APSSAQRLYGFGLamide	*17
Cav-AST-1 (Leu-callatostatin 1)	*Calliphora vomitoria*	DPLNEERRANRYGFGLamide	*77
Cav-AST-2 (Leu-callatostatin 2)	*C vomitoria*	LNEERRANRYGFGLamide	*77
Cav-AST-3 (Leu-callatostatin 3)	*C vomitoria*	ANRYGFGLamide	*77
Cav-AST-4 (Leu-callatostatin 4)	*C vomitoria*	NRPYSFGLamide	*77
Cav-AST-5 (Met-callatostatin 5)	*C vomitoria*	GPPYDFGMamide	*77
Cav-AST-6 ([Hyp]^3-Met-callatostatin)	*C vomitoria*	GPXYDFGMamide X = hydroxyproline	*75
Cav-AST-7 ([Hyp]^2-Met-callatostatin)	*C vomitoria*	GXPYDFGMamide	*76
Cav-AST-8 (des-G-P-Met-callatostatin)	*C vomitoria*	PYDFGMamide	*76
Grb-AST-A1	*G bimaculatus*	AQHQYSFGLamide	*261
Grb-AST-A2	*G bimaculatus*	AGGRQYGFGLamide	*261
Grb-AST-B1	*G bimaculatus*	GWQDLNGGWamide	*260
Grb-AST-B2	*G bimaculatus*	GWRDLNGGWamide	*260
Grb-AST-B3	*G bimaculatus*	AWRDLSGGWamide	*260
Grb-AST-B4	*G bimaculatus*	AWERFHGSWamide	*260

truncated fragments suggested that the amino acids 6–13, an octapeptide, are the biologically active core. Interestingly, the synthetic compound was not active in the biosynthesis of juvenile hormone during other developmental stages (larval, pupal) of *M. sexta*. Furthermore, corpora allata from the beetle, *Tenebrio molitor*, the grasshopper, *Schistocerca nitens*, and the cockroach, *Periplaneta americana*, were not activated by the synthetic allatotropin, whereas the corpora allata of the noctuid moth, *Heliothis virescens*, were stimulated, suggesting order-specificity.

3.2.2.2. Allatostatins

During isolation of the allatostatins, the same bioassay as described above (Sect. 3.2.2.1) was used, but here the inhibition of juvenile hormone biosynthesis was monitored. Either the corpora allata of virgin females (*480*) or the glands from 10-day old pregnant females were incubated *in vitro* (*353*); both research groups obtained the material from the viviparous cockroach *Diploptera punctata*. Brains or brains/retrocerebral complexes of this cockroach comprised the starting material for purification in both studies. Purification was achieved in various steps by reversed-phase HPLC using C-18 and C-8 supports leading to apparent homogeneity of four peaks with allatostatic activity, allatostatins I to IV (*480*); or purification was successful with inclusion of pre-purification steps on C-18 Sep-Pak followed by Diol Sep-Pak which separated two types of allatostatins: one with a lower molecular mass, designated type A allatostatins, and the other with a higher molecular mass, designated type B allatostatins (*352, 353*). Later, two further allatostatins, VI and VII, were isolated from this cockroach (*479*). Both research groups employed Edman degradation sequencing techniques and mass spectrometry for structure elucidation. It became clear that the six allatostatins (I, II, III, IV, VI, and VII or Dip-AST-7, -9, -8, -5, -11, -4; for nomenclature and structure see Table 6) vary between 8 and 13 residues and apparently belong to a family of peptides. This is suggested by the highly conserved sequence at the C-terminus; Arg/Ser-Leu-Tyr-Xaa-Phe-Gly-Leu-NH_2. The larger allatostatin was identified by tandem mass spectrometry as an octadecapeptide (V or Dip-AST-2, Table 6) having an amidated three residue C-terminus identical with the termini of the other allatostatins (*352*).

The synthetic peptides had the same elution times as the native material and inhibition of juvenile hormone synthesis of more than 40% was achieved with 7×10^{-7} M, 10^{-8} M and 10^{-9} M (allatostatin III, II, IV and I respectively; *480*). Allatostatin I also inhibits juvenile hormone synthesis in another, only distantly related cockroach, *Periplaneta americana*; thus there appears that no species specificity exists (*480*).

Cockroaches synthesize juvenile hormone III in their corpora allata. *De novo* synthesis starts from acetyl CoA through the classical isoprenoid pathway to farnesyl pyrophosphate (see *417*). Studies by PRATT *et al.* (*351*, *352*, *353*) revealed that allatostatins I and V were totally ineffective in the presence of 200 μM farnesol, indicating that the action of allatostatins must be located at the beginning of the biosynthetic pathway. The same conclusion was drawn from experiments using allatostatins IV and VII (*479*).

Structure-activity studies showed that allatostatins lacking the C-terminal amide produce no detectable inhibition of juvenile hormone biosynthesis (*373*, *480*). When allatostatin IV was truncated by either the first two or three residues from the N-terminus, the products had progressively reduced activity when compared with the parent molecule (*441*). Using the tridecapeptide allatostatin I PRATT *et al.* (*354*) found no activity at all when changes were made at the C-terminus: Gly^6 (instead of Phe), Ala^{13} (instead of Leu), shortening of the peptide by the last two amino acids (des-Gly^{12} and Leu^{13}), an extra Ala (amidated or not). All these results suggest that the C-terminal part of the molecule is important in signal transmission. However, when Lys^7 or D-Arg^7 (instead of L-Arg) were bioassayed, the affinity was only marginally less than that of the unchanged peptide. A lower binding strength was observed, but the magnitude of the response was not reduced. Two N-terminally truncated analogues of allatostatin I, a decapeptide (= allatostatin I 4–13) and an octapeptide (= allatostatin I 6–13) showed substantially lower affinities, but still the magnitude of the response (> 85% juvenile hormone inhibition at concentrations of 1 μM or lower) was identical with that produced by the intact molecule indicating that the message segment in these peptides is still intact.

Some structure-activity studies were also performed with allatostatin V, the octadecapeptide. An N-terminal nona- or undecapeptide amide (allatostatin V 1–9 or 1–11) is completely inactive as is a peptide missing the Leu^{18} residue. These data indicate that the nine residue N-terminus of allatostatin V has no independent action on the corpora allata (*352*). This is interesting because this peptide shows a potential dibasic (Lys^9–Arg^{10}) cleavage site. The C-terminal fragments (allatostatin V 9–18, 10–18 or 11–18) give full responses at high concentrations, but they are less potent than the intact molecule; this again shows that the message is encoded at the C-terminus (*352*). Thus, the current idea is that the N-terminus is important for high affinity binding to the allatostatin receptor and, given the N-terminal differences in the various allatostatins, that each one may bind to a different receptor subtype (*354*).

For one allatostatin (IV or Dip-AST-5) analogues have been synthesized in which either single residues were substituted by replacement with Ala, to study the importance of side chains, or the native L-amino acid at each position was replaced by its D-amino acid counterpart (*157*). Whereas replacement of Tyr^4, Phe^6, Gly^7 or Leu^8 with L-Ala reduced the biological potency of the analogues quite dramatically, replacements of Asp^1, Arg^2 and Leu^3 were less effective and Ser^5 had almost no effect. These data are quite consistent with the fact that the C-terminal pentapeptide is characteristic for this peptide family and that the position of Ser (in Dip-AST-5) is the position which is quite variable in the allatostatins (see Table 6). Substitution with D-amino acids again resulted in significant loss of biological potency, particularly for the residues which form the C-terminal pentapeptide. Since replacement by D-amino acids will also distort the structure of the peptide by reversal of symmetry of either the backbone or the side chain, such studies are of aid in assessing which residues are likely to be necessary for receptor interaction. The data were interpreted from a conformational point of view in the following way: the N-terminal region is either charged or polar and may have an α-helical structure, whereas the C-terminal pentapeptide region is hydrophobic and may have a β-strand structure. Moreover, there is a strong suggestion that residues Phe^6, Gly^7 and Leu^8 form a type II β-turn. More precise information, however, can only be gathered when the allatostatin receptors have been isolated.

Polyclonal antibodies were raised in mice against allatostatin I (Dip-AST-7) coupled to bovine serum albumin. The presence of allatostatin in the corpora allata was shown by binding of these allatostatin antibodies to corpus cardiacum/corpus allatum tissue.

Specifically, immunocytochemistry identified allatostatin-positive axons which transverse the corpus cardiacum and branch extensively in the corpora allata (*444*). This result supports the hypothesis that the allatostatins are synthesized in neurosecretory cells of the brain and transported axonally to the corpora allata. Recent studies, therefore, attempted the isolation and purification of allatostatins from corpora allata instead of brains (*444*). The successful isolation of the same four allatostatins I to IV previously sequenced from the brain was reported after work-up of 6000 glands; identification was achieved by showing that the retention times were identical with those of the synthetic allatostatins in HPLC and by bioassays. No sequencing was reported. These results suggest the transport of peptidergic neurosecretory brain material to the corpora allata to inhibit the rate of juvenile hormone synthesis. Such a process is analogous to the release of hypothalamic peptidergic factors in vertebrates into the portal system and transport to the anterior pituitary.

Recently, a bioactive radioiodinated analogue of allatostatin I (Dip-AST-7) with a N-terminal azidosalicylamide group was synthesized. Such an analogue can be used for photoaffinity labelling (*62*). It was shown that membranes of corpora allata from virgin females of *D. punctata*, when incubated with this analogue and irradiated, contained two protein bands of 59 and 39 kDa after SDS gel electrophoresis which were specifically labelled; thus, these proteins are thought to be the putative receptor proteins for allatostatin. Very recently, an *in vitro* binding assay and a photoaffinity labelling assay were developed and the presence of receptors for allatostatins demonstrated in brain and corpora allata of *D. punctata* (*488*).

By isolation and sequencing methods not only were the seven allatostatins from the cockroach, *D. punctata*, determined, but also 2 resp. 4 allatostatins in the cockroaches, *P. americana* (*474*), and, *Blattella germanica* (*17*), as well as eight allatostatins (four Leu-, and four Met-callatostatins) in the blowfly, *Calliphora vomitoria* (*75, 76, 77*), six allatostatins in the cricket, *Gryllus bimaculatus* (*260, 261*), and one in the tobacco hornworm, *Manduca sexta* (*240*) (see Table 6).

The two allatostatins of *P. americana* are novel members of the allatostatin family, but molecular cloning led to the isolation of cDNA encoding for a total of 14 putative allatostatins (*vide infra*). Two of the four allatostatins of *B. germanica* are identical with isolated or cDNA-inferred allatostatins from *D. punctata* (see Table 6). Whereas the effective dose of *P. americana* allatostatins required to inhibit JH synthesis in this cockroach is similar to the dose required in *D. punctata* (*474*), the peptides from *B. germanica* are at least two orders of magnitude less effective in *B. germanica* (maximal inhibition at about 10^{-5} M) (*17*). The allatostatins of *C. vomitoria* are all unique members of the family, but despite having an inhibitory effect on JH synthesis in cockroaches, they do not affect the synthesis of JH bisepoxide, the endogenous JH of the blowfly itself. They are, however, potent inhibitors of gut motility in the blowfly (*75, 82*). There is also immunocytochemical evidence that immunopositive neurons from the abdominal ganglion project into certain areas of the hindgut, but there are no neural pathways from the brain to the corpus allatum (*75, 82*).

Two cricket allatostatins are novel members of the family. The effective concentration to inhibit JH synthesis in isolated corpora allata of crickets is somewhat higher when compared with the effect of the allatostatin of *D. punctata* in this species, but this can be explained by the different arrangement used for the assay procedure (*261*). The other four allatostatic neuropeptides of the cricket do not contain the highly conserved C-terminus found in all other allatostatins (*260*). These peptides

Diploptera punctata

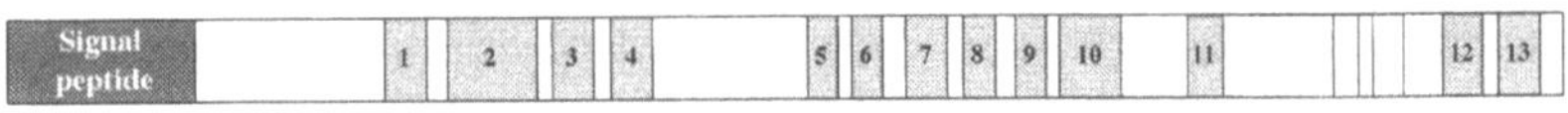

Fig. 5. Schematic diagram of the precursor of the allatostatins from *Diploptera punctata*. Structures of peptides Dip-AST-1-13 are given in Table 6. Modified after (*71*)

consistently have the C-terminal amino acid sequence of G-X-W-amide (X = G or S; see Table 6).

The primary structure of the allatostatin of *M. sexta* does not contain the family-characteristic pentapeptide. This molecule is very effective in inhibiting JH synthesis in the tobacco hornworm and shows cross-reactivity in another moth, *H. virescens*. The corpora cardiaca of adult females of the beetle, *Tenebrio molitor*, the grasshopper, *Melanoplus sanguinipes*, or the cockroach, *P. americana*, are not affected (*240*).

Recently, the sequence of a cDNA encoding the 370 amino acid long preproallatostatin polypeptide has been determined in *D. punctata* (*71*). The sequence deduced for this precursor confirms the identity of the seven previously isolated and sequenced allatostatins of this cockroach. Moreover, the existence of six new allatostatic peptides is predicted (see Table 6 and Fig. 5). Some of these predicted peptides contain the well-known pentapeptide motif Y-X-F-G-L amide, but in three (Dip-AST 10, 11, and 12) Tyr is substituted by Phe, and in Dip-AST-13 the C-terminal Leu is replaced by Ile. The polypeptide precursor also contains three acidic spacer regions (see Fig. 5) and in the third region sequences of two potential peptides with a C-terminal Ile occur. However, there is no indication that these peptides are amidated; since amidation is essential for allatostatic bioactivity, it is highly unlikely that these peptides belong to the allatostatin family.

Similar results have been obtained from a gene sequence of *P. americana* (see *443*). The allatostatin precursor is 379 amino acids long and shares 71% amino acid identity with *D. punctata*. The coding regions of the two allatostatin genes are remarkably similar in structure and organization.

The precursor of *P. americana* contains 14 potential allatostatins, including the two which have been isolated and sequenced (*474*), which are also separated by acidic spacer regions. Five putative peptides of *P. americana* are identical in structure with those of *D. punctata* (see Table 6).

Southern blot analyses indicated the presence of a single copy of the gene per haploid genome in both cockroaches. *In situ* hybridization of brains from native female *D. punctata* and *P. americana* with their respective allatostatin gene showed that the allatostatin mRNA is strongly expressed by two pairs of large medial cells in the pars intercerebralis of the protocerebrum and some weaker signals have been found in other structures like lateral cells, for example.

3.2.3. Prothoracicotropic Hormone, Bombyxin and Other Insulin-Related Neuropeptides

Since the studies of KOPEČ (*239*) which demonstrated that the brain of the larval gypsy moth, *Lymantria dispar*, released a factor that induced pupation, the pivotal role of the brain in the control of moulting and metamorphosis has been well established. This so-called "brain hormone" of KOPEČ is now generally referred to as prothoracicotropic hormone (PTTH) because it stimulates the paired prothoracic glands to synthesize and release ecdysone.

At the beginning of the research to purify PTTH, heads of the easily accessible silkworm, *Bombyx mori*, were used as the source for extraction and the heterologous moth species, *Samia cynthia ricini*, served as the bioassay animal. When pupae of *S. cynthia* were debrained shortly after pupation, adult development stopped. When these debrained pupae were implanted with brains of *B. mori* or injected with *B. mori* brain extracts, the *Samia* pupae restarted their adult development. The same was true when debrained dormant pupae of *B. mori* were injected with brain extracts of *B. mori* or received implanted *Bombyx* brains (*191*). It was thus assumed that the "PTTH" from *B. mori* was not species specific and, because of technical advantages, brainless pupae of *S. cynthia* were first used to assay "PTTH" during purification of *B. mori* heads/brains. When after years of purification efforts an apparently pure form of "PTTH" was obtained (*313*), it could be established that the material was not active on brainless pupae of *B. mori*, but only on *S. cynthia*. Since the crude extract was active in both systems, a re-examination of the bioassay potencies during various purification steps revealed that the brain extract from *B. mori* contained two types of molecules: one, with a molecular weight of about 5 kDa, was active only on debrained *S. cynthia* pupae, while the other of about 30 kDa, was active on brainless *B. mori* pupae but not on those from *S. cynthia* (*189*). The smaller molecule is now called bombyxin and the 30 kDa peptide is the genuine or true PTTH.

3.2.3.1. Prothoracicotropic Hormone

After heroic efforts a 16-step purification scheme was adopted for the isolation of PTTH from 5×10^5 (= 3.7 kg) *B. mori* heads (*211*) which yielded only 15 µg pure material. The N-terminus (amino acids 1 to 13) was sequenced from this material, but another batch of 3×10^6 heads had to be used for purification to get most of the information for the primary structure, including the dimeric state of the molecule (*210*). Peptide sequencing of the purified PTTH and its enzymatic fragments resulted in a monomeric peptide of at least 104 amino acid residues (position 41 was unclear), but also showed microheterogeneity at the amino-terminus (apparently truncation of 6 and 7 residues) and similar slight variations at the carboxy-terminus (*210*).

An antibody raised against a synthetic peptide comprising the amino acids 1 to 15 of the N-terminus of PTTH (*285*) was used for screening an expression cDNA library which was constructed from mRNA of larval brains of *B. mori* (*216*). The amino acid sequence deduced from the nucleotide sequence revealed a *B. mori* PTTH hormone consisting of 109 amino acids; thus the 104 amino acids previously found by direct sequencing and 5 additional residues (R-Y-N-N-N) at the carboxy-terminus (for structure, see Table 7). The previously unidentified residue 41 turned out to be N, which in conjunction with the presence of T at position 43, a typical motif (N-X-T) for asparagine N-glycosylation, suggest that a carbohydrate moiety is linked to the side chain of N^{41}. Therefore, it is very likely that PTTH is a glycoprotein, but the carbohydrate moiety is not yet known. The cDNA work also revealed that PTTH is first synthesized as a large precursor, the prepro-PTTH (see Fig. 6) consisting of 224 amino acids. The cDNA encodes for a signal peptide (29 amino acids) followed by a typical (K-R-K) processing site, then for two smaller peptides (21 amino acids = p2k and 57 amino acids = p6K) which are separated by and end with a proteolytic cleavage site (K-R and R-K-R) and whose functions are not known, followed by the PTTH subunit (109 amino acids). There are seven Cys residues present in the PTTH monomer and it is suggested that there exists one disulfide bridge between the monomers and three intrasubunit disulfide bonds to form the mature PTTH. When a portion of cDNA encoding the PTTH monomer was inserted into a plasmid vector and introduced into *Escherichia coli*, an active peptide that was indistinguishable from natural PTTH was expressed (*190, 216, 311*). This provided good evidence that the cloned cDNA indeed encodes PTTH of *B. mori*, that a dimer was apparently formed, and that glycosylation was not essential for biological activity. Recently, two allelic PTTH genes were cloned from a *B. mori* genomic

Table 7. *Amino acid sequence of prothoracicotropic hormone monomer of Bombyx mori as deduced from cDNA*

Code Name	Sequence	Reference
Bom-PTTH:	GNIQVENQAIPDPPCTCKYKKEIEDLGENSVPRFIETRNCNKTQQPTCR PPYICKESLYSITILKRRETKSQESLEIPNELKYRWVAESHPVSVACLCT RDYQLRYNNN N^{41}: Glycosylated?	*216*

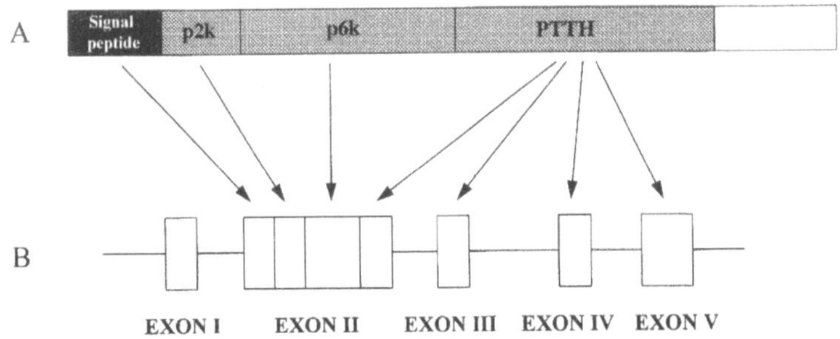

Fig. 6. Schematic diagram of the precursor of Bom-PTTH (prothoracicotropic hormone). p2K and p6K = peptides with mass of 2 or 6 kDa. B: Schematic representation of the Bom-prepro-PTTH-subunit gene. Modified after (*190*)

DNA library using the PTTH cDNA as a probe (3). The genes encode a precursor protein for the PTTH monomer and consist of five exons (see Fig. 6): Exon II contains regions encoding for the signal peptide, the p2k and p6k peptides and the first part of PTTH; the remaining part is encoded in exons III, IV and V. A single copy of the PTTH gene is found in the haploid genome of *B. mori* as evidenced by Southern hybridization experiments, indicating that the microheterogeneities found during peptide sequencing of PTTH have resulted either from post-translational processing or are some sort of artefacts produced during purification steps or products of denaturing conditions during storage.

The monoclonal antibody raised against the N-terminus (1–15) of PTTH was also used for immunocytochemical studies on brain-corpora cardiaca-corpora allata complexes of *B. mori*. Two pairs of dorsolateral neurosecretory cells in the brain were immunostained. Furthermore, immunoreactive-material was also detected in the axons of those neurosecretory cells which run to the corpora allata, a finding which indicates these structures as a possible release site (*285*). The same two pairs of dorsolateral neurosecretory cells in the brain contained mRNA for PTTH as shown by *in situ* hybridization with the PTTH cDNA probe (*216*). In the other moth species which is well known for its PTTH, the tobacco hornworm, *Manduca sexta*, very similar immunohistochemical results were achieved. A monoclonal antibody, very specific for *M. sexta* "big" PTTH (ca. 25.5 kDa), immunostained all four cells (two pairs) of the so-called L-NSC-III cells (neurosecretory cells located dorsally in each hemisphere of the protocerebrum); the axons of these cells traverse medially through the protocerebrum to the contralateral lobe and then pass posteriorly, via the nervi corporis cardiaci I and II, through the corpora cardiaca without branching to the corpora allata where the axon terminals form a typical neurohaemal release site (*330*). Previously, only one of the cells each of the L-NSC-II pair was recognized as producing PTTH when a revolutionary new bioassay was used to measure the amount of PTTH activity (monitored as ecdysone production by *in vitro* incubation of prothoracic glands), in individual somata (*5*), but with this method the corpora allata were already identified as the release site for PTTH (4). In *M. sexta*, PTTHs appear to exist as two different size groups (similar to the "real" PTTH and bombyxin in *B. mori*): a "big" PTTH with different variants of about 25.5 kDa and a "small" heterogenous PTTH of about 7 kDa; however, both forms directly stimulate prothoracicotropic glands of *M. sexta in vitro* (*24*). The "big" PTTH has been isolated from *M. sexta* brains using immunoaffinity chromatography (making use of the previously produced specific monoclonal antibody) and characterized by SDS-polyacrylamide gel electrophoresis, Western blot and partial sequencing (*291*). The mature PTTH is apparently a homodimer consisting of monomers of 16.5 kDa. Trypsin digestion of the monomer and isolation of these fragments on HPLC produced four peptides in sufficient quantities for sequencing. None of these sequences was similar to the PTTH sequence of *B. mori*. Furthermore, isolectric focusing performed on crude "big" PTTH from *M. sexta* yielded a pI of 5.2, while the PTTH of *B. mori* is a basic peptide (see *211*, *216*). Isolated *B. mori* PTTH also showed no biological activity in the *in vitro* prothoracic gland assay of *M. sexta*. Thus, *B. mori* PTTH, which is apparently present in *M. sexta*, as evidenced by 9% sequence similarity by independent PCR of genomic DNA and a L-NSC-III cDNA library (*153*), does not act as a prothoracicotropin in the tobacco hornworm. At the moment it is unclear what the function in *M. sexta* is, but because the *Bombyx*-like PTTH peptide and *Manduca* "big" PTTH are coexpressed in the L-NSC-III cells of *M. sexta* local release into the CNS (or into the haemolymph) and action as neuromodulators have been hypothesized (*153*).

3.2.3.2. Bombyxin

The function of the "small" PTTH is also not well understood. Although this molecule from *B. mori*, now called bombyxin, can induce adult development in brainless pupae of the saturniid moth, *Samia cynthia* and also stimulates *in vitro* the production of ecdysone in prothoracic glands of *S. cynthia*, adult development of a debrained pupae of *B. mori* is not induced (*189*, *313*). After years of work a 15-step purification scheme succeeded in isolating a pure form of bombyxin but with indications of more than one molecular form (*313*). Further studies revealed that at least five molecular forms (bombyxin I to V) could be isolated and more are still to be discovered (*204, 269, 311, 314*). When the N-terminal 19 amino acids were sequenced, it became clear that the bombyxins are homologous to insulin (*314*). After sequencing it was shown that the molecule is a heterodimer and that the A-chain consists of 20 amino acid residues with about 50% homology to insulin, whereas the B-chain, a mixture of at least four microheterogeneous peptides, consists of 28 or 26 residues with about 30% homology to insulin (Table 8) (*315*). Bombyxin contains 6 Cys residues which are distributed as in insulin; they form one intra-(Cys $A^6 \rightarrow$ Cys A^{11}) and two interchain (Cys $A^7 \rightarrow$ Cys B^{10} and Cys $A^{20} \rightarrow$ Cys B^{22}) disulfide bonds (*269*). Using interactive computer graphics and energy minimization techniques, and assuming homology with porcine insulin, a three-dimensional model of bombyxin II has been constructed (*204*). The model proposes two important characteristics: Bombyxin can assume an insulin-like tertiary structure, mostly because the important hydrophobic core residues are identical in bombyxin and insulin, and, when this globular structure is formed, the surface residues in bombyxin are quite different from those in insulin which accounts for the inability of bombyxin to bind anti-insulin antibodies or insulin receptors.

After the structure of some forms of bombyxin were known, studies focused on the chemical synthesis of bombyxins. This faced difficult problems to find the conditions which would induce the formation of the disulfide bonds. The first attempts gave only low yields (*270, 318*), but recently, by stepwise, regio-selective formation of the three disulfide bonds, yields of 50–60% have been achieved (*271*). The synthetic peptides had the same biological activity as the natural bombyxins.

Having established a sequence for bombyxin-II, oligonucleotide probes were designed and a genomic library screened, resulting in the isolation of a genomic DNA encoding for the precursor preprobombyxin (*197*). The organization of the preprobombyxin gene is thus to code for a signal peptide, B-chain followed by dibasic processing site, C-peptide followed by dibasic processing site and A-chain; this overall structure is exactly the same as that of the preproinsulin genes (*16*); however, in contrast to the insulin gene family, the bombyxin gene has no intron. It is predicted – by homology to insulin – that the mature bombyxin is generated in the following way: translation of the preprobombyxin, cleaving off of the signal peptide, generating of the disulfide bridges and, finally, cutting off of the C-peptide.

Using a synthetic oligonucleotide 51-mer of the antisense DNA for the bombyxin-II A-chain, a cDNA library constructed from larval brains of *B. mori* was screened and a clone with the complete coding region for preprobombyxin as given above was isolated (*2*). The *B. mori* genome

Table 8 *Primary structures of bombyxins-II and -IV*, Locusta migratoria *insulin-related peptide (Lom-IRP) and human insulin*

Code Name	Species	Sequence	Reference(s)
A-chain			
Bom-Bombyxin-II	*B mori*	GIVDECCLRPCSVDVLLSYC	*315, 318*
Bom-Bombyxin-IV	*B mori*	GVVDECCIQPCTLDVLATYC	*269*
Lom-IRP	*L migratoria*	GVFDECCRKSCSISELQTYCG	*165*
Human insulin		GIVEQCCTSICSLYQLENYCN	
B-chain			
Bom-Bombyxin-II	*B mori*	pQQPQAVHTYCGRHLARTLADLCWEAGVD	
Bom-Bombyxin-IV	*B mori*	pQEANVAHHYCGRHLANTLADLCWDTSVE	
Lom-IRP	*L migratoria*	SGAPQPVARYCGEKLSWALKLVCRGNYNTMF	
Human insulin		FVNQHLCGSHLVEALYLVCGERGFFYTPKT	

contains multiple copies of the bombyxin gene which contrasts strongly with vertebrate insulin genes (either a single or 2 copies found per haploid genome). Further studies on bombyxin genes revealed the presence of up to 30 gene copies (*190*, *283*). These have been classified into the A, B and C families according to their sequence similarities (*196*). In some cases it was shown that four genes form a cluster in which two genes belonging to different families (A or B) are closely apposed with an opposite transcriptional orientation (*215*). Whether this unique spatial organization has a functional significance for coordinate and differential expression of the bombyxin genes is not yet known. Together with the lack of introns, it shows that differences exist among other members of the insulin gene family of vertebrates and, thus, that there are greater evolutionary distances between these insulin genes.

Knowledge of the primary structures of the bombyxins was also a prerequisite for producing antibodies to study the localization of bombyxin at the cell level. A monoclonal antibody against a synthetic bombyxin fragment corresponding to the N-terminus 1–10 of the A-chain of bombyxin-I was used for immunohistochemical studies (*284*). Four pairs of large dorsomedial neurosecretory cells in the brain of *B. mori* were stained as well as their axons, which traversed to the contralateral lobe of the brain to enter the retrocerebral nerve. This nerve connects the brain with the corpus cardiacum (CC), but the stained axons passed through the CC to the corpora allata (CA) where they arborized and their terminals were preferentially located at the periphery of the CA. Thus, these neuroanatomical studies suggest that eight medial neurosecretory cells produce bombyxin, which is then transported to and released from the CA (*284*). The same cells also contain bombyxin mRNA as shown by *in situ* hybridization (*311*). So far bombyxin transcripts (as analyzed by Northern hybridization experiments) were only found in brain tissue of *B. mori*, but not in the suboesophageal ganglion, fat body, silk gland, Malpighian tubule, ovary or testis (*215*).

As to the putative function of bombyxin, the development of a radioimmunoassay (RIA) using monoclonal antibodies against natural bombyxin-II was very helpful (*283*, *390*). Interestingly, peak levels of ecdysteroids in the haemolymph before larval/larval and larval/pupal ecdysis were accompanied by increases in the titre of bombyxin-immunoreactive material suggesting that bombyxin has some, as yet not clearly defined, physiological role to play during development. Moreover, other experiments showed that bombyxin-immunoreactive material was released when feeding was used as a stimulus. Together with the observation that bombyxin immunoreactive material was released from the brain when glucose was injected into starved larvae, these results, comparable to post-prandial release of insulin by a high glucose titre, indicate a role for bombyxin in regulating carbohydrate metabolism. The levels of trehalose, the major blood sugar of *B. mori*, were indeed decreased by injection of bombyxin into the haemolymph of neck-ligated larvae; but this hypo-trehalosaemic effect was significant only 6 to 9 h after injection. Midgut trehalase, the enzyme that catalyzes trehalose to glucose, of larvae which were injected with bombyxin increased by 40% compared with controls. However, this effect was only present 6h after injection, but not after 3 h.

3.2.3.3. *Locusta* Insulin-Related Peptide

During the search for developmental neurohormones in *Locusta migratoria*, a peptide was isolated from the neurosecretory (storage) part of the corpora cardiaca whose primary structure, as determined by automated sequencing of V8 protease and trypsin fragments and by liquid secondary-ion mass spectrometry, suggested that it was a spacer peptide (*166*). The sequence was used to design oligonucleotide probes with which a cDNA library prepared from mRNA of the pars intercerebralis of the locust brain was screened and several clones encoding a polypeptide of 145 amino acids were isolated (*246*).

This polypeptide serves as a precursor for a molecule with strong sequence similarity to mammalian insulins; its overall organization is signal peptide/B-chain/C-peptide/A-chain. There are seven cysteines in the A- and two in the B-chain as in other insulins and the Cys residues have identical positions as in other insulins. Moreover, most of the hydrophobic core residues are in positions similar to those in other members of the insulin family.

Using a more vigorous extraction procedure than previously (either with 1 M acetic acid or with 75% ethanol containing 0.2 M HCl compared to previous conditions of extraction in deionized water at pH 5.5), crude extracts of neurohaemal parts of locust CC were prepurified on C18 Sep-Pak cartridges. Subsequently fractions of molecular mass between 1 and 15 kDa were obtained on a ProteinPak I-125 gel-permeation column and this material separated on C8-RP-HPLC with an acetonitrile/TFA gradient (*165*). A peptide, here called Lom-IRP (*Locusta migratoria* insulin-related peptide) was characterized, after cleaving the disulfide bridges, the A- and B-chains sequenced by Edman degradation and masses confirmed by plasma-desorption mass spectrometry (see Table 8). These results, in conjunction with the previous cDNA cloning studies (*246*), led to the conclusion that the 145-residue insulin precursor is posttranslationally processed into a 21-residue A chain, a 31-residue

B-chain and 50-residue C-peptide. Furthermore, in contrast to the situation in *B. mori* (see above), there is only a single insulin present in *L. migratoria* and about 5 pmol (thus 10 times more than in *B. mori*) can be extracted from a single corpus cardiacum. The successful cloning of the Lom-IRP gene (*242*) showed that the gene is present as a single copy per haploid genome and consists of three exons separated by two introns, which is remarkably similar to the organization of the gene in vertebrates, but differs dramatically from the situation in *B. mori* (about 30 intronless genes/haploid genome; see above). Northern blot analyses revealed the presence of insulin transcripts in other tissues and organs (fat body, epidermis, midgut, mature oocytes, embryos) than the brain (*241*). After the finding of two transcripts of Lom-IRP, namely T1 and T2 which differ in their 5′ untranslated region, it is proposed that these are produced by alternative usage of two different promoters (*242*). It is clear at least that T1 and T2 are differentially expressed in the various tissues analyzed so far in *L. migratoria*: T1 is the specific one that is massively expressed in the brain, while T2 is found at low levels in all other tissues (*242*).

In vitro production of ecdysone by the prothoracic glands of *L. migratoria* is not increased by natural Lom-IRP; thus no physiological function for this peptide is known. It is speculated that this molecule, as insulin in vertebrates, has a role to play in anabolic processes leading to storage of energy (*242*).

3.2.4. Eclosion Hormones

Insect growth and metamorphosis are characterized by a series of moults in the course of which a new cuticle is produced. A neuropeptide that is secreted by neurosecretory cells in the brain and stored in the neurohaemal corpora cardiaca-corpora allata complex causes the shedding of the old cuticle at ecdysis and is therefore called eclosion hormone (*386*, *459*). The hormone controls the ecdysis behavior not only in adult eclosion, but also in embryonic, larval and pupal ecdyses (*460*). Although its cellular targets and actions are diverse, not only triggering the aforementioned behavior, but also causing cuticle plasticization during the moult and even initiating programmed degeneration of certain intersegmental muscles which are not needed by the imago, the primary target of this peptide appears to be the central nervous tissue (*425*).

The physiology and biochemistry of eclosion hormone has been studied mainly in two lepidopteran moth species, the tobacco hornworm *Manduca sexta* and the silkworm *Bombyx mori*. Eclosion hormone was first isolated from pharate adult heads of *B. mori* by a complex purification scheme and the sequence of the 13 N-terminal amino acid residues was determined (*312*). Later it was found that the N-terminus of eclosion hormone is heterologous (*319*). Purification of eclosion hormone to homogeneity from *B. mori* was achieved from 777 000 pharate adult heads (12 kg fresh weight!); this resulted in isolation of four molecular species of eclosion hormone which were called EH I–IV in the order of elution from reversed-phase HPLC and are here called Bom-EH I–IV (*235*). Although aliquots of each EH were subjected to automated Edman degradation, the amounts were too small to derive complete sequences. Therefore, the whole sequences of these eclosion hormones were constructed by combining sequence data. It appeared that two elosion hormones had 61 amino acid residues, whereas the other two showed a truncation of the two N-terminal residues Ser-Pro (see Table 9).

At the same time two other research groups had isolated eclosion hormone from *Manduca sexta*. Whereas SCHOOLEY's group used 10 000 trimmed heads of pharate adults (*213*), TRUMAN's group dissected brain neurohaemal organs from the heads of over 17 000 pharate adults for extraction (*268*). Fractions from each purification step were injected into pharate adult *Heliothis virescens* moths 7 h before normal eclosion should occur. When eclosion took place within 3 h of injection, this fraction was judged as giving a positive response (*213*). Using different purification schemes both groups detected the same primary structure, a 62-mer peptide, determined by sequence analyses of the intact peptide and/or fragment peptides generated by various proteases or cyanogenbromide cleavage (Table 9). MARTI *et al.* (*268*) found that 20% of their preparation contained a peptide which lacked the N-terminal dipeptide Asn-Pro. Both studies thus confirmed a 62-amino acid peptide for *M. sexta*, whose C-terminus is a free acid and has an extra Leu residue which was not detected in *B. mori*. However, subsequent studies on *B. mori* were successful in cloning the eclosion hormone gene; its nucleotide sequence indicated a 62-mer containing a Leu at the C-terminus (*453*). When the eclosion hormone-encoding gene of *M. sexta* was isolated, it became clear that there is only one gene and eclosion hormone is the only product from the precursor molecule (*183*). The gene contains 7.8 kilobases and consists of three exons. Whereas exon I is non-translated, exon II contains a signal peptide (26-mer) and the four N-terminal amino acids of eclosion hormone and exon III encodes the remainder of the peptide. Experiments using *in situ* hybridization showed expression of the eclosion hormone gene in two pairs of ventromedial neurosecretory cells of the brain of both larvae and developing adults only (*183*). Using a monoclonal antibody against a synthetic C-terminal fragment of *B. mori* eclosion hormone (EH

Table 9. *Primary structures of eclosion hormones*

Code Name	Species	Sequence	Reference(s)
Bom-EH-IV	*B. mori*	SPAIASSYDAMEICIENCAQCKKMFGPWFEGSLCAESCIKARGKDIPECESFASISPFLNKL	*453, 236*
Bom-EH-I	*B. mori*	3–61	
Bom-EH-II	*B. mori*	1–61	
Bom-EH-III	*B. mori*	3–62	
Mas-EH	*M. sexta*	NPAIATGYDPMEICIENCAQCKKMLGAWFEGPLCAESCIKFKGKLIPECEDFASIAPFLNKL	*213, 268*

49–61), immunohistochemistry revealed also two pairs of median neurosecretory cells of the brain in this species which produce eclosion hormone (*234*). The results were confirmed when the cDNA encoding *B. mori* eclosion hormone was isolated and sequenced (*207*). The pre-eclosion hormone molecule contains a 26-mer signal peptide and the 62-mer eclosion hormone; furthermore, *in situ* hybridization showed expression of the eclosion hormone gene in two pairs of neurosecretory cells of the brain of fifth instar larvae. Of interest is a comparison of the data on *B. mori* and *M. sexta*:

1. Primary sequences of eclosion hormones differ by 12 residues, thus 80% sequence homology;
2. DNA sequence encoding eclosion hormone again shows about 80% homology;
3. in contrast, DNA sequences encoding the signal peptide (26-mer) and the non-translated region have less than 50% homology (*207*).

The gene encoding *B. mori* eclosion hormone (1–62) was chemically synthesized, inserted into a secretion vector and expressed in *Escherichia coli*, where it produced biologically active eclosion hormone (*237*). Recent studies on this recombinant eclosion hormone (*237*) and on native eclosion hormone from *M. sexta* (*209*) assigned the location of three disulfide bonds between Cys^{14}–Cys^{38}, Cys^{18}–Cys^{34}, and Cys^{21}–Cys^{49}. These results are consistent with the fact that, although Bom-EH and Mas-EH differ by 12 residues, the six cysteine residues and the residues before and after them are conserved in the two species (see Table 9). Additionally, biological activity was abolished by reductive alkylation, thus disulfide bridges are necessary for activity. Since both eclosion hormones are active in the heterologous species, arrangement of disulfide bonds was anticipated to be identical in both molecules, because it is apparently essential for correct receptor-binding.

Lastly, with the help of the recombinant Bom-EH it was shown that the molecular species I and II (3–61) (1–61) are very likely produced artefactually from EH III (3–62) and IV (1–62), possibly by digestion by a carboxypeptidase A-like enzyme present in the extract during purification (*236*).

3.2.5. *Peptides Affecting Gonad Activity*

Reproduction in insects is a very precisely regulated process in which hormones are involved (*87*). The key players of hormonal regulation are the true epithelial hormones, the ecdysteroids and juvenile hormones. In the adult stage these hormones do not interact with moulting, but control the synthesis and uptake of yolk protein (vitellogenesis), the maturation of ovaries, and the development of eggs (oogenesis). Although the prothoracic glands degenerate at metamorphosis, other tissues in the adult insect (mostly the gonads, but fat body and integumental tissue as well) are the main ecdysteroid producers. Endocrine regulation of reproduction is very complex because different species have developed different physiological mechanics. In most insects the synthesis of vitellogenin is stimulated by JH, but it may not be the only stimulator. In many dipterans a pulse of ecdysteroids is needed to trigger vitellogenesis. Most Diptera have to take a proteinaceous meal before oogenesis as well; the amino acids of the ingested proteins are the precursors for the vitellogenin. However, also the stretching of the abdomen by the blood meal in the blood-sucking bug, *Rhodnius prolixus*, provides the physiological stimulus which initiates endocrine events.

In locusts, for example, growth of the oocytes is synchronized. Thus gonotrophic cycles occur which follow immediately upon each other; thus locusts produce and lay their eggs in batches. Both vitellogenin synthesis and the uptake of yolk proteins by the oocytes are stimulated by JH. However, new data have shown that peptide hormones play a role as well.

3.2.5.1. Ovary Maturating Peptide and Neuroparsin of *Locusta migratoria*

In the migratory locust it was shown that a factor residing in the nervous (neurosecretory) part of the corpus cardiacum leads to premature oocyte development when injected into young adult females (*59*). For purification, 2000 nervous parts of corpora cardiaca were extracted with 70% methanol and isolated via a Pharmacia Mono Q anion-exchanger; the active material was desalted and further purified by C-8 RP-HPLC (*142*, *144*): Sequence determination was achieved by a combination of Edman degradation of the N-terminal intact peptide and various fragments produced by 2-iodobenzoic acid or tryptic digestion and quadrupole electrospray mass spectrometric measurements. The peptide, code-named Lom-OMP (*Locusta migratoria* ovary maturating parsin; Table 10), consists of 65 amino acids, does not contain any cystein residues (thus there is no possibility to form dimers via disulfide bridges), but has a polyalanine sequence (8 Ala residues at positions 43 to 50). The existence of two isopeptides, due to a point mutation at position 26 (Ala or Ser), was noted by sequencing and mass spectrometry. A polyclonal antibody was raised and, by immunocytochemical staining, 250 acidophilic cells in the pars intercerebralis-corpora cardiaca gave a positive response. Injection

Table 10 *Primary structures of ovary maturating peptide (OMP) and neuroparsin of* Locusta migratoria

Code Name	Species	Sequence	Reference(s)
Lom-OMP	*L migratoria*	YYEAPPDGRHLLLQPAPAAPAVAPA(S/A)PASWPHQQRRQALDEFAAAAAAAD AQFQDEEEDGGRRV	*144*
Lom-neuroparsin	*L migratoria*	NPISRSCEGANCVVDLTRCEYGDVTDFFGRKVCAKGPGDKCGGPYELHGKCGVG MDCRCGLCSGCSLHNLQCFFFEGGLPSSC	*142, 167*
γ		1–83	
β		3–83 (= Neuroparsin B)	
α		6–83 (= Neuroparsin A)	

of the immune serum into young adult females blocks the rapid growth of oocytes. Although JH produced in the corpora allata stimulates the synthesis of vitellogenin in normal fat body, implantation of supplementary corpora allata does not induce normal oocyte growth when immunoneutralization had taken place prior to this treatment. Other preliminary experiments showed that Lom-OMP apparently does not stimulate the incorporation of vitellogenin into the oocytes, but rather, like JH and very likely synergistically, induces the expression of the vitellogenin genes.

The "counterpart" of Lom-OMP is a large peptide which causes an antigonadotropic effect (but also stimulates fluid reabsorption and elevates trehalose and lipids in the haemolymph) which is called Lom-neuroparsin (*100*, *141*, *289*). Its name was coined because it was isolated from the neurosecretory part of the locust's corpus cardiacum and is produced by the A 1 type of the protocerebral median neurosecretory cells (*141*). The isolation procedure for this compound was as described for Lom-OMP; during the anion-exchange run two fractions, called neuroparsins A and B, were found which were further purified and characterized (*142*, *143*). Whereas it was assumed initially that the neuroparsins were dimers containing 12 Cys residues and being microheterogenous at the N-terminus, it appeared later that they are monomers containing 6 intramolecular disulfide bridges (*167*). The latter authors fractionated the crude extract of nervous lobes of corpora cardiaca on C-18 RP-HPLC and analysed the purified peaks by liquid secondary-ion and electrospray mass spectrometry. According to HIETTER *et al.* (*167*), three main peptides could be found (see Table 10): the longest one consists of 83 amino acids (compound γ), whereas compounds β and α are two and five amino acids shorter from the N-terminus, respectively. In the terminology of GIRARDIE *et al.* (*143*) the 83- and 81-mers are neuroparsin A and the 78-mer is neuroparsin B. It is not yet known whether compounds α and β result from proteolytic cleavage by aminopeptidases from compound γ or, alternatively, whether all three peptides are synthesized in the corpus cardiacum and have each a different function.

When a cDNA library, prepared from mRNA of *L. migratoria* brains, was screened with appropriate probes, a cDNA encoding a precursor protein (107 amino acids) consisting of a signal peptide (22 amino acids), two processing sites (3 amino acids) and neuroparsin A (83 amino acids) was isolated and sequenced. The deduced amino acid sequence shows complete identity between residues 25 and 107 with the peptide sequence for neuroparsin A determined previously by Edman degradation (*143*).

3.2.5.2. Oostatic Hormones of Diptera

Substances that inhibit egg development and are therefore called antigonadotropins or oostatic hormones have been found in a number of insect species (*27*). Most of the research has been done on flies, mosquitoes and the blood-sucking bug, *Rhodnius prolixus*. Until very recently the only very limited information on the chemical nature of these factors has been reported for the latter two groups (*25*, *259*). New data have been compiled for the mosquito *Aedes aegypti* culminating in the elucidation of the primary structure (*27*). Therefore, the present status of research of oostatic hormones is described in some detail only for this species.

Egg development in anautogenous (insects which need a blood/protein meal to produce eggs) mosquitoes like *A. aegypti* depends on digestion of ingested blood as well as on the release of an until now not well-characterized egg development neurosecretory hormone (EDNH) also called ovarian ecdysteroidogenic hormone (OEH) (*152*, *249*, *274*). This EDNH is apparently produced in medial neurosecretory cells of the brain and stored in the corpus cardiacum (*250*). The peak of release of this neuropeptide is brief, its action is to stimulate the ovarian follicular cells to secrete ecdysone. In contrast, the ovary of the mosquito controls its own growth and development, because an oostatic hormone fraction highly purified from ovaries was able to inhibit egg development and the biosynthesis of vitellogenin, the main yolk protein precursor in insects (*25*). Recently, one peptide hormone has been purified from 30000 ovaries of female *A. aegypti* by using low and high pressure liquid chromatography (*28*), Fourier transform mass spectrometry was critical for the proper characterization of the minute amounts available (*29*). The last step of isolation was on a RP-HPLC-C18 column, where a single peak with bioactivity was eluted. Amino acid analysis combined with tandem quadrupole mass spectrometry with ion cyclotron resonance revealed the primary structure of an unblocked decapeptide with the rather unusual C-terminal sequence of 6 proline residues (see Table 11). Surprisingly, a computer search found significant structural homology to mammalian, plant and several viral proteins that are either synthesized by double

Table 11. *Primary structures of oostatic peptides (= trypsin modulating oostatic factors, TMOFs)*

Code Name	Species	Sequence	Reference(s)
Aea-TMOF	*Aedes aegypti*	YDPAPPPPPP	*28*, *29*
Neb-TMOF	*Neobellieria bullata*	NPTNLH	*37*
Neb-colloostatin	*N. bullata*	SIVPLGLPVPIGPIVVGPR	*38*

stranded DNA viruses (Epstein Barr virus; *Herpes simplex* virus) or single stranded RNA viruses (Abelson murine leukemia viruses, and HIV-2, for example). The peptide directly modulates trypsin biosynthesis in the gut and indirectly regulates egg development and is therefore now called Trypsin Modulatory Oostatic Factor or Aea-TMOF.

Synthesis occurs in the follicular epithelium of the ovary and active secretion by the ovary, as shown by immunocytochemistry and *in vitro* incubations of ovaries (*29*) and by RIA and ELISA (*31*), is reported 24 to 48 hours after the blood meal. It is proposed that it is then bound to a receptor on the midgut epithelium cell where it acts, possibly via a repressor, by inhibiting trypsin synthesis. Trypsin biosynthesis, which was initiated during the first 24 hours after the blood meal, is then stopped, blood digestion cannot progress after some time, no more amino acids are transported to the fat body, and therefore no more vitellogenin can be made which will lead in the end to an arrest of egg development (*30*).

TMOF is rapidly hydrolyzed in intact mosquito having a half-life of about 1.6 h (*29*). The source of hormonal inactivation is suggested to reside in the thorax: when ligated abdomens were injected with synthetic Aea-TMOF, lower concentrations than injected into intact animals inhibited 90% of the trypsin-like enzyme biosynthesis (*29*). Some structure-activity studies using the test on ligated abdomens, where inactivation was not crucial, revealed the following: when the left-handed helix at the C-terminus was abolished by removing four or two of the Pro residues, the ED_{50} values were increased, thus the molecule was less active. Similarly, changing Tyr at position 1 with Asp at position 2 also increased the ED_{50}; thus, N- and C-termini are apparently important for biological activity (*29*).

Since natural or synthetic TMOFs are not species specific (*26, 28*), it was proposed that sequence-related TMOFs control trypsin biosynthesis in other insect species as well (*29*).

A peptide that inhibits trypsin-like synthesis by the midgut of liver-fed female flies of the species *Neobellieria* (*Sarcophaga*) *bullata* was purified from 10 000 ovaries of late vitellogenic state by using five HPLC steps and identified as a hexapeptide called Neb-TMOF (see Table 11) (*37*). Despite the difference in sequence to Aea-TMOF cross-reactivity was noted: Neb-TMOF is 6-fold more active in the fly and Aea-TMOF is 5-fold more active in the mosquito. This may probably be attributed to some properties of the physico-chemical structure; in both molecules an aromatic amino acid (Tyr in Aea-TMOF and His in Neb-TMOF) sticks out of the molecular axis (*37*). In contrast, the six C-terminal Pro residues in Aea-TMOF were predicted by computer modelling and NMR (*28, 60*) to form an α-helix, which is absent in Neb-TMOF.

Recently, a second peptide which displays oostatic activity has been purified from whole abdomens of adult *N. bullata* (*38*). This 19-mer peptide is called Neb-colloostatin because it has a striking structural resemblance to a particular part of the sequence of preprocollagen of *Drosophila* (Table 11). Its effect is different from that of Neb-TMOF; trypsin biosynthesis is not inhibited. It is more likely that this peptide prevents yolk deposition in the penultimate oocytes.

3.2.6. Diapause Hormones

Diapause is defined as a spontaneous developmental arrest which occurs to adapt to changing environmental conditions. It is not restricted to a specific developmental stage in the life cycle; thus insects may enter diapause either in the embryonic, larval, pupal or imaginal stage (*67*, *485*). Diapause is induced by a variety of environmental cues (e.g. temperature, humidity, photoperiod, diet) which are transduced by the neuroendocrine system to result in the various and complex adaptive responses of a physiological, biochemical and endocrinological nature.

In this chapter we are only concerned with embryonic diapause (egg diapause) occurring in the silkworm, *Bombyx mori*. Depending on the strain and environmental conditions, the number of annual generations varies in this species. Generalizing, one can say that a univoltine strain produces a single generation per year and all eggs enter diapause. Bivoltine or quadrivoltine strains which produce two or four generations annually lay eggs that undergo diapause when the female moth is subjected to high temperature (25 °C) and long-day photoperiods (16 h light: 8 h dark) during the embryonic stage; low temperature (15 °C) and short-day photoperiods (12 h light: 12 h dark), however, produce non-diapause eggs (*484*).

Early experiments had already demonstrated that the suboesophageal ganglion is involved in the nature of diapause. Neurosecretory cells of the suboesophageal ganglion secrete a substance which was called diapause hormone and promotes diapause. Attempts to isolate the active compound from huge quantities of dried heads of male adults using several conventional column chromatographic steps led to a highly purified sample of peptidic nature (*192*, *193*, *194*). Recently, the isolation proper and sequence determination was successfully executed (*185*). 55 000 complexes of the suboesophageal ganglion and the first thoracic ganglion were dissected from day 1 old silkworm pupae, homogenized in ethanol and the pellet, after centrifugation, sequentially washed with ethanol, methanol/dichloromethane (1:1), 80% ethanol and 50% 2-propanol. This procedure did not extract the diapause hormone which was

Table 12. *Primary structures of diapause hormones causing egg diapause in* Bombyx mori

Code Name	Species	Sequence	Reference(s)
Bom-DH	*B. mori*	TDMKDESDRGAHSERGALCFGPRLamide	*185*
Bom-DH-I	*B. mori*	TDMKDESDRGAHSERGALWFGPRLamide	*392*

finally extracted by hot water. The aqueous extract was fractionated by reversed-phase HPLC into several broad peaks one of which contained biological activity. The pooled material was applied again to the RP-HPLC including the ion-pairing reagent trifluoroacetic acid (0.5%) into the organic solvent (2-propanol). Diapause hormone bioactivity was mainly found in one sharp peak. The yield of pure peptide was calculated to be less than 500 ng from the 55000 ganglia. Gas-phase sequencing revealed a 24 mer peptide (see Table 12) with one ambiguity at position 19 which was assumed to be Cys. The sequence from Arg^{15} to Gly^{21} was confirmed by sequencing a fragment after digestion with endoproteinase Glu-C. Sequencing showed that the N-terminus was not blocked, but it was not clear whether the C-terminus was amidated or not. Synthesis of both alternatives (free and amidated Leu at the C-terminus) solved this uncertainty: the peptide with the Leu as free acid had no biological activity when injected at the very high dose of 1 µg/pupa, whereas the amidated form eluted exactly at the same retention time as the native diapause hormone and was comparably biologically active.

The information of the primary structure of diapause hormone (DH) made it possible to isolate cDNA clones coding for DH. A cDNA library constructed from mRNA of suboesophageal ganglia was screened using oligonucleotide probes. Sequence data of the cloned cDNA encoding DH indicated the possibility that a second DH was produced in *B. mori*. Therefore, 110000 suboesophageal ganglia (plus first thoracic ganglia) were excised from day 2 to 3 pupae of a bivoltine race and the material purified as previously described (*185*) yielding a single peptidic compound which was characterized as having a Trp at position 19 (as predicted from the cDNA) instead of Cys which was previously found at this position (*392*). The synthetic Bom-DH (Trp^{19}) in its amidated form had the same retention time as the native molecule; it also has similar dose-response relationship in the biological assay as its Cys^{19} analogue. Thus, it was concluded that this new peptide containing Trp^{19} is a novel DH molecule. It is interesting to note that no cDNA clone was isolated which bears the codon for Cys^{19}.

Interestingly, four out of five amino acids at the C-terminus are identical to those of the pheromone biosynthesis-activating neuropeptide from *Helicoverpa zea* and *Bombyx mori* (see Table 5), and of the locustamyotropins and locusta- and leucopyrokinins (see Table 13).

On a molecular level this was clearly shown when cDNA encoding DH was cloned and sequenced (*217*, *393*): the cDNA encodes a polyprotein precursor from which DH is processed post-translationally together with PBAN and three other, shorter neuropeptides; all of these peptides share the common pentapeptide C-terminal sequence F-X-P-R/K-L-amide (see also Sect. 3.2.1.). Using these molecular tools it was shown that the transcript of the diapause hormone polyprotein precursor was found in the suboesophageal ganglia of pupae and pharate adults, but brains, thoracic and abdominal ganglia had no positive reaction (*391*). *In situ* hybridization revealed 12 cells in the suboesophageal ganglia aggregated in three clusters.

3.3. Peptides Modifying Spontaneous Muscle Contractions: Myotropic Peptides

The majority of insect neuropeptides fully characterized thus far have the property of regulating the contractile activity of visceral and/or skeletal muscles. The first insect neuropeptide which was isolated and whose primary structure elucidated was proctolin (*34*, *439*). The heroic efforts of isolation (11 steps were used starting with 125 kg of whole cockroaches) at a time when only quite insensitive techniques were available have been reported many times (*418*, *177*) (Table 13).

3.3.1. Proctolin and Cardiostimulatory Peptides

Proctolin was present in extracts of the hindgut of the American cockroach *Periplaneta americana* and caused a slow graded contraction of the longitudinal muscles of the hindgut. Since then proctolin has been found, by using RIA, immunocytochemistry and/or HPLC, to be widely distributed among insects and other arthropods (*333*, *426*). At first, proctolin was proposed to be a visceral muscle neurotransmitter (*34*). The pentapeptide, however, not only stimulated visceral muscles but also skeletal muscles (see *334*). Moreover, most of the effects of proctolin can be attributed more to a neuromodulatory role than to the classical effect of a neurotransmitter or of a neurohormone; in an insect neuromuscular junction, proctolin acts as a cotransmitter with a second, conventional (possibly glutamate) neurotransmitter (*334*). Whereas the hindgut assay in

References, pp 97–128

Table 13 Primary structures of various myotropic peptides isolated from insects

Peptide Name or Family Name	Code Name	Species	Sequence	Reference(s)
I Proctolin	Pea-proctolin	*P americana*	RYLPT	*439*
II Cardioacceleratory neuropeptides				
Periplanetin CC-1 (= hypertrehalosaemic peptide I, see Table 1)	Pea-CAH-I	*P americana*	pQVNFSPNWamide	*14, 394, 478* *430*
Periplanetin CC-2 (= hypertrehalosaemic peptide II, see Table 1)	Pea-CAH-II	*P americana*	pQLTFTPNWamide	*394, 478, 430*
Corazonin	Pea-corazonin	*P americana*	pQTFQYSRGWTNamide	*464*
His^7-corazonin	Scg-corazonin	*S gregaria*	pQTFQYSHGWTNamide	*466*
Crustacean cardioactive peptide (= CAP_{2a})	Cama-CCAP	*L migratoria*	PFCNAFTGCamide	*438*
		M sexta		*45, 255*
		T molitor		*102*
		Spodoptera eridania		*102*
Manduca cardioacceleratory peptide (= CAP_{2b})	Mas-CAP	*M sexta*	pQLYAFPAVamide	*184*
III Myokinins				
a) Leucokinins				
Leucokinin I, LK-I	Lem-M-I	*L maderae*	DPAFNSWGamide	*170*
Leucokinin II, LK-II	Lem-M-II	*L maderae*	DPGFSSWGamide	*170*
Leucokinin III, LK-III	Lem-M-III	*L maderae*	DQGFNSWGamide	*171*
Leucokinin IV, LK-IV	Lem-M-IV	*L maderae*	DASFHSWGamide	*171*
Leucokinin V, LK-V	Lem-M-V	*L maderae*	GSGFSSWGamide	*174*
Leucokinin VI, LK-VI	Lem-M-VI	*L maderae*	pQSSFHSWGamide	*174*
Leucokinin VII, LK-VII	Lem-M-VII	*L maderae*	DPAFSSWGamide	*175*
Leucokinin VIII, LK-VIII	Lem-M-VIII	*L maderae*	GADFYSWGamide	*175*
b) Achetakinins				
Achetakinin I, AK-I	Acd-K-I	*A domesticus*	SGADFYPWGamide	*177*
Achetakinin II, AK-II	Acd-K-II	*A domesticus*	AYFSPWGamide	*177*
Achetakinin III, AK-III	Acd-K-III	*A domesticus*	ALPFSSWGamide	*177*
Achetakinin IV, AK-IV	Acd-K-IV	*A domesticus*	NFKFNPWGamide	*177*
Achetakinin V, AK-V	Acd-K-V	*A domesticus*	AFHSWGamide	*177*

c) Locustakinin	Lom-K	*L migratoria*	AFSSWGamide	*413*
d) Aedeskinins				
Aedes leukokinin 1	Aea-K-I	*A aegypti*	NSKYVSKQKFYSWGamide	*467*
Aedes leukokinin 2	Aea-K-II	*A aegypti*	NPFHAWGamide	*467*
Aedes leukokinin 3	Aea-K-III	*A aegypti*	NNPNVFYPWGamide	*467*
e) Culekinins				
Culekinin I, CDP-I	Cus-CDP-I	*Culex salinarius*	NPFHSWGamide	*158*
Culekinin II, CDP-II	Cus-CDP-II	*C salinarius*	NNANVFYPWGamide	*51*
Culekinin III, CDP-III	Cus-CDP-III	*C salinarius*	WKYVSKQFFSWGamide	*51*
f) Helicokinins				
Helicokinin I	Hez-K-I	*H zea*	YFSPWGamide	*22*
Helicokinin II	Hez-K-II	*H zea*	VRFSPWGamide	*22*
Helicokinin III	Hez-K-III	*H zea*	KVKFSAWGamide	*22*
IV Sulfakinins				
a) Leucosulfakinins				
LSK	Lem-SK-I	*L maderae*	EQFEDY(SO_3H)GHMRFamide	*298*
LSK-II	Lem-SK-II	*L maderae*	pQSDDY(SO_3H)GHMRFamide	*295*
		P americana	pQSDDYGHMRFamide	*465*
b) Locustasulfakinin	Lom-SK	*L migratoria*	pQLASDDY(SO_3H)GHMRFamide	*405*
c) Perisulfakinin	Pea-SK	*P americana*	EQFDDY(SO_3H)GHMRFamide	*465*
d) Dipteran sulfakinins				
Neosulfakinin I	Neb-SK-I	*N bullata*	FDDY(SO_3H)GHMRFamide	*98*
(= Drosulfakinin I)	(Drm-SK-I)	*D melanogaster*		*321, 322,324*
(= Callisulfakinin I)	(Cav-SK-I)	*C vomitoria*		*83*
(= Lucisulfakinin I)	(Luc-SK-I)	*L cuprina*		*83*
Neosulfakinin II	Neb-SK-II	*N bullata*	??EEQFDDY(SO_3H)GHMRFamide	*98*
(= Callisulfakinin II)	(Cav-SK-II)	*C vomitoria*	GGEEQFDDY(SO_3H)GHMRFamide	*83*
(= Lucisulfakinin II)	(Luc-SK-II)	*L cuprina*	GGEEQFDDY(?)GHMRFamide	*83*
		C vomitoria		

Table 13 (*continued*)

Peptide Name or Family Name	Code Name	Species	Sequence	Reference(s)
Drosulfakinin II	Drm-SK-II	*D. melanogaster*	GGDDQFDDY(?)GHMRFamide	*324*
V. Myotropins/pyrokinins				
a) Leucopyrokinin, LPK	Lem-PK	*L. maderae*	pQTSFTPRLamide	*172*
b) Locustapyrokinin I	Lom-PK-I	*L. migratoria*	pQDSGDGWPQQPFVPRLamide	*407*
Locustapyrokinin II	Lom-PK-II	*L. migratoria*	pQSVPTFTPRLamide	*411*
c) Locustamyotropin I	Lom-MT-I	*L. migratoria*	GAVPAAQFSPRLamide	*410*
Locustamyotropin II	Lom-MT-II	*L. migratoria*	EGDFTPRLamide	*406*
Locustamyotropin III	Lom-MT-III	*L. migratoria*	RQQPFVPRLamide	*409*
Locustamyotropin IV	Lom-MT-IV	*L. migratoria*	RLHQNGMPFSPRLamide	*409*
d) Helicomyotropin I	Hez-MT-I	*H. zea*	MEFTPRLamide	*64*
Helicomyotropin II	Hez-MT-II	*H. zea*	TMNFSPRLamide	*64*
e) Bommyotropin I (= α-suboesophageal neuropeptide)	Bom-MT-I (Bom-α-SGNP)	*B. mori*	IIFTPKLamide	*393, 217*
Bommyotropin II (= β-suboesophageal neuropeptide)	Bom-MT-II (Bom-β-SGNP)	*B. mori*	SVAKPQTHESLEFIPRLamide	*393, 217*
Putative bommyotropin III (= γ-suboesophageal neuropeptide)	Bom-MT-III	*B. mori*	TMSFSPRLamide	*393, 217*
VI. Tachykinins				
a) Locustatachykinin I	Lom-TK-I	*L. migratoria*	GPSGFYGVRamide	*404*
Locustatachykinin II	Lom-TK-II	*L. migratoria*	APLSGFYGVRamide	*404*
Locustatachykinin III	Lom-TK-III	*L. migratoria*	APQAGFYGVRamide	*403*
Locustatachykinin IV	Lom-TK-IV	*L. migratoria*	APSLGFHGVRamide	*403*
Locustatachykinin V	Lom-TK-V	*L. migratoria*	?PSWFYGVRamide	*415*
b) Callitachykinin I	Cav-TK-I	*C. vomitoria*	APTAFYGVRamide	*263*
Callitachykinin II	Cav-TK-II	*C. vomitoria*	GLGNNAFVGVRamide	*263*
c) Culetachykinin I	Cus-TK-I	*C. salinarius*	APSGFMGMRamide	*51*
Culetachykinin II	Cus-TK-II	*C. salinarius*	APWGFTGMRamide	*51*

References, pp. 97–128

VII Accessory glands- and midgut-myotropins				
a) Male accessory glands myotropin I	Lom-AG-MT-I	*L migratoria*	GFKNVALSTARGFamide	*342*
Male accessory glands myotropin II	Lom-AG-MT-II	*L migratoria*	AHRFAAEDFGALDTAamide	*341*
Female accessory glands myotropin	Mud-AG-MT	*Musca domestica*	LLNALPLDALSSLTGamide	*473*
b) Midgut myotropin I	Mas-MG-MT-I	*M sexta*	AGPYTamide	*487*
Midgut myotropin II	Mas-MG-MT-II	*M sexta*	DIPPRamide	*486*
c) Oviductal motility stimulating head peptide	Led-OVM	*L decemlineata*	IAYKPEamide	*434*
VIII Periviscerokinin	Pea-PVK	*P americana*	GASGLIPVMRNamide	*356*
IX Myoinhibitory peptides and other FMRFamide related peptides (FaRPs)				
a) Locustamyoinhibitory peptide	Lom-MIP	*L migratoria*	AWQDLNAGWamide	*408*
Manducamyoinhibitory peptide I	Mas-MIP-I	*M sexta*	AWQDLNSAWamide	*21*
Manducamyoinhibitory peptide II	Mas-MIP-II	*M sexta*	GWQDLNSAWamide	*21*
Locustamyoinhibin	Lom-MIH	*L migratoria*	pQ?Y?KQSAFNAVSamide	*416*
b) Myosuppressin and FaRPs				
Leucomyosuppressin	Lem-MS	*L maderae*	pQDVDHVFLRFamide	*173*
SchistoFLRFamide	Scg-FLRFamide	*S gregaria*	PDVDHVFLRFamide	*389, 412, 345*
		L migratoria		*248*
LocustaFaRP (locustamyosuppressin)	Lom-MS	*L migratoria*	ADVGHVFLRFamide	*345, 248*
ManducaFLRFamide	Mas-FLRFamide	*M sexta*	*pQDVVHSFLRFamide*	*228*
Neomyosuppressin	Neb-MS	*N bullata*	TDVDHVFLRFamide	*97, 323*
		D melanogaster		
LocustaFaRPs	Lom-FaRP-I	*L migratoria*	GQERNFLRFamide	*345, 248*
	Lom-FaRP-II	*L migratoria*	A??RNFIRFamide	*248*
	Lom-FaRP-III	*L migratoria*	AFIRFamide	*248*
Aedes head peptide I	Aea-HP-I	*A aegypti*	pQRPHypSLKTRFamide	*273*
Aedes head peptide II	Aea-HP-II	*A aegypti*	TRFamide	*273*
CalliFMRFamides*				
CalliFMRFamide-I	Cav-FMRF-NH_2-I	*C vomitoria*	SVQDNFIRFamide	*78*
		L cuprina		*74*

Table 13 (*continued*)

Peptide Name or Family Name	Code Name	Species	Sequence	Reference(s)
CalliFMRFamide-II	Cav-FMRF-NH_2-II	*C vomitoria*	GDNFMRFamide	*78*
		L cuprina		*74*
CalliFMRFamide-III(= 9)	Cav-FMRF-NH_2-III	*C vomitoria*	SVNTKDNFMRFamide	*78, 74*
CalliFMRFamide-IIIa	Cav-FMRF-NH_2-IIIa	*L cuprina*	SANTKDNFMRFamide	*74*
CalliFMRFamide-IV(= 8)	Cav-FMRF-NH_2-IV	*C vomitoria*	GANDFMRFamide	*78, 74*
CalliFMRFamide-IVa	Cav-FMRF-NH_2-IVa	*L cuprina*	GGNDFMRFamide	*74*
CalliFMRFamide-V(= 3)	Cav-FMRF-NH_2-V	*C vomitoria*	SPSQDFMRFamide	*78*
		L cuprina		*74*
CalliFMRFamide-Va	Cav-FMRF-NH_2-Va	*L cuprina*	SPTQDFMRFamide	*74*
CalliFMRFamide-VI(= 12)	Cav-FMRF-NH_2-VI	*C vomitoria*	AAGTDNFMRFamide	*78, 74*
CalliFMRFamide-VIa	Cav-FMRF-NH_2-VIa	*L cuprina*	AAASDNFMRFamide	*74*
CalliFMRFamide-VII	Cav-FMRF-NH_2-VII	*C vomitoria*	QASQDFMRFamide	*74*
CalliFMRFamide-VIIa	Cav-FMRF-NH_2-VIIa	*L cuprina*	QANQDFMRFamide	*74*
CalliFMRFamide-VIII(= 5)	Cav-FMRF-NH_2-VIII	*C vomitoria*	APGQDFMRFamide	*78, 74*
CalliFMRFamide-VIIIa	Cav-FMRF-NH_2-VIIIa	*L cuprina*	AAGQDFMRFamide	*74*
CalliFMRFamide-IX(= 10)	Cav-FMRF-NH_2-IX	*C vomitoria*	TPNRDFMRFamide	*78, 74*
CalliFMRFamide-X(= 2)	Cav-FMRF-NH_2-X	*C vomitoria*	TPSQDFMRFamide	*78, 74*
CalliFMRFamide-XI(= 5)	Cav-FMRF-NH_2-XI	*C vomitoria*	APGQDFMRFamide	*78, 74*
CalliFMRFamide-XIa	Cav-FMRF-NH_2-XIa	*L cuprina*	APSQDFMRFamide	*74*
CalliFMRFamide-XII(= 6)	Cav-FMRF-NH_2-XII	*C vomitoria*	ASGQDFMRFamide	*78, 74*
CalliFMRFamide-XIIa	Cav-FMRF-NH_2-XIIa	*L cuprina*	AGQDNFMRFamide	*74*
CalliFMRFamide-XIII(= 13)	Cav-FMRF-NH_2-XIII	*C vomitoria*	AGQDGFMRFamide	*78, 74*
CalliFMRFamide-XIIIa	Cav-FMRF-NH_2-XIIIa	*L cuprina*	NPQQDFMRFamide	*74*
CalliFMRFamide-XIV(= 1)	Cav-FMRF-NH_2-XIV	*C vomitoria*	TPQQDDFMRFamide	*78, 74*
		L cuprina		*74*
CalliFMRFamide-XV(= 11)	Cav-FMRF-NH_2-XV	*C vomitoria*	PDNFMRFamide	*78, 74*
		L cuprina		*74*
		D melanogaster		*320, 402*
		D virilis		*454*

CalliFMRFamide-XVI(= 14)	Cav-FMRF-NH_2-XVI	*C vomitoria*	APPQPSDNFIRFamide	*78, 74*
CalliFMRFamide-XVIa	Cav-FMRF-NH_2-XVIa	*L curprina*	TPPQPSDNFIRFamide	*74*
DroFMRFamides				
DroFMRFamide-I	Drm-FMRFNH_2-I	*D melanogaster*	SVKQDFMHFamide	*454*
DroFMRFamide-Ia	Drm-FMRFNH_2-Ia	*D virilis*	SLKQDFMHFamide	*454*
DroFMRFamide-II	Drm-FMRFNH_2-II	*D melanogaster* *D virilis*	SVKQDFMRFamide	*454*
DroFMRFamide-III	Drm-FMRFNH_2-III	*D melanogaster*	TPAEDFMRFamide	*454*
DroFMRFamide-IV	Drm-FMRFNH_2-IV	*D melanogaster* *D virilis*	SDNFMRFamide	*454*
DroFMRFamide-V	Drm-FMRFNH_2-V	*D melanogaster* *D virilis*	SPKQDFMRFamide	*454*
DroFMRFamide-VI	Drm-FMRFNH_2-VI	*D melanogaster*	SAPQDVRSamide	*454*
DroFMRFamide-VIa	Drm-FMRFNH_2-VIa	*D virilis*	SAPTEFERNamide	*454*
DroFMRFamide-VII	Drm-FMRFNH_2-VII	*D melanogaster*	MDSNFIRFamide	*454*
DroFMRFamide-VIIa	Drm-FMRFNH_2-VIIa	*D virilis*	MDSNFMRFamide	*454*
DroFMRFamide-VIII	Drm-FMRFNH_2-VIII	*D virilis*	APPSDFMRFamide	*454*
DroFMRFamide-IX	Drm-FMRFNH_2-IX	*D virilis*	APSDFMRFamide	*454*
DroFMRFamide-X	Drm-FMRFNH_2-X	*D virilis*	DPSQDFMRFamide	*454*
(DroFMRFamide-XI see neomyosuppressin)	(Drm-FMRFNH_2-XI = Neb-MS)	*D melanogaster*	TDVDHVFLRFamide	*323*

* Arabic numerals are assigned calliFMRFamide members in (*78*)

P. americana was (first) used to detect proctolin, later the far more sensitive (picomolar range) tests using the locust extensor tibia or the locust oviduct bioassays were implemented (*247, 334*). Using the latter preparation or the locust ovipositor muscles, a skeletal muscle preparation, proctolin's role as a contransmitter was clearly shown (*333*). There is no direct evidence for a role of proctolin as a neurohormone in insects, but immunocytochemistry shows proctolin-like immunoreactive neurons in the blowfly, with endings terminating outside the neural sheath (*308*), in the corpora cardiaca of a beetle and in moths (*65, 472*) or in the corpora allata of a moth (*180*).

Studies on the pharmacology of the proctolin receptor have been carried out by several groups who determined the myotropic effects of various proctolin analogues in different bioassay systems like the cockroach hindgut (*440, 452*), the desert locust foregut (*154*), the migratory locust oviduct (*363*) and the heart of the cockroach and the mealworm, *Tenebrio molitor* (*238*). Although species- or bioassay-specific responses occur, it can be generalized that activity depends on the full pentapeptide while the amino acids have to have the L-configuration. C- or N-terminally truncated analogues (di-, tri- or tetrapeptides) were inactive and slight modifications at a single position resulted mostly in a complete loss of activity. Some analogues, however, for example [Ala^4]-proline, had substantial activity in a particular bioassay, causing locust oviduct contraction in this case, but were inactive in others. A supra-analogue, which had twice the potency of proctolin in the cockroach hindgut assay and was 4- and 1.5-fold more active in the cockroach and mealworm heart assay respectively, contained a methoxygroup instead of the hydroxylgroup at the *p*-position of the aromatic side chain of Tyr^2. Phe^2, however, showed little or no activity, whereas analogues substituting the hydroxylgroup of Tyr^2 with various nitrogen containing groups (Phe(*p*-NH_2); Phe(*p*-NMe_2); Phe(*p*-NO_2)) were all more active than proctolin in the cardiostimulatory assay. This was also true for the Phe(*p*-fluoro)-analogue in the locust foregut assay. In this system the Tyr(3′mono-iodo)-analogue had reasonable potency and, if still active when ^{125}I-labelled, such a compound could be extremely useful for receptor binding studies. Another useful tool may be the Tyr(α-methyl)-analogue which reduced the maximum response of the locust foregut by 88% at a concentration of 10^{-6} M and thus is an antagonist. Another antagonist was the tripeptide Arg-Tyr-Thr, but, at higher concentrations (10^{-5} M), the reduction in the maximal response to proctolin was smaller. It is speculated that the tripeptide in high concentration reduces the rate of proctolin inactivation by enzymes either by competing with proctolin for the active sites of the proteolytic enzymes or by exerting end product inhibition. From a number of studies on proctolin degradation using cockroach haemolymph (*441*, *442*), cockroach tissue homogenates (*364*), membranes of desert locust synatosomes (*186*) and membrane preparations of migratory locust hindgut and ovary (*360*), the presence of aminopeptidase, carboxypeptidase and endopeptidase activity is known. Depending on the pH it became clear that, at pH 6, either a carboxypeptidase followed by an endopeptidase cleaves the Tyr-Leu bond or that immediately the endopeptidase degrades proctolin to yield Arg – Tyr + Leu – Pro + Thr, whereas, at pH 8, an aminopeptidase is apparently favoured which produces Arg + Tyr-Leu-Pro-Thr and later Arg + Tyr + Leu-Pro + Thr. The effects discussed above for the tripeptide Arg – Tyr – Thr, which were detected *in vitro*, could thus relate to the situation *in vivo* by the degradation product Arg + Tyr.

Tritiated proctolin, [^{3}H] proctolin, was used to investigate binding to locust hindgut- and oviduct membranes and specific binding was shown (*361*, *362*).

Besides proctolin, whose effect on the insect heart has been mentioned above, two peptides isolated from the corpus cardiacum of *P. americana* (Pea-CAH-I and II) have cardioacceleratory activity. These peptides, which belong to the adipokinetic hormone/red pigment-concentrating hormone family, also have hypertrehalosaemic activity in cockroaches and have been dealt with in Sect. 3.1.1.

The most potent cardiostimulatory peptide in *P. americana* is Pea-corazonin. It was isolated from corpora cardiaca of this cockroach by RP-HPLC on a C-18 support using a water/acetonitrile gradient with TFA or HFBA as ion pairing agents and, after deblocking with pyroglutamate aminopeptidase, was shown to be a blocked undecapeptide (*464*) (Table 13). Subsequently, using an ELISA to monitor the presence of corazonin, the same molecule was shown by retention time on HPLC and amino acid composition to be present in the corpora cardiaca of the cockroach, *Nauphoeta cinerea*, and the tobacco hawk moth, *Manduca sexta*. The primary structure of a bioanalogue, [His^7] corazonin instead of Arg^7, was determined for the material isolated from the corpus cardiacum of the desert locust, *Schistocerca americana* (*466*) (Table 13). Because of its isolation from corpora cardiaca and its distribution, as shown by immunocytochemistry with antisera specific to Pea-corazonin, in neurosecretory cells of the protocerebrum and their axon terminals in the storage part of the corpus cardiacum (*355*, *469*), it is suggested that this peptide is released from the corpus cardiacum and acts as a neurohormone to control heart beat. Moreover, immunoreactivity was also found in interneurons of the brain and segmental body ganglia of *P. americana* (*469*). A similar distribution of Pea-corazonin immunoreactivity was found in another cockroach, *Leucophaea maderae* (*355*). In the blowfly

Phormia terraenovae, two cell groups (lateral and median) with immunoreactivity were found in the protocerebrum of all postembryonic stages and a large plexus of varicose fibres located in the wall of the aorta, a possible release site, was shown to contain peripheral processes as well (*41*). When brain-corpora cardiaca-aorta complexes of *P. terraenovae* were extracted, the material was identified by a Pea-corazonin specific ELISA to co-elute with authentic Pea-corazonin. This suggests that *P. terraenovae* also contains Pea-corazonin. Synthetic Pea-corazonin was also able to stimulate contraction of the hyperneural muscle of *P. americana*, but neither the oviduct nor the proctodeum. Interestingly, only the hyperneural muscle of *P. americana* is stimulated in a very sensitive way, but not those of other cockroaches such as *Blatta orientalis*, *Blattella germanica* (weakly), *Blaberus craniifer*, *Blabtica dubia*, *Pycnoscelus surinamensis*, *Leucophaea maderae* (weakly), *Gromphadorhina portentosa* and *Nauphoeta cinerea* (*355*). Since Pea-corazonin appears to be present in some of the species, another, as yet not discovered, target tissue and possibly another function has to be postulated for Pea-corazonin.

The heart of the moth, *Manduca sexta*, is modulated by a number of neuropeptides called cardioacceleratory peptides (CAPs) of which two groups (CAP_1 and CAP_2) with at least two and three members, respectively, exist (*461*). It is believed that these peptides stimulate the heart immediately after adult emergence, facilitating wing inflation and are also active during flight to achieve adequate haemolymph circulation between abdomen and thorax. In larvae, the hindgut may be the primary target (*461*). Isolation of one of the CAP_2 peptides was achieved by dissecting 6000 ventral abdominal nerve cords from pharate adult moths and, after heat-treatment, extraction in 0.5 M acetic acid. Pre-purification was on Sep-Pak (C-18 support), followed by a 6-step HPLC procedure (*45*), resulting finally in a pure peptide as judged by Edman sequencing and mass spectral analysis. The primary structure yielded an amidated nonapeptide containing cystein residues at positions 3 and 9 (*45*) (Table 13). An identical peptide had earlier been isolated and sequenced from 800 locust brain-suboesophageal ganglia-ventral nerve cord complexes; it was shown to have a potent myotropic effect on the locust hindgut (*438*). Purification had been achieved by antibody affinity chromatography followed by RP-HPLC using a RIA developed for the detection and quantification of the crustacean cardioactive peptide (Cama[1]-CCAP) (*437*). Thus, *L. migratoria* and *M. sexta* were shown to contain authentic Cama-CCAP in their nervous tissue. This peptide was subsequently also

[1] A four letter code is used for this crustacean peptide to distinguish between Cam = *Carausius morosus* (a stick insect) and Cama = *Carcinus maenas* (a crab). See also Table 4.

sequenced in *M. sexta* (*255*), the southern armyworm, *Spodoptera eridania*, and the mealworm, *Tenebrio molitor* (*102*). The presence of Cama-CCAP in the latter beetle species was not too surprising, since CCAP-immunoreactive neurons in the ventral nerve cord and the brain (lateral neurosecretory cell) had been demonstrated previously (*33*). In the locust, the antiserum stained efferent and intersegmental neuronal systems in the ventral nerve cord, some of which are recognized as release sites (*70*). In the blowfly, only four cells in the fused thoracic-abdominal ganglion are immunopositive. Axons of these cells reach the hindgut (*306*) and the peptide may be involved in modulating hindgut myotropic activity. Recently, another cardioacceleratory peptide, CAP_{2b}, of *M. sexta* has been fully structurally elucidated (*184*). This N- and C-terminally blocked octapeptide has no sequence homology to CAP_{2a} (=Cama-CCAP) or any other insect neuropeptide (see Table 13).

During 1987/1988 12 novel myotropic peptides were isolated and characterized from head extracts of the cockroach, *Leucophaea maderae*, using the hindgut bioassay (see Sect. 2.1.2) and a four-step HPLC purification procedure (see Sect. 2.2) (*176*, *178*). The same purification and bioassay system was used for the identification of five myotropic peptides from head extracts of *Acheta domesticus* (*176*, *178*) and a very similar procedure yielded 21 novel neuropeptides from brain-corpora cardiaca-corpora allata-suboesophageal ganglion complexes of *Locusta migratoria* (*415*). The peptides are now placed in distinct peptide families because of their structural similarities; additional members of these families have been elucidated in the meantime from other insect species and are all listed in Table 13.

3.3.2. Myokinins

To date eight myokinins from *L. maderae*, five from *A. domesticus*, three each from *Aedes aegypti*, *Culex salinarius* and *Helicoverpa zea* are known and a single one from *Locusta migratoria* (for original references see Table 13). They all share a common C-terminal pentapeptide sequence. FX^1X^2WGamide (where X^1 = H, S, N, Y, F and X^2 = S, P, A). The leucokinins stimulate the hindgut most potently by increasing the frequency and amplitude of spontaneous phasic contractions at lower concentrations and with a tonic component at higher concentrations (threshold concentration: 0.3 to 2.0×10^{-10} M for the various peptides) (*294*). Their effects on stimulating the muscles of the foregut and oviduct are about 100- and 1000-fold less (*57*, *58*).

An antiserum against leucokinin I and synthetic leucokinin I labelled with ^{125}I-Bolton-Hunter reagent were used to develop a sensitive RIA.

Low levels of immunoreactive material were measured in the ventral nerve cord of *L. maderae*, but high values (1.9 pmol) in the brain and largest amounts (6.6 pmol per tissue) in the corpora cardiaca-corpora allata complexes, whereas the titre in the haemolymph was in the nanomolar range (*292*). High-potassium depolarization combined with Ca^{2+}-induced release of about 2% of the stored material suggested that the leucokinins may act via the circulation as neurohormones.

The achetakinins are almost as potent on the cockroach hindgut (*178*), but the locustakinin is inactive on hindgut and oviduct of the locust (*415*). Achetakinins also exhibit an adipokinetic effect and cause inhibition of protein synthesis in the fat body of crickets and locusts, actions well known for peptides from the AKH/RPCH family (see Sect. 3.1.1).

Achetakinins have a diuretic effect in the cricket (*54*, *436*) and, after raising antibodies in rabbits for immunocytochemistry and establishing a RIA, achetakinin-like immunoreactive material was found in brain and other nervous tissues. Activity was mainly in the retrocerebral complex of crickets (CC, CA and hypocerebral ganglion), and was detected in the haemolymph, where it increased 10-fold in starved crickets (*47*). Moreover, achetakinin binding sites on the membranes of Malpighian tubules of the cricket have been studied by using a biologically active ^{125}I-labelled analogue and specific binding sites have been demonstrated (*48*). Leucokinins are also known for their effect on Malpighian tubules of the mosquito, *Aedes aegypti*, where they cause a depolarization of the transepithelial potential (*161*). This bioassay, in conjunction with the cockroach hindgut myotropic assay, was actually used to monitor the separation of the myokinins from *Culex salinarius* (*158*). Diuretic and antidiuretic effects of locustakinin on locust tubules and rectum, respectively, have been shown as well (*415*). The novel kinins from *H. zea*, the helicokinins, were isolated from the abdominal ventral nerve cord and stimulated fluid secretion of the Malpighian tubules at concentrations below 10^{-11} M (*22*).

Antisera raised against Lem–M–I recognized about 160 immunoreactive cell bodies from mainly interneurons and neurosecretory cells in the protocerebrum and optic lobes of *L. maderae* (*307*); neurosecretory cells in the protocerebrum have also been stained in the blowfly brain (*304*), and abdominal ganglia also contained immunoreactive neurons (*40*). In larvae of the lepidopteran species *Agrotis segetum*, immunoreactive fibers innervate the perisympathetic organ, which are known release sites (*39*). These data, in conjunction with the failure to detect immunoreactivity to leucokinin I in fibers directly innervating the hindgut of *L. maderae*, indicate that leucokinin may act as a neurohormone (*307*).

References, pp. 97–128

Structure-activity studies on Lem-M-VIII showed that truncated analogues, Lem-M-VIII (1–7) or Lem-M-VIII (5–8) are totally inactive; the core pentapeptide (FYSWG-amide) is as active as the parent molecule, but not its free acid (*294*). Replacements of Phe^1 or Trp^4 by Ala resulted in inactive analogues, but Trp^1 and Phe^4 are tolerated, thus aromatic groups are needed at these positions. While Ala analogues at positions 2, 3 or 5 had reasonable activity, the D-Ala^2 analogue is inactive. A β-turn is predicted for the C-terminal region of the leucokinins. A conformationally-constrained analogue of the core region, *cyclo*-[CFYSWCamide], retains activity although the threshold activity is now in the range of 9×10^{-7} M instead of 0.2×10^{-10} M (*294*). Since bioanalogues (naturally occurring peptides) tolerated various substitutions at positions 2 of the pentapeptide core region, a pseudopeptide analogue containing a reduced amide bond linkage (–CH_2–NH– instead of –C(O)NH–) between residues 1 and 2 was synthesized, FΨ[CH_2–NH_2]FSWGamide (*299*). The biological activity of this pseudopeptide is 1% when compared to its amide bond-containing counterpart (FFSWGamide). Thus it retains activity, but, most importantly, the pseudopeptide is stable to proteolytic digestion by aminopeptidase M, whereas its natural conterpart is not (*299*). This experiment proves that peptide mimetics, which may be extremely useful as potential insect pest control agents, are active and have an improved half life.

3.3.3. *Sulfakinins*

To date two sulfakinins each from *L. maderae* and *Neobellieria bullata*, and one each from *L. migratoria* and *P. americana* have been isolated and sequenced (Table 13). Moreover, the non-sulfated Lem-SK-II molecule has been sequenced from *P. americana* (*465*); Sulfakinins in Diptera (*Drosophila melanogaster*, *Calliphora vomitoria*, and *Lucilia cuprina*) have been deduced from cloning and sequencing the respective genes. It still has to be demonstrated that they are expressed in these species, but since identical or very similar peptides have been sequenced in another dipteran insect, *N. bullata* (*98*), expression in the other dipterans is very likely. In fact, very recently, the peptide from *Calliphora vomitoria* has been isolated from heads of this fly (*83*). The sulfakinin insect family is characterized by high conservation of the C-terminal decapeptide sequence: $X^1X^2DY(SO_3H)$GHMRFamide (where X^1 = F, S and X^2 = E, D). They share sequence near-identity of the C-terminus with the human gastrin and the vertebrate hormone cholecystokinin (CCK):

gastrin II: . . . $Y(SO_3H)$GWMDFamide

CCK_8: $DY(SO_3H)$MGWMDFamide

However, these vertebrate molecules were inactive on the cockroach hindgut, but introduction of Arg instead of Asp transformed them into active analogues in this bioassay (*296*). The structural homology between these vertebrate hormones and their insect counterparts and their analogous myotropic actions (gastrin and CCK also stimulate smooth muscle contractions in the intestine (see *295*)) point to a long evolutionary history. It is also interesting to note in this context that the sulfakinins share sequence similarities with the so-called FMRFamide related peptides (FaRPs) which are dealt with later (see Sect. 3.3.8).

Structure-activity studies demonstrated that non-sulfated analogues were inactive, and the C-terminal hexapeptide is the smallest fragment ("active core") possessing about 10% of the myotropic activity of the parent molecule. Full activity requires the C-terminal octapeptide (*294*). The relative importance of amino acid residues within the active core region was established by synthesizing and bioassaying single (by Ala) replacement octapeptide analogues (*297*). All contractile activity on the hindgut was lost when the last (Phe), − 1(Arg) and − 3(His) positions were replaced. These and additional experiments suggested that aromaticity (− 3 and last position) and basicity (− 1 position) are critical for interaction with the putative receptor. Furthermore, although the presence of a sulfate group is required for biological activity, the position is less critical; it can be moved by one position towards the C-terminus without complete loss of activity (0.3% of parent molecule) and by one (still 38% active) or up to five (about 0.2% active) positions to the N-terminus (*294*).

3.3.4. Pyrokinins/Myotropins

This family, characterized by the carboxy-terminal sequence FXPRL (where X = T, V, S), consists of the myotropins (Lom-MT-I to IV) and two pyrokinins (Lom-PK-I and II) from *L. migratoria*, the pyrokinin (Lem-PK) from *L. maderae*, which was the first member of this family fully elucidated, as well as some peptides from *H. zea* and *B. mori*, which were deduced from cDNA work (see Table 13 for references). Lem-PK has the highest concentration (1.4 pmol/head) of all myotropic peptides in *L. maderae*, but had surprisingly weak activity on the hindgut (threshold concentration: 0.6 nM). However, it was active on the cockroach foregut and oviduct (*178*). The locustapyrokinins and -myotropins were all monitored during isolation by their effect on the cockroach hindgut, but the synthetic peptides were also shown to stimulate the oviduct of *L. migratoria* (*415*).

The structural requirements for Lem-PK were assessed by synthesizing a series of octapeptide analogues (*293*). Analogues with substitutions

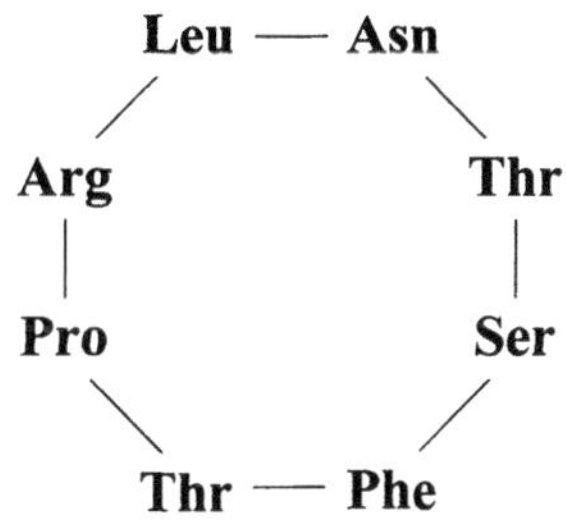

Fig. 7. Structure of the cyclic analogue of Lem-PK (Leucopyrokinin)

of Thr^2 by Leu^2 or Ser^3 by Thr^3 retained most of their activity. This was not surprising since even a peptide truncated by the N-terminal tripeptide (pGlu-Thr-Ser) still had 30% of the parent molecule's activity; surprisingly the des-pGlu-analogue was even 40% more active than the intact Lem-PK. At the C-terminus the amide was essential and replacement of Pro^6 by Gly^6 or D-Ala^6 or Arg^7 by Lys^7 resulted in very weak activity (all at least 1000-fold less active).

Conformational information was gained by studying a cyclic, biologically active, Lem-PK analogue (see Fig. 7) in which the N- and C-termini are linked by an amide bond (*301*). Analyzing data from circular dichroism, nuclear magnetic resonance and molecular dynamics, the presence of a type 1 β-turn in the active core region formed by residues Thr-Pro-Arg-Leu was established for this conformationally restricted analogue; the biological activity is about 4% of the linear molecule suggesting that its C-terminal β-turn is the active pyrokinin conformation recognized by the specific receptor.

Additional members of this family containing the pentapeptide FXPRLamide sequence at their C-terminus are the insect hormones pheromone biosynthesis activating neuropeptides (PBAN) isolated from *Heliothis zea*, *Bombyx mori*, *Pseudaletia separata*, and *Lymantria dispar* (see Sect. 3.2.1) and the diapause hormone from *B. mori* (see Sect. 3.2.6). Cloning of the PBAN and PBAN/DH genes of *H. zea* and *B. mori* (*64, 217, 393*) led to the deduction of other peptides with the above C-terminal sequence (see Table 13); only one of these putative Bom "myotropins" contained Lys as the penultimate amino acid instead of Arg. Since structure-activity studies had revealed that the pentapeptide sequence is sufficient to elicit myotropic (*293*) and pheromonotropic (*371*) activity, it was not surprising to find that leucopyrokinin (*294*) and the locustamyotropins (*96, 244*) have considerable cross-activity in the pheromonotropic assay of the silkworm, *B. mori*. Lom-MT-II, for example, was even

100-fold more active in this assay than the 33-mer Bom-PBAN-I (*244*). Locustamyotropins also stimulate pheromone biosynthesis in *Spodoptera litura* (*96*). Furthermore, PBAN is also able to stimulate visceral muscle contractions in *L. maderae* and *L. migratoria* (*176, 415*). Since Bom-PBAN is the same molecule as the hormone responsible for cuticular melanization and epidermal reddish brown pigmentation, the so-called MRCH (*277*), and a similar "pheromonotropic peptide" with MRCH activity was isolated from *P. separata* (*275*), it was again in keeping with the "active core theory" that locustamyotropins induced larval cuticular melanization in *P. separata* (*276*).

Lastly, Lem-PK, Lom-PK-I and II, and Lom-MT-I and II all elicit significant diapause-inducing activity in *B. mori* (*300*). Lom-PK-I was even 3-fold more potent than the native Bom-DH-I. Conversely, Bom-DHs elicited contraction of the hindgut, but were several orders of magnitude less active as native Lem-PK. All these results clearly show cross-reactivity for this peptide family in different physiological processes, myotropic, pheromonotropic, diapause inducing and cuticular melanization, suggesting homologous features of the receptor sites.

Antisera raised against Lom-MT-I and II and Hez-PBAN were used for studying the distribution of these immunoreactivities in the nervous system of *L. migratoria* and various other insects (*414, 456*). Since the antisera cross-react with all peptides of this family, interpretation of the results is difficult and, therefore, no further comments are given here, but the interested reader is referred to the literature (see above) or a recent review (*415*).

3.3.5. Tachykinins

To date nine members comprise the insect tachykinin family which is characterized by the C-terminal pentapeptide sequence FX^1GX^2Ramide (where X^1 is mostly Y but in one member each H and T; X^2 is V except M in 2 members; see Table 13). The "true" tachykinins from vertebrates, of which the undecapeptide substance P is the most well-known member, contain the pentapeptide C-terminus of FXGLMamide (X = F, Y, I, V). Substance P (RPKPQQFFGLMamide), for example, has been identified in mammals to act on many systems – as an excitatory neurotransmitter and as a modulator involved in regulating such diverse functions as sensory processing, control of movement, gastric motility, vasodilation and salination (*164, 337*). Because of some structural homology with the tachykinins, especially with the physalaemin subfamily of tachykinins (see *415*), and because the first members were discovered to stimulate the hindgut of *L. maderae* these peptides were grouped together into the insect tachykinin family (*404*). The Lom-TKs were also shown to stimulate the visceral muscles of locust foregut and oviduct (*415*). Moreover, they stimulate the slow excitatory motor neurons of the locust extensor tibiae (*415*) and display some pheromonotropic activity in *B. mori* (*95*).

Antisera against Lom-TK-I have been raised and applied to nervous tissue of various insect species, including *L. migratoria*, *L. maderae*, *C. vomitoria* and *D. melanogaster* (see *303*, *305*), to determine the cellular localization of tachykinins. Most of the immunoreactive neurons are interneurons. In *L. migratoria* immunoreactive neurons project to the intrinsic neurosecretory cells in the corpus cardiacum and make synapses there with these cells known to synthesize adipokinetic hormone (*303*); this is also corroborated by immunocytochemical studies on the electron microscopical level (*309*). The suggestion that the Lom-TK immunoreactive cells may be interneurons regulating Lom-AKH release is substantiated by the demonstration of release of Lom-AKH-I *in vitro* from isolated corpora cardiaca by authentic Lom-TK-I (*309*). In the blowfly, *C. vomitoria*, those neurons reacting to the Lom-TK-I antiserum were identical with those which were immunoreactive with antisera against kassinin, a member of the tachykinin family in frogs (*263*). This is explained by the structure of the native tachykinins in *C. vomitoria*, Cav-TK-I and II (see Table 13); whereas the C-terminal pentapeptide of Cav-TK-I is identical to Lom-TK-I, the C-terminus of Cav-TK-II is similar to kassinin.

Interestingly, peptides which were isolated from salivary glands of the mosquito, *A. aegypti*, and therefore called sialokinins I and II (I: NTGDKFYGLM; II: DTGDKFYGLM), contain the "true" tachykinin C-terminal pentapeptide FXGLM (*43*). It is not yet known whether they are produced in neurons.

3.3.6. Periviscerokinin

The perisympathetic organs of insects, first discovered in stick insects (*365*), have been identified as a major neurosecretory storage and release site of the ventral nerve cord. Using these organs as starting material for isolation, a peptide was purified from extracts of 1000 abdominal perisympathetic organs of male American cockroaches by a 3-step HPLC procedure. This peptide had an excitatory action on the hyperneural muscle of *P. americana* (*356*). After Edman degradation and mass spectral analysis, the structure of a unique undecapeptide, called periviscerokinin (Pea-PVK), was elucidated (Table 13). The synthetic amidated form, but not the free acid, was biologically active. Since this compound was isolated from a neurohaemal site and is active on the isolated hyperneural muscle at low concentrations (10^{-9} M), it is believed that periviscerokinin has a physiological role. Immunocytochemical studies revealed Pea-PVK-like immunoreactivity in three cell clusters of the abdominal ganglia. These neurons project into the perivisceral organs (*85*).

3.3.7. Accessory Glands- and Midgut-Myotropins and Others

Peptides which stimulate the spontaneous contractions of the oviduct have been isolated by several (in the case of *L. migratoria*) or a single (in the case of *M. domestica*) HPLC step(s) from either male accessory reproductive glands of the migratory locust (Lom-AG-MT-I, II; *341*, *342*) or from female accessory sex glands of the house fly (Mud-AG-MT; *473*) (Table 13). Lom-AG-MT-I resembles in structure the juvenile hormone biosynthesis stimulating peptide allatotropin from *M. sexta* (Mas-AT; see Sect. 3.2.2.1), but this compound had no allatotropic effect on the corpora allata of the desert locust (*212*). It is not known yet whether Lom-AG-MT-I stimulates the biosynthesis of juvenile hormone in locusts. The neuropeptide status of the Mud-AG-MT is not established, but Lom-AG-MTs immunoreactive cells, stained with polyclonal antibody raised against each of the peptides (*340*), were not only found in the tubules of the glands, but also in cell bodies of proto- and deuterocerebrum, optic lobes, frontal ganglion, thoracic and the last abdominal ganglion (for Lom-AG-MT-I). The antiserum against Lom-AG-MT-II also stained cells of the central nervous systems, but double staining revealed the presence of Lom-AG-MT-I and II immunoreactive materials in distinct cell population and nerve fibres (*340*, *415*).

It is well known that endocrine cells are present in the insect gut. Recently, two myoactive peptides isolated from the midgut of *M. sexta* have been sequenced (Mas-MG-MT-I and II), but, again, it is unclear whether they are synthesized in neurons (*486*, *487*). The same is true of a peptide that stimulates the contraction of the oviducts of *L. migratoria* and was isolated from 10000 heads of the Colorado potato beetle, *L. decemlineata*, by a 4-step HPLC procedure. After prepurification on Sep-Pak, a phenyl support, followed by C-1 and C-8 RP and subsequently normal phase Protein Pak 125 columns were used to achieve purification to homogeneity (*434*). Edman degradation resulted in the sequence of an amidated hexapeptide code-named Led-OVM (Table 13). The peptide had no influence on the contraction of the beetle's hindgut.

3.3.8. *Myoinhibitory Peptides and Other FMRFamide Related Peptides (FaRPs)*

The purification of 9000 brain complexes of *L. migratoria* led not only to the isolation of the contracting-stimulatory peptides (see previous section), but some fractions were also found which inhibited the contractions of the cockroach hindgut. Further purification lead to the isolation and identification of three myoinhibiting peptides which have structurally nothing in common with each other.

Locustamyoinhibitory peptide (Lom-MIP; see Table 13) is a blocked nonapeptide; the C-terminal tripeptide sequence,... AGWamide, is identical with that of the locust adipokinetic hormone Lom-AKH-II (*408*). Immunocytochemical studies found immunoreactivity in neurons innervating the heart and oviduct of the locusts (*415*). This pattern corresponds well with the functional aspect of Lom-MIP, which was shown to suppress the spontaneous contractions of the hindgut and oviduct of *L. migratoria* as well. The same tissues seem to be targets for the partially sequenced tridecapeptide locustamyoinhibin (Lom-MIH) which is blocked at both termini (*416*) (Table 13). Two peptides structurally related to Lom-MIP have been isolated and sequenced from the ventral nerve cord of adult *M. sexta* (*21*). These nonapeptides, Mas-MIP-I and II (Table 13), significantly reduced the rate of peristalsis of the isolated anterior hindgut (ileum) of *M. sexta* at low concentrations (10^{-9} M).

The other myosuppressins belong to the large family of FMRFamide related peptides (FaRPs), which is characterized by at least an RFamide sequence at the C-terminus; but mostly by an FLRFamide.

We have already discussed one of the "FaRPs" of insects – the sulfakinins which consistently contain the C-terminal sequence HMRF-amide (see Sect. 3.3.3.). Myosuppressins (FLRFamides), which are structurally closely related, have been found in *L. maderae*, *S. gregaria*, *L. migratoria*, *M. sexta* and *N. bullata*/*D. melanogaster* (see Table 13). During isolation most of them were detected by monitoring HPLC fractions via a immunoassay using an FMRFamide antiserum. Functionally diverse actions were found. For example, Mas-FLRFamide may be involved in flight behavior patterns, since it increases the force of contraction of dorsal longitudinal flight muscles in *M. sexta* (*228*), whereas Scg-FLRFamide inhibits the heart rhythm, but also potentiates twitch tension in the extensor tibiae muscles of *S. gregaria* (*389*) and inhibits spontaneous contraction of the oviduct of *L. migratoria* (*248*, *345*, *412*; Table 13).

Three further FaRPs, here code-named Lom-FaRP I to III, have been isolated from ventral nerve cords of *L. migratoria*. Two of them, one not yet fully sequenced, contain a FIRFamide C-terminus, whereas the other one has the known FLRFamide C-terminal sequence (*248*). These peptides had excitatory actions on the locust oviduct, indicating that the N-terminus of such FaRPs is important as well. Moreover, it is evident that a number of FaRPs exist in one species. This was very clearly shown for some dipteran species, where not only the peptides but the genes are known as well.

It had been shown by immunocytochemical studies that ventral neurosecretory cells of the thoracic ganglion of *Calliphora vomitoria* projecting axons into a neurohaemal area were immunoreactive against the vertebrate C-terminally extended enkephalin (YGGMRF; *81*), against vertebrate gastrin/cholecystokinin which has the C-terminus WMDFamide (*80*) and against FMRFamide (*264*). YGGMRF and FMRF, but not the amidated forms, were active in inducing saliva excretion from isolated salivary glands (*79*), and this was true using partially purified extracts of the thoracic ganglia, which have been shown to contain YGGFMRF-immunoreactive material. Processing thoracic ganglia from the blowfly in a 5-step HPLC procedure and using radioimmunoassays against YGGMRF and RFamide for monitoring the fractions, thirteen neuropeptides of varying length (7 to 11 residues) and ending C-terminally in FMRFamides, designated calliFMRFamides, and one (a dodecapeptide) ending in IRFamide, were isolated and sequenced (*78*). By cloning and sequencing a genomic DNA fragment encoding the FMRFamide prohormone it became clear that the prohormone contains 16 copies of potential FMRFpeptides and additionally two copies of FIRF peptides (*74*) (Figure 8, Table 13). Potential amidation (a Gly residue at the C-terminus of the putative peptide sequence) and cleavage sites (mostly single Arg residues) were found as well. This organization of the prohormone precursor divided into signal peptide, acidic spacer region, first FMRFamide peptide, spacer region and then a high amount of more FMRFamide-related peptides without spacers is very similar in all dipteran species investigated, but there are species-specific differences in the putative FMRFamide peptides in the precursor from *C. vomitoria, Lucilia cuprina* (see *74*) and *D. melanogaster* and *D. virilis* (*46, 320, 402, 454*) (Fig. 8). Only one peptide, PDNFMRF-amide, is present in all four species. Five peptides are shared between *C. vomitoria* and *L. cuprina* and four between the two *Drosophila* species (Table 13). In *Drosophila* another precursor of the FaRPs has been isolated; it contains two copies of FaRPs, the drosulfakinins I and II (*322*, see 3.3.3.). However, recently another FaRP, TDVDHVFLRFamide, was isolated and sequenced (*323*), which is not encoded on the two known precursors (Table 13). Thus, a third precursor appears to be present in *Drosophila.*

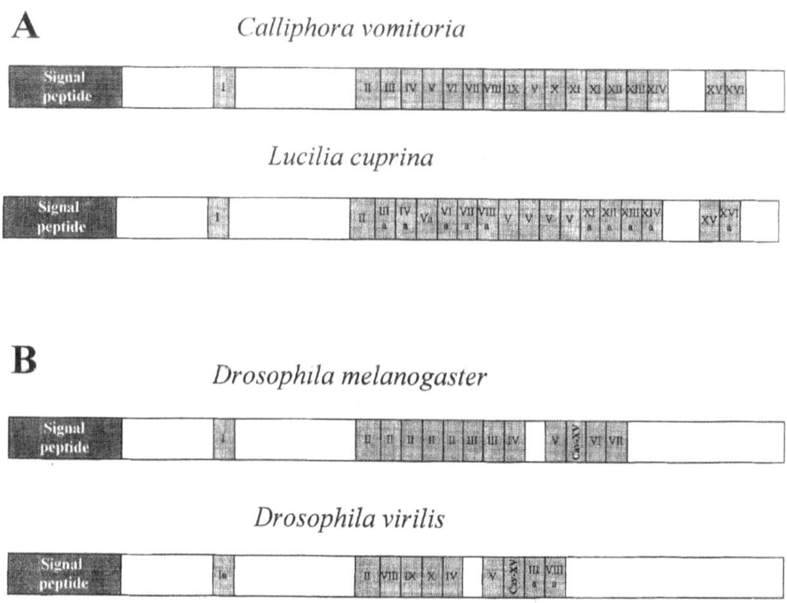

Fig. 8. Schematic diagrams of the precursor peptides for FMRF amide-related peptides from various Diptera. A The precursors of *C. vomitoria* and *L. cuprina*. Roman numbers correspond to the Cav-FMRF amide peptides given in Table 13. Modified after (*74*). B The precursors of *D. melanogaster* and *D. virilis*. Roman numbers correspond to the Drm-FMRF amide peptides given in Table 13. Modified after (*454*)

What does this molecular diversity mean? At the moment it is not known whether all the deduced peptides are expressed, but the studies on *C. vomitoria* show that at least 13 of 16 peptides are and, thus, it may be true for the remaining peptides in this and other species as well. It seems unlikely that of this array of peptides each has a different task, which would also mean a multiplicity of receptors. However, there is at least some evidence in the blowfly that certain calliFMRFamides are active secretagogues for the salivary glands, whereas others are inactive. In contrast, only two of these peptides are active on the heart of the blowfly increasing either the frequency alone or frequency and amplitude of the heartbeat (*73*).

In the mosquito *Aedes aegypti* two FaRPs were isolated and characterized from whole heads and designated Aea-HP-I and II (*273*, Table 13). Recent studies suggest that Aea-HP-I inhibits the host-seeking behavior (*35*). This behavior is employed by female mosquitoes to locate a vertebrate host for taking a blood meal, which, in turn, triggers the onset of oogenesis. After initiation of oogenesis the female does not engage in host-seeking. Synthetic Aea-HP-I injected into non-oogenic females, which actively seek a host, inhibited this behavior. Based on RIA determinations the haemolymph titre of Aea-HP-I in females that had ingested a blood meal was increased.

3.4. Chromatotropic Factors in Insects

A true color change in insects within one developmental stage is rare. This is especially true for the physiological color change resulting from pigment movement, since most terrestrial insects have developed a robust cuticle to prevent water loss, and thus pigment movement in the underlying epidermal cells, even if it takes place, is not so obviously noticeable. Morphological color change is characterized by pigment concentration and mostly occurs during specific developmental stages such as moulting. Although studies have shown that hormonal regulation is involved in color change in some species, here only those examples where molecules have been structurally identified either by controlling the color change in insects or in crustaceans are briefly reported.

The only structural knowledge of a true neuropeptide regulating insect pigmentation is for the melanization and reddish coloration hormone (MRCH) from the silkmoth *Bombyx mori* (*227*). The penultimate instar larvae of *Spodoptera separata* served as bioassay animals. In this species cuticular melanization and epidermal reddish-brown pigmentation in morphological color change is regulated hormonally. The sequence analysis revealed that Bom-MRCH was the same molecule as Bom-PBAN (see Sect. 3.2.1.).

It was found around 1940 that extracts from insect nervous tissue caused body blanching in prawns and shrimps due to concentration of pigments in the chromatophores of these crustaceans; furthermore, extracts from heads of insects also caused dispersion of pigments in crabs (*155*). When the locust adipokinetic hormone I (Lom-AKH-I) was structurally characterized (*450*) and its similarity to the crustacean red pigment-concentrating hormone (*92*) was noted, it became clear that the substances from insects causing "blanching" in crustaceans are the various members of the AKH/RPCH family (see Sect. 3.1.1.).

Using eyestalkless (the eyestalks are the source for synthesis and storage of endogenous crustacean neuropeptides) fiddler crabs, *Uca pugilator*, as bioassay animals by monitoring the dispersion of pigment in epidermal melanophores, pigment-dispersing factors were purified from

Table 14 *Primary structures of crustacean pigment-dispersing hormones (α- and β-PDHs) and insect pigment-dispersing factors (PDFs)*

Code Name (Alternative Designations)	Species	Sequence	Reference(s)
Pab-PDH(α-PDH)	*Pandalus borealis*	NSGMINSILGIPRVMTEAamide	*91*
	P jordani		*379*
Paj-PDH-I (= K^{13}, A^{16}, D^{17}-α-PDH)	*P jordani*	NSGMINSILGIPKVMADAamide	*379*
Ucp-PDH (β-PDH)	*Uca pugilator*	NSELINSILGLPKVMNDAamide	*379*
	Cancer magister		
	Callinectes sapidus		
	Pacifastacus leniusculus		
Prc-PDH (= E^{17}-β-PDH)	*Procambarus clarkii*	NSELINSILGLPKVMNEAamide	*379*
	Orconectes immunis		
Peaz-PDH (= L^{8}, I^{11}-β-PDH)	*Penaeus aztecus*	NSELINSLLGIPKVMNDAamide	*379*
Paj-PDH (= L^{8}, T^{16}-β-PDH)	*P jordani*	NSELINSLLGLPKVMTDAamide	*379*
Arv-PDH (Pillbug-PDH)	*Armadillidium vulgare*	NSELINSLLGAPRVLNNAamide	*379*
Acd-PDF (Acheta-PDF)	*A domesticus*	NSEIINSLLGLPKVMTDAamide	*379*
Rom-PDF (Romalea-PDF)	*R microptera*	NSEIINSLLGLPKLLNDAamide	*378*
Pea-PDF (Periplaneta-PDF)	*P americana*	NSELINSLLGLPKVLNDAamide	*287*
Cam-PDF (Carausius-PDF)	*C morosus*	NSELINSLLALPKVLNDAamide	*286*

whole heads of the cricket, *Acheta domesticus* (*379*), the grasshopper, *Romalea microptera* (*378*), and the American cockroach, *Periplaneta americana* (*287*) (see Table 14). The isolation procedure was very complex and used many chromatographic steps including partition-, gel filtration- and ion exchange chromatography. The result for each species was a compound which was characterized by protein chemical analysis to be an octadecapeptide, as was shown previously for the pigment dispersing hormones from crustaceans itself, the α- and β-PDHs of *Pandalus borealis* and *Uca pugilator*, respectively (*91,380*). Antisera were raised either against crustacean β-PDH or against the *Romalea*-PDF. Immunocytochemical studies showed prominent PDH or PDF-immunoreactive neurons which are associated with the visual system in a variety of insects (*181, 310, 377*) leading to the conclusion that the PDFs in insects have probably something to do with a circadian pacemaker system. At least their function in insects is not, as in crustaceans, to regulate pigment dispersion.

4. Conclusions

The last decade or so has seen an explosion of structural data on insect neuropeptides. This is well-documented in this review. Mainly this was possible because techniques for isolation and acquiring sequence information have been improved. It became clear that an array of methods, including Edman degradation sequencing, mass spectrometry and cDNA work, has to be used for arriving at the correct structures since post-translational modifications occur quite often. Using one method alone would, in most cases, not have been sufficient for structure elucidation. Immunocytochemistry was also a helpful tool for localizing the site of peptide synthesis in the cells/tissues, especially when the starting materials for isolation were whole animals or whole heads. In this context we have to acknowledge that for the majority of neuropeptides we still do not know exactly whether they are true hormones or not. One way to demonstrate this would be to show their production or storage in known neurohaemal organs from which they can easily be released; another way would be to show an increased neuropeptide concentration in the haemolymph upon some physiological stimulus.

Because quite a few neuropeptides were pleiotropic, thus had different biological activities, future research will possibly reveal that the primary effect of some so-called myotropic peptides, for example, will be different from that described in the bioassay used for isolation purposes. The pyrokinins, for example, share the C-terminal sequence with the PBANs and the diapause hormones and cross-reactivity occurs (*1*). Thus, are the

References, pp. 97–128

pyrokinins also involved in pheromone production *in vivo*? Interestingly, a single mRNA in *B. mori* encodes for a large precursor protein from which the diapause hormone, PBAN and three putative pyrokinins can be produced (*482*). Moreover, expression of this gene was regulated by temperature which leads to the induction of diapause. Temperature-independent, but stage-dependent regulation seems to be related to the production of pheromone (*483*). With the current interest of molecular biologists in insect endocrinology, much more of this type of research will occur in the future.

Another area that will be investigated quite actively during the next decade will involve characterization of receptors. Since only a few receptor molecules are probably present, protein purification methodology alone will not be successful and, again, molecular biological techniques will have to be used.

One aspect of great interest in insect neuropeptide research, which has not been dealt with in this review, is the exploration of alternative strategies to combat insect pests. This is very well outlined in a review by KEELEY and HAYES (*222*). Among other strategies, one is to synthesize peptidomimetics, *i.e.*, substances in which at least some of the peptide bonds susceptible to degradation by exo- or endo-peptidases in the insect's gut or haemolymph, have been replaced. NACHMAN and co-workers are very active in this field. Recently, they have synthesized a pseudodipeptide analogue of the C-terminal core pentapeptide of the pyrokinins/ PBANs/diapause hormones which had almost the same biological activity in the myotropic assay (cockroach hindgut) as the pentapeptide itself (*302*). This line of research will surely be intensified once pharmaceutical companies become fully convinced that insect neuropeptides may be useful as insecticides.

Acknowledgments

The author is indebted to A Lancha for her support during the initial phase of writing this review, to my colleagues Drs K -H Hoffmann (University of Bayreuth, Germany) and J H Spring (University of Southwestern Louisiana, Lafayette, LA, U S A) for reading and correcting earlier drafts, to L Auerswald, M P -E Janssens and H G Marco for their help with typing and drawing of figures Financial support for the recent work was provided by the Foundation for Research Development and a University of Cape Town staff award

References

1. ABERNATHY, R.L , R J. NACHMAN, P E A TEAL, O YAMASHITA, and J H TUMLISON Pheromonotropic Activity of Naturally Occurring Pyrokinin Insect Neuropeptides (FXPRLamide) in *Helicoverpa zea*. Peptides **16**, 215 (1995)

2 ADACHI, T , S TAKIYA, Y SUZUKI, M IWAMI, A KAWAKAMI, S Y TAKAHASHI, H ISHIZAKI, H NAGASAWA, and A SUZUKI cDNA structure and expression of bombyxin, an insulin-like brain secretory peptide of the silkmoth *Bombyx mori* J Biol Chem **264**, 7681 (1989)

3 ADACHI-YAMADA T , M IWAMI, H KATAOKA, A SUZUKI, and H ISHIZAKI Structure and Expression of the Gene for the Prothoracicotropic Hormone of the Silkmoth *Bombyx mori* Eur J Biochem **220**, 633 (1994)

4 AGUI, N , W E BOLLENBACHER, N A GRANGER, and L I GILBERT Corpus Allatum is Release Site for Insect Prothoracicotropic Hormone Nature **285**, 669 (1980)

5 AGUI, N , N A GRANGER, L I GILBERT, and W E BOLLENBACHER Cellular Localization of the Insect Prothoracicotropic hormone *In vitro* Assay of a Single Neurosecretory Cell Proc Natl Acad Sci U S A **76**, 5694 (1979)

6 ALTSTEIN, M , O BEN-AZIZ, and Y GAZIT Pheromone Biosynthesis Activating Neuropeptide (PBAN) and Colour Polymorphism an Immunochemical Study in *Spodoptera littoralis* J Insect Physiol **40**, 303 (1994)

7 ARIMA, R , K TAKAHARA, T KADOSHIMA, F NUMAZAKI, T ANDO, M UCHIYAMA, H NAGASAWA, A KITAMURA, and A SUZUKI Hormonal Regulation of Pheromone Biosynthesis in the Silkworm Moth, *Bombyx mori* Appl Ent Zool **26**, 137 (1991)

8 AUDSLEY, N , G M COAST, and D A SCHOOLEY The Effects of *Manduca sexta* Diuretic Hormone on Fluid Transport by the Malpighian Tubules and Cryptonephric Complex of *Manduca sexta* J Exp Biol **178**, 231 (1993)

9 AUDSLEY, N , I KAY, T K HAYES, and G M COAST Cross Reactivity Studies of CRF-Related Peptides on Insect Malpighian Tubules Comp Biochem Physiol **110A**, 87 (1995)

10 AUDSLEY, N , C MCINTOSH, and J E PHILLIPS Isolation of a Neuropeptide from Locust Corpus Cardiacum which Influences Ileal Transport J Exp Biol **173**, 261 (1992)

11 AUDSLEY, N , C MCINTOSH, J E PHILLIPS, D A SCHOOLEY, and G M COAST Neuropeptide Regulation of Ion and Fluid Reabsorption in the Insect Excretory System In Perspectives in Comparative Endocrinology (K G Davey, R E Peter, and S S Tobe, eds), pp 74–80 Ottawa National Research Council of Canada 1994

12 BAHR, U , M KARAS, and F HILLENKAMP Analysis of Biopolymers by Metrix-Assisted Laser Desorption/Ionization (MALDI) Mass Spectrometry In Microcharacterization of Proteins (R KELLNER, F LOTTSPEICH, and H E MEYER, eds), pp 149–166 Weinheim VCH Verlagsgesellschaft 1994

13 BAUMANN, E, G GADE, and H PENZLIN Structure-Function Studies on Neurohormone D Activity of Naturally-Occurring Hormone Analogues J Comp Physiol **160B**, 423 (1990)

14 BAUMANN, E , and H PENZLIN Sequence Analysis of Neurohormone D, a Neuropeptide of an Insect, *Periplaneta americana* Biomed Biochim Acta **43**, K13 (1984)

15 BEENAKKERS, A M T The Influence of Corpus Allatum and Corpus Cardiacum on Lipid Metabolism in *Locusta migratoria* Gen Comp Endocrinol **13**, 492 abstr 12 (1969)

16 BELL, G I , W F SWAIN, R PICTET, B CORDELL, H M GOODMAN, and W J RUTTER Nucleotide Sequence of a cDNA Clone Encoding Human Preproinsulin Nature **282**, 525 (1979)

17 BELLES, X , J -L MAESTRO, M -D PIULACHS, A H JOHNSEN, H DUVE, and A THORPE Allatostatic Neuropeptides from the Cockroach *Blattella germanica* (L) (Dictyoptera, Blattellidae) Identification, Immunolocalization and Activity Regul Pept **53**, 237 (1994)

18 BEYENBACH, K W Mechanism and Regulation of Electrolyte Transport in Malpighian Tubules J Insect Physiol **41**, 197 (1995)

19 BLACKBURN, M B , T G KINGAN, W BODNAR, J SHABANOWITZ, D F HUNT, T KEMPE, R M WAGNER, A K RAINA, M E SCHNEE, and M C MA Isolation and Identification of a new Diuretic Peptide from the Tobacco Hornworm, *Manduca sexta* Biochem Biophys Res Comm **181**, 927 (1991)

20 BLACKBURN, M B and M C MA Diuretic Activity of Mas-DP II, an Identified Neuropeptide from *Manduca sexta* an *in vivo* and *in vitro* Examination in the Adult Moth Archs Insect Biochem Physiol **27**, 3 (1994)

21 BLACKBURN, M B , R M WAGNER, J P KOCHANSKY, D J HARRISON, P THOMAS-LAEMONT, and A K RAINA The Identification of two Myoinhibitory Peptides, with Sequence Similarities to the Galanins, Isolated from the Ventral Nerve Cord of *Manduca sexta* Regul Pept **57**, 213 (1995)

22 BLACKBURN, M B , R M WAGNER, J SHABANOWITZ, J P KOCHANSKY, D F HUNT, and A K RAINA The Isolation and Identification of Three Diuretic Kinins from the Abdominal Ventral Nerve Cord of Adult *Helicoverpa zea* J Insect Physiol **41**, 723 (1995)

23 BOGERD, J , F P KOOIMAN, M A P PIJNENBURG, L H P HEKKING, R C H M OUDEJANS, and D J VAN DER HORST Molecular Cloning of Three Distinct cDNAs, Each Encoding a Different Adipokinetic Hormone Precursor, of the Migratory Locust, *Locusta migratoria* Differential Expression of the Distinct Adipokinetic Hormone Precursor Genes During Flight Activity J Biol Chem **29**, 23038 (1995)

24 BOLLENBACHER, W E , E J KATAHIRA, and M A O'BRIEN Insect Prothoracicotropic Hormone Evidence of Two Molecular Forms Science **224**, 1243 (1984)

25 BOROVSKY, D Isolation and Characterization of Highly Purified Mosquito Oostatic Hormone Arch Insect Biochem Physiol **2**, 333 (1985)

26 BOROVSKY, D Oostatic Hormone Inhibits Biosynthesis of Midgut Proteolytic Enzymes and Egg Development in Mosquitoes Arch Insect Biochem Physiol **7**, 187 (1988)

27 BOROVSKY, D , and D A CARLSON The Role of Mosquito Oostatic Hormone in the Regulation of Midgut Serine Proteases In Insect Neurochemistry and Neurophysiology 1989 (A B BORKOVEC and E P MASLER, eds) pp 251 254 Clifton, N J The Humana Press Inc 1990

28 BOROVSKY, D , D A CARLSON, P R GRIFFIN, J SHABANOWITZ, and D F HUNT Mosquito Oostatic Factor a Novel Decapeptide Modulating Trypsin-like Enzyme Biosynthesis in the Midgut The FASEB Journal **4**, 3015 (1990)

29 BOROVSKY, D , D A CARLSON, P R GRIFFIN, J SHABANOWITZ, and D F HUNT Mass Spectrometry and Characterization of *Aedes aegypti* Trypsin Modulating Oostatic Factor (TMOF) and its Analogs Insect Biochem Molec Biol **23**, 703 (1993)

30 BOROVSKY, D , D A CARLSON and D F HUNT Mosquito Oostatic Hormone A Trypsin-modulating Oostatic Factor In Insect Neuropeptides Chemistry, Biology, and Action ACS Symposium Series No 453 (J J MENN, T J KELLY, and E P MASLER, (eds), pp 133–142 Washington, D C American Chemical Society Books 1991

31 BOROVSKY, D , C A POWEL, and D A CARLSON Development of Specific RIA and ELISA to Study Trypsin Modulating Oostatic Factor in Mosquitoes Arch Insect Biochem Physiol **21**, 13 (1992)

32 BRADFIELD, J Y , and L L KEELEY Adipokinetic Hormone Gene Sequence from *Manduca sexta* J Biol Chem **264**, 12791 (1989)

33 BREIDBACH, O , and H DIRCKSEN Crustacean Cardioactive Peptide-immunoreactive Neurons in the Ventral Nerve Cord and the Brain of the Meal Beetle *Tenebrio molitor* During Postembryonic Development Cell Tissue Res **265**, 129 (1991)

34 BROWN, B E, and A N STARRATT Isolation of Proctolin, a Myotropic Peptide, from *Periplaneta americana* J Insect Physiol **23**, 1879 (1975)

35 BROWN, M R, M J KLOWDEN, J W CRIM, L YOUNG, L A SHROUDER, and A O LEA Endogenous Regulation of Mosquito Host-seeking Behavior by a Neuropeptide J Insect Physiol **40**, 399 (1994)

36 BUTENANDT, A, R BECKMANN, D STAMM, and E HECKER Uber den Sexuallockstoff des Seidenspinners *Bombyx mori* Reindarstellung und Konstitution Z Naturforsch **14b**, 283 (1959)

37 BYLEMANS, D, D BOROVSKY, D F HUNT, J SHABANOWITZ, L GRAUWELS, and A DE LOOF Sequencing and Characterization of Trypsin Modulating Oostatic Factor (TMOF) from the Ovaries of the Grey Fleshfly, *Neobellieria (Sarcophaga) bullata* Regul Pept **50**, 61 (1994)

38 BYLEMANS, D, P PROOST, B SAMYN, D BOROVSKY, L GRAUWELS, R HUYBRECHTS, J VAN DAMME, J VAN BEEUMEN, and A DE LOOF Neb-colloostatin, a second Folliculostatin of the Grey Fleshfly *Neobellieria bullata* Eur J Biochem **228**, 45 (1995)

39 CANTERA, R, B S HANSSON, E HALLBERG, and D R NASSEL Postembryonic Development of Leucokinin I Immunoreactive Neurons Innervating a Neurohemal Organ in the Turnip Moth *Agrotis segetum* Cell Tissue Res **269**, 65 (1992)

40 CANTERA, R, and D R NASSEL Segmental Peptidergic Innervation of Abdominal Targets in Larval and Adult Dipteran Insects Revealed with an Antiserum Against Leucokinin I Cell Tissue Res **269**, 459 (1992)

41 CANTERA, R, J A VEENSTRA, and D R NASSEL Postembryonic Development of Corazonin-containing Neurons and Neurosecretory Cells in the Blowfly, *Phormia terraenovae* J Comp Neurol **350**, 559 (1994)

42 CARLISLE, J, and B G LOUGHTON The Inhibition of Protein Synthesis in *Locusta migratoria* by Adipokinetic Hormone J Insect Physiol **32**, 573 (1986)

43 CHAMPAGNE, D E, and J M C RIBEIRO Sialokinin I and II Vasodilatory Tachykinins from the Yellow Fever Mosquito *Aedes aegypti* Proc Natl Acad Sci U S A **91**, 138 (1994)

44 CHEN, R, K A LEWIS, M H PERRIN, and W W VALE Expression Cloning of a Human Corticotropin-releasing-factor Receptor Proc Natl Acad Sci U S A **90**, 8967 (1993)

45 CHEUNG, C C, P K LOI, A W SYLWESTER, T D LEE, and N J TUBLITZ Primary Structure of a Cardioactive Neuropeptide from the Tobacco Hawkmoth, *Manduca sexta* FEBS Lett **313**, 165 (1992)

46 CHIN, A C, E R REYNOLDS, and R H SCHELLER Organization and Expression of the *Drosophila* FMRFamide-related Prohormone Gene DNA Cell Biol **9**, 263 (1990)

47 CHUNG, J S, G J GOLDSWORTHY, and G M COAST Haemolymph and Tissue Titres of Achetakinins in the House Cricket *Acheta domesticus* Effect of Starvation and Dehydration J exp Biol **193**, 307 (1994)

48 CHUNG, J S, C H WHEELER, G J GOLDSWORTHY, and G M COAST Properties of Achetakinin Bindings Sites on Malpighian Tubule Membranes from the House Cricket, *Acheta domesticus* Peptides **16**, 375 (1995)

49 CLOTTENS, F, G GADE, R HUYBRECHTS, and A DE LOOF Immunohistochemical Localisation of the Hypertrehalosaemic Hormone II (Cam-HrTH-II) and Related Peptides in the Nervous System of *Carausius morosus* and *Sarcophaga bullata* Cell Tissue Res **258**, 631 (1989)

50 CLOTTENS, F L, G M HOLMAN, G M COAST, N F TOTTY, T K HAYES, I KAY, A I MALLET, M S WRIGHT J-S CHUNG, O TRUONG, and D L BULL Isolation and

Characterization of a Diuretic Peptide Common to the House Fly and Stable Fly Peptides **15**, 971 (1994)

51 CLOTTENS, F L, S M MEOLA, G M COAST, T K HAYES, M S WRIGHT, R J NACHMAN, and G M HOLMAN Characterization of an Antiserum against an Achetakinin-I analog and its use for the Localization of Culekinin Depolarizing Peptide-II in the Mosquito, *Culex salinurius* Regul Pept **49**, 145 (1993)

52 COAST, G M, J -S CHUNG, G J GOLDSWORTHY, M PATEL, T K HAYES, and I KAY Corticotropin Releasing Factor Related Diuretic Peptides in Insects In Perspectives in Comparative Endocrinology (K G DAVEY, R E PETER, and S S TOBE, eds), pp 67–73 Ottawa National Research Council of Canada 1994

53 COAST, G M, T K HAYES, I KAY, and J -S CHUNG Effect of *Manduca sexta* Diuretic Hormone and Related Peptides on Isolated Malpighian Tubules of the House Cricket *Acheta domesticus* (L) J Exp Biol **162**, 331 (1992)

54 COAST, G M, G M HOLMAN, and R J NACHMAN The Diuretic Activity of a Series of Cephalomyotropic Neuropeptides, the Achetakinins, on Isolated Malpighian Tubules of the House Cricket, *Acheta domesticus* J Insect Physiol **36**, 481 (1990)

55 COAST, G M, I KAY, and C H WHEELER Diuretic Peptides in the House Cricket, *Acheta domesticus* (L) a Possible Dual Control of Malpighian Tubules In Molecular Comparative Physiology (K W Beyenbach, ed), Vol 12, pp 38–66 Basel Karger 1993

56 COAST, G M, R C RAYNE, T K HAYES, A I MALLET, K S J THOMPSON, and J P BACON A Comparison of the Effects of two Putative Diuretic Hormones from *Locusta migratoria* on Isolated Locust Malpighian Tubules J Exp Biol **175**, 1 (1993)

57 COOK, B J, G M HOLMAN, R M WAGNER, and R J NACHMAN Pharmacological Actions of a New Class of Neuropeptides, the Leucokinins I–IV, on the Visceral Muscles of *Leucophaea maderae* Comp Biochem Physiol **93C**, 257 (1989)

58 COOK B J, G M HOLMAN, R M WAGNER, and R J NACHMAN Comparative Pharmacological Actions of Leucokinins V–VIII on the Visceral Muscles of *Leucophaea maderae* Comp Biochem Physiol **95C**, 19 (1990)

59 COUILLAUD, F, A GIRARDIE, and J GIRARDIE Identification of Gonadotropic and Antigonadotropic Factors from the Nervous Part of the Corpora Cardiaca in the African Locust Invert Reprod Develop **16**, 17 (1989)

60 CURTO, E V, M A JARPE, J E BLALOCK, D BOROVSKY, and N R KRISHNA Solution Structure of Trypsin Modulating Oostatic Factor is a Left-handed Helix Biochem Biophys Res Comm **193**, 688 (1993)

61 CUSINATO, O, C H WHEELER, and G J GOLDSWORTHY The Identity and Physiological Actions of an Adipokinetic Hormone in *Acheta domesticus* J Insect Physiol **37**, 461 (1991)

62 CUSSON, M, G D PRESTWICH, B STAY, and S S TOBE Photoaffinity Labeling of Allatostatin Receptor Proteins in the Corpora Allata of the Cockroach, *Diploptera punctata* Biochem Biophys Res Comm **181**, 736 (1991)

63 DAVEY, K G The Mode of Action of the Corpus Cardiacum on the Hindgut in *Periplaneta americana* J Exp Biol **39**, 319 (1962)

64 DAVIS, M -T B, V N VAKHARIA, J HENRY, T G KEMPE, and A K RAINA Molecular Cloning of the Pheromone Biosynthesis-activating Neuropeptide in *Helicoverpa zea* Proc Natl Acad Sci U S A **89**, 142 (1992)

65 DAVIS, N T, S G VELLEMAN, T G KINGAN, and H KESHISHIAN Identification and Distribution of a Proctolin-like Neuropeptide in the Nervous System of the Gypsy Moth, *Lymantria dispar*, and in Other Lepidoptera J Comp Neurol **283**, 71 (1989)

66 DELBECQUE, J -P, K WEIDNER, and K H HOFFMANN Alternative Sites for Ecdysteroid Production in Insects Inv Reprod Dev **18**, 29 (1990)

67 DENLINGER, D L Hormonal Control of Diapause In Comprehensive Insect Physiology, Biochemistry and Pharmacology (G A KERKUT and L I GILBERT, eds), Vol 8, pp 353–412 Pergamon Press, Oxford 1985

68 DIEDEREN, J H B, H A MAAS, H J PEL, H SCHOONEVELD, W F JANSEN, and H G B VULLINGS Co-localization of the Adipokinetic Hormones I and II in the Same Glandular Cells and in the Same Secretory Granules of Corpus Cardiacum of *Locusta migratoria* and *Schistocerca gregaria* Cell Tissue Res **249**, 379 (1987)

69 DIGAN, M E, D N ROBERTS, F E ENDERLIN, A R WOODWORTH, and S J KRAMER Characterization of the Precursor for *Manduca sexta* Diuretic Hormone Mas-DH Proc Natl Acad Sci U S A **89**, 11074 (1992)

70 DIRCKSEN, H, A MULLER, and R KELLER Crustacean Cardioactive Peptide in the Nervous System of the Locust *Locusta migratoria* an Immunocytochemical Study of the Ventral Nerve Cord and Peripheral Innervation Cell Tissue Res **263**, 439 (1991)

71 DONLY, B C, Q DING, S S TOBE, and W G BENDENA Molecular Cloning of the Gene for the Allatostatin Family of Neuropeptides from the Cockroach *Diploptera punctata* Proc Natl Acad Sci U S A **90**, 8807 (1993)

72 DOW, J A T Countercurrent Flows, Water Movements and Nutrient Absorption in the Locust Midgut J Insect Physiol **27**, 579 (1981)

73 DUVE, H, A J ELIA, I ORCHARD, A JOHNSEN, and A THORPE The Effects of CalliFMRFamides and Other FMRFamide-related Neuropeptides on the Activity of the Heart of the Blowfly *Calliphora vomitoria* J Insect Physiol **39**, 31 (1993)

74 DUVE, H, A H JOHNSEN, P EAST, and A THORPE Comparative Aspects of the FMRFamides of Blowflies Isolation of the Peptides, Genes, and Functions In Perspectives in Comparative Endocrinology (K G Davey, R E Peter, and S S Tobe, eds), pp 91–96 Ottawa National Research Council of Canada 1994

75 DUVE, H, A H JOHNSEN, A G SCOTT, P EAST, and A THORPE [Hyp^3] Met-Callatostatin Identification and Biological Properties of a Novel Neuropeptide from the Blowfly *Calliphora vomitoria* J Biol Chem **269**, 21059 (1994)

76 DUVE, H, A H JOHNSEN, A G SCOTT, and A THORPE Isolation, Identification and Functional Significance of [Hyp^2] Met-Callatostatin and Des Gly-Pro Met-Callatostatin, two Further Post-translational Modifications of the Blowfly Neuropeptide Met-Callatostatin Regul Pept **57**, 237 (1995)

77 DUVE, H, A H JOHNSEN, A G SCOTT, C G YU, K J YAGI, S S TOBE, and A THORPE Callatostatins Neuropeptides from the Blowfly *Calliphora vomitoria* with Sequence Homology to Cockroach Allatostatins Proc Natl Acad Sci U S A **90**, 2456 (1993)

78 DUVE, H, A H JOHNSEN, J C SEWELL, A G SCOTT, I ORCHARD, J F REHFELD, and A THORPE Isolation, Structure, and Activity of-Phe-Met-Arg-Phe-NH_2 Neuropeptides (Designated calliFMRFamides) from the Blowfly *Calliphora vomitoria* Proc Natl Acad Sci USA **89**, 2326 (1992)

79 DUVE, H, J C SEWELL, A G SCOTT, and A THORPE Chromatographic Characterisation and Biological Activity of Neuropeptides Immunoreactive to Antisera Against Met^5-Enkephalin-Arg^6-Phe^7 (YGGFMRF) Extracted from the Blowfly *Calliphora vomitoria* (Diptera) Regul Pept **35**, 145 (1991)

80 DUVE, H, and A THORPE Mapping of Enkephalin-Related Peptides in the Nervous System of the Blowfly *Calliphora vomitoria* and their Colocalization with Cholecystokinin (CCK) and Pancreatic Polypeptide (PP)-like Peptides Cell Tissue Res **251**, 399 (1988)

81 DUVE, H, and A THORPE Distribution and Functional Significance of Met-Enkephalin-Arg^{6}-Phe^{7}- and Met-Enkephalin-Arg^{6}-Gly^{7}-Leu^{8}-like Peptides in the Blowfly *Calliphora vomitoria* II Immunocytochemical Mapping of Neuronal Pathways in the Retrocerebral Complex and Thoracic Ganglion Cell Tissue Res **259**, 147 (1990)
82 DUVE, H, and A THORPE Distribution and Functional Significance of Leu-Callatostatins in the Blowfly *Calliphora vomitoria* Cell Tissue Res **276**, 367 (1994)
83 DUVE, H, A THORPE, A G SCOTT, A H JOHNSON, J F REHFELD, E HINES, and P D EAST The Sulfakinins of the Blowfly *Calliphora vomitoria* Peptide Isolation, Gene Cloning and Expression Studies Eur J Biochem **232**, 633 (1995)
84 ECKERT, M, J GABRIEL, H BIRKENBELL, G GREINER, J RAPUS, and G GADE A Comparative Immunocytochemical Study Using an Antiserum Against a Synthetic Analogue of the Corpora Cardiaca Peptide Pea-CAH-I (MI, Neurohormone D) of *Periplaneta americana* Cell Tissue Res 284, 401 (1996)
85 ECKERT, M, R PREDEL, J RAPUS, D LINDE, and H PENZLIN Immunocytochemical Localization of Periviscerokinin, a New Myotropic Neuropeptide from the Perisympathetic Organs of the American Cockroach In Learning and Memory (N ELSNER and R MENZEL, eds), 23rd Neurobiol Conf Goettingen, p 668 Stuttgart, New York Thieme Verlag 1995
86 ELDRIDGE, R, F M HORODYSKI, D B MORTON, D R O'REILLY, J W TRUMAN, L M RIDDIFORD, and L K MILLER Expression of an Eclosion Hormone in Insect Cells Using Baculovirus Vectors Insect Biochem **21**, 341 (1991)
87 ENGELMANN, F The Physiology of Insect Reproduction Oxford Pergamon 1970
88 ENGELMANN, F Vitellogenesis Controlled by Juvenile Hormone In Endocrinology of Insects (R G H DOWNER and H LAUFER, eds), pp 259–270 New York A Liss Inc 1983
89 ENGELMANN, F Hormonal Control of Arthropod Reproduction In Progress in Comparative Endocrinology (A EPPLE, C G SCANES, and M H STETSON, eds), pp 357–364 New York Wiley-Liss 1990
90 ESCH, F S, N C LING, and P BOHLEN Microisolation of Neuropeptides Methods Enzymol **103**, 72 (1983)
91 FERNLUND, P Structure of a Light-Adapting Hormone from the Shrimp, *Pandalus borealis* Biochem Biophys Acta **439**, 17 (1976)
92 FERNLUND, P, and L JOSEFSSON Crustacean Color-change Hormone Amino Acid Sequence and Chemical Synthesis Science **177**, 173 (1972)
93 FEYEREISEN, R Regulation of Juvenile Hormone Titre Synthesis In Comprehensive Insect Physiology, Biochemistry and Pharmacology (G A KERKUT and L I GILBERT, eds), Vol 7, pp 391–429 Oxford Pergamon Press 1985
94 FEYEREISEN, R, and S S TOBE A rapid Partition Assay for Routine Analysis of Juvenile Hormone Release by Insect Corpora Allata Anal Biochem **111**, 372 (1981)
95 FONAGY, A, S MATSUMOTO, L SCHOOFS, A DE LOOF, and T MITSUI *In vivo* and *in vitro* Pheromonotropic Activity of Two Locustatachykinin Peptides in *Bombyx mori* Biosci Biotechnol Biochem **56**, 1692 (1992)
96 FONAGY A, L SCHOOFS, S MATSUMOTO, A DE LOOF and T MITSUI Functional Cross-Reactivities of Some Locustamyotropins and *Bombyx* Pheromone Biosynthesis Activating Neuropeptide J Insect Physiol **38**, 651 (1992)
97 FONAGY, A, L SCHOOFS, P PROOST, J VAN DAMME, H BUEDS, and A DE LOOF Isolation, Primary Structure and Synthesis of Neomyosuppressin, a Myoinhibiting Neuropeptide from the Grey Fleshfly, *Neobellieria bullata* Comp Biochem Physiol **102C**, 239 (1992)

98 FONAGY, A, L SCHOOFS, P PROOST, J VAN DAMME, and A DE LOOF Isolation and Primary Structure of Two Sulfakinin-Like Peptides from the Fleshfly, *Neobellieria bullata* Comp Biochem Physiol **103C**, 135 (1992)

99 FORD, M M, T K HAYES, and L L KEELEY Structure-activity Relationships for Insect Hypertrehalosaemic Hormone The Importance of Side Chains and Termini In Peptides, Chemistry and Biology (G R Marshall, ed), pp 653–655 Leiden Escom Press 1988

100 FOURNIER, B, and J GIRARDIE A New Function for the Locust Neuroparsins Stimulation of Water Reabsorption J Insect Physiol **34**, 309 (1988)

101 FOX, A M, and S E REYNOLDS The Pharmacology of the Lipid-Mobilizing Response to Adipokinetic Hormone Family Peptides in the Moth, *Manduca sexta* J Insect Physiol **37**, 373 (1991)

102 FURUYA, K, S LIAO, S E REYNOLDS, R B OTA, M HACKETT, and D A SCHOOLEY Isolation and Identification of a Cardioactive Peptide from *Tenebrio molitor* and *Spodoptera eridania* Biol Chem Hoppe-Seyler **374**, 1065 (1993)

103 FURUYA, K, R B OTA, S E REYNOLDS, D A SCHOOLEY, R TROETSCHLER, and S J KRAMER Development of an Enzyme-Linked Immunosorbent Assay for a Diuretic Hormone of *Manduca sexta* Proc 2nd Int Symp Mol Insect Sci, Flagstaff, p 65 (1993)

104 GADE, G Relative Hypertrehalosaemic Activities of Naturally Occurring Neuropeptides from the AKH/RPCH Family Z Naturforsch **41c**, 315 (1986)

105 GADE, G Isolation, Physiological Characterization, Release and Sequence Elucidation of a Hypertrehalosaemic Neuropeptide from the Corpus Cardiacum of the Stick Insect, *Sipyloidea sipylus* Physiol Entomol **14**, 405 (1989)

106 GADE, G The Adipokinetic Hormone/Red Pigment-Concentrating Hormone Peptide Family Structures, Interrelationships and Functions J Insect Physiol **36**, 1 (1990)

107 GADE, G Extraction, Purification and Sequencing of Adipokinetic/Red Pigment-Concentrating Hormone-Family Peptides In Chromatography and Isolation of Insect Hormones and Pheromones (A R MCCAFFERY and I D WILSON, eds), pp 165–182 New York Plenum Press, 1990

108 GADE, G The Putative Ancestral Peptide of the Adipokinetic/Red-Pigment-Concentrating Hormone Family Isolated and Sequenced from a Dragonfly Biol Chem Hoppe-Seyler **371**, 475 (1990)

109 GADE, G Structure-Function Studies on Hypertrehalosaemic and Adipokinetic Hormones Activity of Naturally Occurring Analogues and Some N- and C-Terminally Modified Analogues Physiol Entomol **15**, 299 (1990)

110 GADE, G The Adipokinetic Neuropeptide of Mantodea Sequence Elucidation and Evolutionary Relationships Biol Chem Hoppe-Seyler **372**, 193 (1991)

111 GADE, G A Unique Charged Tyrosine-Containing Member of the Adipokinetic Hormone/Red-Pigment-Concentrating Hormone Peptide Family Isolated and Sequenced from Two Beetle Species Biochem J **275**, 671 (1991)

112 GADE, G The Hormonal Integration of Insect Flight Metabolism Zool Jb Physiol **96**, 211 (1992)

113 GADE, G Isolation and Structure Elucidation of Neuropeptides of the AKH/RPCH family in Long-Horned Grasshoppers (Ensifera) Biol Chem Hoppe-Seyler **373**, 1169 (1992)

114 GADE, G Structure-Activity Relationships for the Carbohydrate-Mobilizing Action of Further Bioanalogues of the Adipokinetic Hormone/Red Pigment-Concentrating Hormone Family of Peptides J Insect Physiol **38**, 259 (1992)

115 GADE, G Structure-Activity Relationships for the Lipid-Mobilizing Action of Further

Bioanalogues of the Adipokinetic Hormone/Red Pigment-Concentrating Hormone Family of Peptides J Insect Physiol **39**, 375 (1993)

116 GADE, G Isolation and Structure Elucidation of a Neuropeptide from Three Species of Namib Desert Tenebrionid Beetles S Afr J Zool **29**, 11 (1994)

117 GADE, G Functional and Evolutionary Aspects of Peptides of the AKH/RPCH Family the Odonata and Dictyoptera Story In Insects Chemical, Physiological and Environmental Aspects (D KONOPINSKA, ed), pp 28–34 Wroclaw Wroclaw University Press 1995

118 GADE, G Isolation and Identification of AKH/RPCH Family Peptides in Blister Beetles (Meloidae) Physiol Entomol **20**, 45 (1995)

119 GADE, G , G J GOLDSWORTHY, G KEGEL, and R KELLER Single Step Purification of Locust Adipokinetic Hormones I and II by Reversed-Phase High-Performance Liquid Chromatography and the Amino-Acid Composition of the Hormone II Hoppe-Seyler's Z Physiol Chem **365**, 393 (1984)

120 GADE, G , G J GOLDSWORTHY, M H SCHAFFER, J C COOK and K L RINEHART, JR Sequence Analyses of Adipokinetic Hormones II from Corpora Cardiaca of *Schistocerca nitans*, *Schistocerca gregaria*, and *Locusta migratoria* by Fast Atom Bombardment Mass Spectrometry Biochem Biophys Res Comm **134**, 723 (1986)

121 GADE, G , and T K HAYES Structure-activity Relationships for *Periplaneta americana* Hypertrehalosaemic Hormone I the Importance of Side Chains and Termini Peptides **16**, 1173 (1995)

122 GADE, G , C HILBICH, K BEYREUTHER and K L RINEHART Sequence Analyses of Two Neuropeptides of the AKP/RPCH–Family from the Lubber Grasshopper, *Romalea microptera* Peptides **9**, 681 (1988)

123 GADE, G , and M P -E JANSSENS Cicadas Contain Novel Members of the AKH/RPCH Family Peptides with Hypertrehalosaemic Activity Biol Chem Hoppe-Seyler **375**, 803 (1994)

124 GADE, G , M P -E JANSSENS and R KELLNER A Novel Peptide in the AKH/RPCH Family Isolated from the Corpora Cardiaca of the Emperor Dragonfly, *Anax imperator* Peptides **15**, 1 (1994)

125 GADE, G , and R KELLNER The Metabolic Neuropeptides of the Corpus Cardiacum from the Potato Beetle and the American Cockroach are Identical Peptides **10**, 1287 (1989)

126 GADE, G , and R KELLNER Primary Structures of the Hypertrehalosemic Peptides from Corpora Cardiaca of the Primitive Cockroach *Polyphaga aegyptiaca* Gen Comp Endocrin **86**, 119 (1992)

127 GADE, G , and R KELLNER Isolation and Primary Structure of a Novel Adipokinetic Peptide from the Pyrgomorphid grasshopper, *Phymateus leprosus* Regul Pept **57**, 247 (1995)

128 GADE, G , R KELLNER, K L RINEHART, and M L PROEFKE A Tryptophan-substituted Member of the AKH/RPCH Family Isolated from a Stick Insect Corpus Cardiacum Biochem Biophys Res Comm **189**, 1303 (1992)

129 GADE, G, A LOPATA, R KELLNER, and K L RINEHART Primary Structures of Neuropeptides Isolated from the Corpora Cardiaca of Various Cetonid Beetle Species Determined by Pulsed-liquid Phase Sequencing and Tandem Fast Atom Bombardment Mass Spectrometry Biol Chem Hoppe-Seyler **373**, 133 (1992)

130 GADE, G , S E REYNOLDS, and J R BEECHING Molecular Evolution of Peptides of the AKH/RPCH family In Perspectives in Comparative Endocrinology (K G DAVEY, R E PETER, and S S TOBE, eds), pp 119–128 Ottawa National Research Council of Canada 1994

131 GADE, G, and K L RINEHART, JR Amino Acid Sequence of a Hypertrehalosaemic Neuropeptide from the Corpus Cardiacum of the Cockroach, *Nauphoeta cinerea* Biochem Biophys Res Comm **141**, 774 (1986)

132 GADE, G, and K L RINEHART Primary Sequence Analysis by Fast Atom Bombardment Mass Spectrometry of a Peptide with Adipokinetic Activity from the Corpora Cardiaca of the Cricket *Gryllus bimaculatus* Biochem Biophys Res Comm **149**, 908 (1987)

133 GADE, G, and K L RINEHART, JR Primary Structure of the Hypertrehalosaemic Factor II from the Corpus Cardiacum of the Indian Stick Insect, *Carausius morosus*, Determined by Fast Atom Bombardment Mass Spectrometry Biol Chem Hoppe-Seyler **368**, 67 (1987)

134 GADE, G, and K L RINEHART Primary Structures of Hypertrehalosaemic Neuropeptides Isolated from the Corpora Cardiaca of the Cockroaches *Leucophaea maderae*, *Gromphadorhina portentosa*, *Blattella germanica* and *Blatta orientalis* and of the Stick Insect *Extatosoma tiaratum* Assigned by Tandem Fast Atom Bombardment Mass Spectrometry Biol Chem Hoppe-Seyler **371**, 345 (1990)

135 GADE, G, and G ROSINSKI The Primary Structure of the Hypertrehalosaemic Neuropeptide from Tenebrionid Beetles a Novel Member of the AKH/RPCH-Family Peptides **11**, 455 (1990)

136 GADE, G, H WILPS, and R KELLNER Isolation and Structure of a Novel Charged Member of the Red-Pigment-Concentrating Hormone-Adipokinetic Hormone Family of Peptides Isolated from the Corpora Cardiaca of the Blowfly *Phormia terraenovae* (Diptera) Biochem J **269**, 309 (1990)

137 GAUS, G, L H KLEINHOLZ, G KEGEL, and R KELLER Isolation and Characterization of Red Pigment-Concentrating Hormone (RPCH) from Six Crustacean Species J Comp Physiol **160B**, 373 (1990)

138 GAZIT, Y, E DUNKELBLUM, O BEN-AZIZ, and M ALTSTEIN Immunochemical and Biological Analysis of Pheromone Biosynthesis Activating Neuropeptide in *Heliothis peltigera* Arch Insect Biochem Physiol **19**, 247 (1992)

139 GILBERT, L I, and T A MILLER (eds) Immunological Techniques in Insect Biology New York Springer 1988

140 GIRARDIE, J Molecular Approaches to Study Invertebrate Hormones, with Particular Reference to Insects Netherlands J Zool **45**, 10 (1995)

141 GIRARDIE, J, D BOURÊME, F COUILLARD, M TAMARELLE, and A GIRARDIE Anti-Juvenile Effect of Neuroparsin-A, a Neuroprotein Isolated from Locust *corpora cardiaca* Insect Biochem **17**, 977 (1987)

142 GIRARDIE, J, A GIRARDIE, J-C HUET, and J-C PERNOLLET Amino Acid Sequence of Locust Neuroparsins FEBS Letters **245**, 4 (1989)

143 GIRARDIE, J, J-C HUET, and J-C PERNOLLET The Locust Neuroparsin A Sequence and Similarities with Vertebrate and Insect Polypeptide Hormones Insect Biochem **20**, 659 (1990)

144 GIRARDIE, J, O RICHARD, J-C HUET, C NESPOULOUS, A VAN DORSSELAER, and J-C PERNOLLET Physical Characterization and Sequence Identification of the Ovary Maturating Parsin A New Neurohormone Purified from the Nervous Corpora Cardiaca of the African Locust (*Locusta migratoria migratorioides*) Eur J Biochem **202**, 1121 (1991)

145 GOKULDAS, M, P A HUNT, and D J CANDY The Inhibition of Lipid Synthesis *in vitro* in the Locust *Schistocerca gregaria* by Factors from the Corpora Cardiaca Physiol Entomol **13**, 43 (1988)

146 GOLDSWORTHY, G J The Endocrine Control of Flight Metabolism in Locusts

In Advances in Insect Physiology (M J BERRIDGE, J E TREHERNE, and V B WIGGLESWORTH, eds), pp 149–204 New York Academic Press 1983

147 GOLDSWORTHY, G J Hormonal Control of Flight Metabolism in Locusts In Biology of Grasshoppers (R F CHAPMAN and A JOERN, eds), pp 205–225 New York John Wiley & Sons, Inc 1990

148 GOLDSWORTHY, G J Adipokinetic Hormones of Insects are they the Insect Glucagons? In Perspectives in Comparative Endocrinology (K G DAVEY, R E PETER, and S S TOBE, eds), pp 486–492 Ottawa National Research Council of Canada 1994

149 GOLDSWORTHY, G J , G M COAST, C H WHEELER, O CUSINATO, I KAY, and B KHAMBAY The Structure and Functional Activity of Neuropeptides In Proceedings of Royal Entomological Society Symposium on Insect Molecular Science (J M Crampton and P Eggleston, eds), pp 205–225 London and San Diego Academic Press 1992

150 GOLDSWORTHY, G J , K MALLISON, and C H WHEELER The Relative Potencies of Two Known Locust Adipokinetic Hormones J Insect Physiol **32**, 95 (1986)

151 GOLDSWORTHY, G J , K MALLISON, C H WHEELER, and G GADE Relative Adipokinetic Activities of Members of the Adipokinetic Hormone/Red Pigment Concentrating Hormone Family J Insect Physiol **32**, 433 (1986)

152 GOODING, R H Physiological Aspects of Digestion of the Blood Meal by *Aedes aegypti* (L) and *Culex fatigans* Wiedemann J Med Entomol **3**, 53 (1966)

153 GRAY, R S , D P MUEHLEISEN, E J KATAHIRA, and W E BOLLENBACHER The Prothoracicotropic Hormone (PTTH) of the Commercial Silkmoth, *Bombyx mori*, in the CNS of the Tobacco Hornworm, *Manduca sexta* Peptides **15**, 777 (1994)

154 GRAY, A S , R H OSBORNE, and P J JEWESS Pharmacology of Proctolin Receptors in the Isolated Foregut of the Locust *Schistocerca gregaria*–Identification of [α-Methyl-L-Tyrosine^2]-Proctolin as a Potent Receptor Antagonist J Insect Physiol **40**, 595 (1994)

155 HANSTROM, B Hormones in Invertebrates Oxford Oxford University Press 1939

156 HAYASHI, H , M NAKANO, Y SHIBANAKA, and N FUJITA Expression of a Silkworm Eclosion Hormone Gene in Yeast Biochem Biophys Res Comm **173**, 1065 (1990)

157 HAYES, T K , X -C GUAN, V JOHNSON, A STREY, and S S TOBE Structure-Activity Studies of Allatostatin 4 on the Inhibition of Juvenile Hormone Biosynthesis by Corpora Allata the Importance of Individual Side Chains and Stereochemistry Peptides **15**, 1165 (1994)

158 HAYES, T K , G M HOLMAN, T L PANNABECKER, M S WRIGHT, A A STREY, R J NACHMAN, D F HOEL, J K OLSON, and K W BEYENBACH Culekinin Depolarizing Peptide a Mosquito Leucokinin-Like Peptide that Influences Insect Malpighian Tubule Ion Transport Regul Pept **52**, 235 (1994)

159 HAYES, T K , and L L KEELEY Structure-Activity Relationships on Hyperglycemia by Representatives of the Adipokinetic/Hyperglycemic Hormone Family in *Blaberus discoidalis* Cockroaches J Comp Physiol **160B**, 187 (1990)

160 HAYES, T K , L L KEELEY, and D W KNIGHT Insect Hypertrehalosemic Hormone Isolation and Primary Structure from *Blaberus discoidalis* Cockroaches Biochem Biophys Res Comm **140**, 674 (1986)

161 HAYES, T K , T L PANNABECKER, D J HINCKLEY, G M HOLMAN, R J NACHMAN, D H PETZEL, and K W BEYENBACH Leucokinins, a New Family of Ion Transport Stimulators and Inhibitors in Insect Malpighian Tubules Life Sci **44**, 1259 (1989)

162 HEKIMI, S , and M O'SHEA Antisera Against AKHs and AKH Precursors for Experimental Studies of an Insect Neurosecretory System Insect Biochem **19**, 79 (1989)

163 HERMODSON, M A A Short History of Protein Sequence Analysis In Laboratory Methodology in Biochemistry Amino Acid Analysis and Protein Sequencing (C FINI, A FLORIDI, V N FINELLI, and B WITTMAN-LIEBOLD, eds), pp 1–8 Boca Raton, Florida, U S A CRC Press Inc 1990

164 HERSHEY, A D , L POLENZANI, R M WOODWARD, R MILLEDI, and J E KRAUSE Molecular and Genetic Characterization, Functional Expression, and mRNA Expression Patterns of a Rat Substance P Receptor Ann NY Acad Sci **632**, 63 (1991)

165 HETRU, C , K W LI, P BULET, M LAGUEUX, and J A HOFFMANN Isolation and Structural Characterization of an Insulin-Related Molecule, a Predominant Neuropeptide from *Locusta migratoria* Eur J Biochem **201**, 495 (1991)

166 HIETTER, H , A VAN DORSSELAER, B GREEN, L DENOROY, J HOFFMAN, and B LUU Isolation and Structure Elucidation of a Novel 5-kDa Peptide from Neurohaemal Lobes of the *corpora cardiaca* of *Locusta migratoria* (Insecta, Orthoptera) Eur J Biochem **187**, 241 (1990)

167 HIETTER, H , A VAN DORSSELAER, and B LUU Characterization of three Structurally Related 8–9 kDa Monomeric Peptides Present in the *corpora cardiaca* of *Locusta* a Revised Structure for the Neuroparsins Insect Biochem **21**, 259 (1991)

168 HOFSTEENGE, J , D R MULLER, T DE BEER, A LOFFLER, W J RICHTER, and J F G VLIEGENTHART New Type of Linkage between a Carbohydrate and a Protein C-Glycosylation of a Specific Tryptophan Residue in Human RNase U_s Biochemistry **33**, 13524 (1994)

169 HOLMAN, G M , and B J COOK Pharmacological Properties of Excitatory Neuromuscular Transmission in the Hindgut of the Cockroach, *Leucophaea maderae* J Insect Physiol **16**, 1891 (1970)

170 HOLMAN, G M , B J COOK, and R J NACHMAN Isolation, Primary Structure and Synthesis of Two Neuropeptides from *Leucophaea maderae* Members of a New Family of Cephalomyotropins Comp Biochem Physiol **84C**, 205 (1986)

171 HOLMAN, G M , B J COOK, and R J NACHMAN Isolation, Primary Structure and Synthesis of Two Additional Neuropeptides from *Leucophaea maderae* Members of a New Family of Cephalomyotropins Comp Biochem Physiol **84C**, 271 (1986)

172 HOLMAN, G M , B J COOK, and R J NACHMAN Isolation, Primary Structure and Synthesis of a Blocked Myotropic Neuropeptide Isolated from the Cockroach, *Leucophaea maderae* Comp Biochem Physiol **85C**, 219 (1986)

173 HOLMAN, G M , B J COOK, and R J NACHMAN Isolation, Primary Structure and Synthesis of Leucomyosuppressin, an Insect Neuropeptide which Inhibits Spontaneous Contractions of the Cockroach Hindgut Comp Biochem Physiol **85C**, 329 (1986)

174 HOLMAN, G M , B J COOK, and R J NACHMAN Isolation, Primary Structure and Synthesis of Leucokinins V and VI Myotropic Peptides of *Leucophaea maderae* Comp Biochem Physiol **88C**, 27 (1987)

175 HOLMAN, G M , B J COOK, and R J NACHMAN Isolation, Primary Structure and Synthesis of Leucokinins of VII and VIII the Final Members of this New Family of Cephalomyotropic Peptides Isolated from Head Extracts of *Leucophaea maderae* Comp Biochem Physiol **88C**, 31 (1987)

176 HOLMAN, G M , R J NACHMAN, L SCHOOFS, T K HAYES, M S WRIGHT, and A DE LOOF The *Leucophaea maderae* Hindgut Preparation a Rapid and Sensitive Bioassay Tool for the Isolation of Insect Myotropins of Other Insect Species Insect Biochem **21**, 107 (1991)

177 HOLMAN, G M , R J NACHMAN, and M S WRIGHT A Strategy for the Isolation and Structural Characterization of Certain Insect Myotropic Peptides that Modify the

Spontaneous Contractions of the Isolated Cockroach Hindgut In Chromatography and Isolation of Insect Hormones and Pheromones (A R McCAFFERY and I D WILSON, eds), pp 195–204 New York Plenum Press 1990

178 HOLMAN, G M, R J NACHMAN, M S WRIGHT, L SCHOOFS, T K HAYES, and A DE LOOF Insect Myotropic Peptides Isolation, Structural Characterization, and Biological Activities In Insect Neuropeptides Chemistry, Biology, and Action ACS Symposium Series No 453 (J J MENN, T J KELLY, and E P MASLER, eds), pp 40–50 Washington, D C American Chemical Society Books 1991

179 HOLWERDA, D A, J VAN DOORN, and A M TH BEENAKKERS Characterization of the Adipokinetic and Hyperglycemic Substances from the Locust Corpus Cardiacum Insect Biochem **7**, 151 (1977)

180 HOMBERG, U, N T DAVIS, and J G HILDEBRAND Peptide Immunocytochemistry of Neurosecretory Cells in the Brain and Retrocerebral Complex of the Sphinx Moth *Manduca sexta* J Comp Neurol **303**, 35 (1991)

181 HOMBERG, U, S WURDEN, H DIRCKSEN and K R RAO Comparative Anatomy of Pigment-Dispersing Hormone-Immunoreactive Neurons in the Brain of Orthopteroid Insects Cell Tissue Res **266**, 343 (1991)

182 HORNE, T J, D G DOAK, R C RAYNE, G BALACCO, M O'SHEA, and I D CAMPBELL A Model for the Structure of a Homodimeric Prohormone the Precursor to the Locust Neuropeptide AKH I Proteins Structure, Function and Genetics **20**, 356 (1994)

183 HORODYSKI, F M, L M RIDDIFORD, and J W TRUMAN Isolation and Expression of the Eclosion Hormone Gene from the Tobacco Hornworm, *Manduca sexta* Proc Natl Acad Sci U S A **86**, 8123 (1989)

184 HUESMANN, G R, C C CHEUNG, P K LOI, T D LEE, K M SWIDEREK, and N J TUBLITZ Amino Acid Sequence of CAP_{2b}, an Insect Cardioacceleratory Peptide from the Tobacco Hawkmoth *Manduca sexta* FEBS Lett **371**, 311 (1995)

185 IMAI, K, T KONNO, Y NAKAZAWA, T KOMIYA, M ISOBE, K KOGA, T GOTO, T YAGINUMA, K SAKAKIBARA, K HASEGAWA, and O YAMASHITA Isolation and Structure of Diapause Hormone of the Silkworm, *Bombyx mori*, Proc Japan Acad **67B**, 98 (1991)

186 ISAAC, R E Proctolin Degradation by Membrane Peptidases from Nervous Tissues of the Desert Locust (*Schistocerca gregaria*) Biochem J **245**, 365 (1987)

187 ISAAC, R E Neuropeptide-Degrading Endopeptidase Activity of Locust (*Schistocerca gregaria*) Synaptic Membranes Biochem J **255**, 843 (1988)

188 ISHIBASHI, J, H KATAOKA, H NAGASAWA, A ISOGAI, and A SUZUKI Isolation and Identification of Adipokinetic Hormone of the Silkworm, *Bombyx mori* Biosci Biotech Biochem **56**, 66 (1992)

189 ISHIZAKI, H, A MIZOGUCHI, M FUJISHITA, A SUZUKI, I MORIYA, H O'OKA, H KATAOKA, A ISOGAI, H NAGASAWA, S TAMURA, and A SUZUKI Species Specificity of the Insect Prothoracicotropic Hormone (PTTH) the Presence of *Bombyx*- and *Samia*-Specific PTTHs in the Brain of *Bombyx mori* Dev Growth Diff **25**,593 (1983)

190 ISHIZAKI, H, and A SUZUKI Brain Secretory Peptides of the Silkmoth *Bombyx mori* Prothoracicotropic Hormone and Bombyxin In Progress in Brain Research (J JOOSSE, R M BUIJS, and F J H TILDERS, eds) Vol 92, pp 1–14 Amsterdam Elsevier Science Publishers B V 1992

191 ISHIZAKI, H, A SUZUKI, I MORIYA, A MIZOGUCHI, M FUJISHITA, H O'OKA, H KATAOKA, A ISOGAI, H NAGASAWA, and A SUZUKI Prothoracicotropic Hormone Bioassay Pupal-Adult *Bombyx assay* Dev Growth Diff **25**, 585 (1983)

192 ISOBE, M, and T GOTO Diapause hormone In Neurohormonal Techniques in Insects (T A Miller, ed), pp 216–243 Berlin Springer 1980

193 ISOBE, M , K HASEGAWA, and T GOTO Isolation of the Diapause Hormone from the Silkworm, *Bombyx mori* J Insect Physiol **19**, 1221 (1973)

194 ISOBE, M , K HASEGAWA, and T GOTO Further Characterization of the Silkworm Diapause Hormone A J Insect Physiol **21**, 1917 (1975)

195 IVERSON, L L What next? Trends Neurosci **6**, 293 (1983)

196 IWAMI, M , T ADACHI, H KONDO, A KAWAKAMI, Y SUZUKI, H NAGASAWA, A SUZUKI, and H ISHIZAKI A Novel Family C of the Genes that Encode Bombyxin, an Insulin-Related Brain Secretory Peptide of the Silkmoth, *Bombyx mori* Isolation and Characterization of Gene C-1 Insect Biochem **20**, 295 (1990)

197 IWAMI, M , A KAWAKAMI, H ISHIZAKI, S Y TAKAHASHI, T ADACHI, Y SUZUKI, H NAGASAWA, and A SUZUKI Cloning of a Gene Encoding Bombyxin, an Insulin-Like Brain Secretory Peptide of the Silkmoth *Bombyx mori* with Prothoracicotropic Activity Dev Growth Diff **31**, 31 (1989)

198 JAFFE, H , A K RAINA, C T RILEY, B A FRASER, T G BIRD, C -M TSENG, Y -S ZHANG, and D K HAYES Isolation and Primary Structure of a Neuropeptide Hormone from *Heliothis zea* with Hypertrehalosemic and Adipokinetic Activities Biochem Biophys Res Comm **155**, 344 (1988)

199 JAFFE, H , A K RAINA, C T RILEY, B A FRASER, G M HOLMAN, R M WAGNER, R L RIDGWAY, and D K HAYES Isolation and Primary Structure of a Peptide from the Corpora Cardiaca of *Heliothis zea* with Adipokinetic Activity Biochem Biophys Res Comm **135**, 622 (1986)

200 JAFFE, H , A K RAINA, C T RILEY, B A FRASER, R J NACHMAN, V W VOGEL, Y -S ZHANG, and D K HAYES Primary Structure of Two Neuropeptide Hormones with Adipokinetic and Hypotrehalosemic Activity Isolated from the Corpora Cardiaca of Horse Flies (Diptera) Proc Natl Acad Sci USA **86**, 8161 (1989)

201 JAFFE, H , A K RAINA, R M WAGNER, H M FALES, T G KEMPE, P KEIM, R W BLACKER, and C T RILEY Pheromone Biosynthesis-Activating Neuropeptide Hormone of *Heliothis zea* Isolation and Characterization In Insect Neuropeptides Chemistry, Biology, and Action ACS Symposium Series No 453 (J J MENN, T J KELLY, and E P MASLER, eds), pp 215–225 Washington, D C American Chemical Society Books 1991

202 JANSSENS, M P -E , R KELLNER, and G GADE A Novel Adipokinetic Octapeptide Found in the Damselflies *Pseudagrion inconspicuum* and *Ischnura senegalensis* Biochem J **302**, 539 (1994)

203 JAROS, P P , and G GADE Evidence for a Crustacean Hyperglycemic Hormone-Like Molecule in the Nervous System of the Stick Insect, *Carausius morosus* Cell Tissue Res **227**, 555 (1982)

204 JHOTI, H , A N MCLEOD, T L BLUNDELL, H ISHIZAKI, H NAGASAWA, and A SUZUKI Prothoracicotropic Hormone has an Insulin-Like Tertiary Structure FEBS Lett **219**, 419 (1987)

205 JURENKA, R A , E JACQUIN, and W L ROELOFS Control of the Pheromone Biosynthetic Pathway in *Helicoverpa zea* by the Pheromone Biosynthesis Activating Neuropeptide Arch Insect Biochem Physiol **17**, 81 (1991)

206 KADONO–OKUDA, K , M YAMAMOTO, Y HIGASHINO, K TANIAI, Y KATO, S CHOWDHURY, J XU, S K CHOI, M SUGIYAMA, K NAKASHIMA, S MAEDA, and M YAMAKAWA Baculovirus-Mediated Production of the Human Growth Hormone in Larvae of the Silkworm, *Bombyx mori* Biochem Biophys Res Comm **213**, 389 (1995)

207 KAMITO, T , H TANAKA, B SATO, H NAGASAWA, and A SUZUKI Nucleotide Sequence of cDNA for the Eclosion Hormone of the Silkworm, *Bombyx mori*, and the Expression in a Brain Biochem Biophys Res Comm **182**, 514 (1992)

208 KARLSON, P, and M LUSCHER Pheromones a New Term for a Class of Biologically Active Substances Nature **183**, 55 (1959)

209 KATAOKA, H, J P LI, A S T LUI, S J KRAMER, and D A SCHOOLEY Complete Structure of Eclosion Hormone of *Manduca sexta* Assignment of Disulfide Bond Location Int J Peptide Protein Res **39**, 29 (1992)

210 KATAOKA, H, H NAGASAWA, A ISOGAI, H ISHIZAKI, and A SUZUKI Prothoracicotropic Hormone of the Silkworm, *Bombyx mori* Amino Acid Sequence and Dimeric Structure Agric Biol Chem **55**, 73 (1991)

211 KATAOKA, H, H NAGASAWA, A ISOGAI, S TAMURA, A MIZOGUCHI, Y FUJIWARA, C SUZUKI, H ISHIZAKI, and A SUZUKI Isolation and Partial Characterization of a Prothoracicotropic Hormone of the Silkworm, *Bombyx mori* Agric Biol Chem **51**, 1067 (1987)

212 KATAOKA, H, A TOSCHI, J P LI, R L CARNEY, D A SCHOOLEY, and S J KRAMER Identification of an Allatotropin from Adult *Manduca sexta* Science **243**, 1481 (1989)

213 KATAOKA, H, R G TROETSCHLER, S J KRAMER, B J CESARIN, and D A SCHOOLEY Isolation and Primary Structure of the Eclosion Hormone of the Tobacco Hornworm, *Manduca sexta* Biochem Biophys Res Comm **146**, 746 (1987)

214 KATAOKA, H, R G, TROETSCHLER, J P LI, S J KRAMER, B J CESARIN, and D A SCHOOLEY Isolation and Identification of a Diuretic Hormone from the Tobacco Hornworm moth, *Manduca sexta* Proc Natl Acad Sci U S A **86**, 2976 (1989)

215 KAWAKAMI, A, M IWAMI, H NAGASAWA, A SUZUKI, and H ISHIZAKI Structure and Organization of Four Clustered Genes that Encode Bombyxin, an Insulin-Related Brain Secretory Peptide of the Silkmoth *Bombyx mori* Proc Natl Acad Sci U S A **86**, 6843 (1989)

216 KAWAKAMI, A, H KATAOKA, T OKA, A MIZOGUCHI, M KIMURA-KAWAKAMI, T ADACHI, M IWAMI, H NAGASAWA, A SUZUKI, and H ISHIZAKI Molecular Cloning of the *Bombyx mori* Prothoracicotropic Hormone Science **247**, 1333 (1990)

217 KAWANO, T, H KATAOKA, H NAGASAWA, A ISOGAI, and A SUZUKI cDNA Cloning and Sequence Determination of the Pheromone Biosynthesis Activating Neuropeptide of the Silkworm, *Bombyx mori* Biochem Biophys Res Comm **189** 221 (1992)

218 KAY, I, G M COAST, O CUSINATO, C H WHEELER, N F TOTTY, and G J GOLDSWORTHY Isolation and Characterization of a Diuretic Peptide from *Acheta domesticus* Evidence for a Family of Insect Diuretic Peptides Biol Chem Hoppe-Seyler **372**, 505 (1991)

219 KAY, I, M PATEL, G M COAST, N F TOTTY, A I MALLET, and G J GOLDSWORTHY Isolation, Characterization and Biological Activity of a CRF-Related Diuretic Peptide from *Periplaneta americana* L Regul Pept **42**, 111 (1992)

220 KAY, I, C H WHEELER, G M COAST, N F TOTTY, O CUSINATO, M PATEL, and G J GOLDSWORTHY Characterisation of a Diuretic Peptide from *Locusta migratoria* Biol Chem Hoppe-Seyler **372**, 929 (1991)

221 KEELEY, L L, J Y BRADFIELD, S M SOWA, Y-H LEE, and K-H LU Physiological Actions of Hypertrehalosemic Hormones in Cockroaches In Perspectives in Comparative Endocrinology (K G Davey, R E Peter, and S S Tobe, eds), pp 475–485 Ottawa National Research Council of Canada 1994

222 KEELEY, L L, and T K HAYES Speculations on Biotechnology Applications for Insect Neuroendocrine Research Insect Biochem **17**, 639 (1987)

223 KEELEY, L L, S M SOWA, T K HAYES, and J Y BRADFIELD Neuroendocrine and Juvenile Hormone Effects on Fat Body Protein Synthesis During the Reproductive Cycle in Female *Blaberus discoidalis* Cockroaches Gen Comp Endocrinol **72**, 364 (1988)

224 KEGEL, G, B REICHWEIN, C P TENSON, and R KELLER Amino Acid Sequence of Crustacean Hyperglycaemic Hormone (CHH) from the Crayfish *Orconectes limosus* Emergence of a Novel Neuropeptide Family Neuropeptides **12**, 909 (1991)

225 KEGEL, G, B REICHWEIN, S WEESE, G GAUS, J PETER–KATALINIC, and R KELLER Amino acid Sequence of the Crustacean Hyperglycaemic Hormone (CHH) from the Shore crab, *Carcinus maenas* FEBS Lett **255**, 10 (1989)

226 KELLNER, R Chemical and Enzymatic Fragmentation of Proteins In Microcharacterization of Proteins (R KELLNER, F LOTTSPEICH, and H E MEYER, eds), pp 11–27 Weinheim VCH Verlagsgesellschaft 1994

227 KINGAN, T G, M B BLACKBURN, and A K RAINA The Distribution of Pheromone-Biosynthesis-Activating Neuropeptide (PBAN) Immunoreactivity in the Central Nervous System of the Corn Earworm Moth, *Helicoverpa zea* Cell Tissue Res **270**, 229 (1992)

228 KINGAN, T G, D B TEPLOW, J M PHILLIPS, J P RIEHM, K R RAO, J G HILDEBRAND, U HOMBERG, A E KAMMER, I JARDINE, P R GRIFFIN, and D F HUNT A New Peptide in the FMRFamide Family Isolated from the CNS of the Hawkmoth, *Manduca sexta* Peptides **11**, 849 (1990)

229 KITAMURA, A, H NAGASAWA, H KATAOKA, T ANDO, and A SUZUKI Amino Acid Sequence of Pheromone Biosynthesis Activating Neuropeptide-II (PBAN-II) of the Silkmoth, *Bombyx mori* Agric Biol Chem **54**, 2495 (1990)

230 KITAMURA, A, H NAGASAWA, H KATAOKA, T INOUE, T ANDO, and A SUZUKI Amino Acid Sequence of Pheromone-Biosynthesis-Activating Neuropeptide (PBAN) of the Silkmoth, *Bombyx mori* Biochem Biophys Res Comm **163**, 520 (1989)

231 KLEIN, J M, C J MOHRHERR, F SLEUTELS, N JAENECKE, J P RIEHM, and K R RAO A Highly Conserved Red Pigment-Concentrating Hormone Precursor in the Blue Crab *Callinectes sapidus* Biochem Biophys Res Comm **212**, 151 (1995)

232 KODRIK, D, and G J GOLDSWORTHY Inhibition of RNA Synthesis by Adipokinetic Hormones and Brain Factor(s) in Adult Fat Body of *Locusta migratoria* J Insect Physiol **41**, 127 (1995)

233 KONINGS, P N M, H G B VULLINGS, R SIEBINGA, J H B DIEDEREN, and W F JANSEN Serotonin-Immunoreactive Neurones in the Brain of *Locusta migratoria* Innervating the Corpus Cardiacum Cell Tissue Res **254**, 147 (1988)

234 KONO, T, A MIZOGUCHI, H NAGASAWA, H ISHIZAKI, H FUGO, and A SUZUKI A Monoclonal Antibody Against a Synthetic Carboxyl-Terminal Fragment of the Eclosion Hormone of the Silkworm, *Bombyx mori* Characterization and Application to Immunohistochemistry and Affinity Chromatography Zool Sci **7**, 47 (1990)

235 KONO, T, H NAGASAWA, A ISOGAI, H FUGO, and A SUZUKI Amino Acid Sequence of Eclosion Hormone of the Silkworm, *Bombyx mori* Agric Biol Chem **51**, 2307 (1987)

236 KONO, T, H NAGASAWA, A ISOGAI, H FUGO, and A SUZUKI Isolation and Complete Amino Acid Sequences of Eclosion Hormones of the Silkworm, *Bombyx mori* Insect Biochem **21**, 185 (1991)

237 KONO, T, H NAGASAWA, H KATAOKA, A ISOGAI, H FUGO, and A SUZUKI Eclosion Hormone of the Silkworm *Bombyx mori* Expression in *Escherichia coli* and Location of Disulfide Bonds FEBS Lett **263**, 358 (1990)

238 KONOPINSKA, D, G ROSINSKI, and W SOBOTKA Insect Peptide Hormones, an Overview of the Present Literature Int J Peptide Protein Res **39**, 1 (1992)

239 KOPEČ, S Studies on the Necessity of the Brain for the Inception of Insect Metamorphosis Biol Bull **42**, 323 (1922)

240 Kramer, S J, A Toschi, C A Miller, H Kataoka, G B Quistad, J P Li, R L Carney, and D A Schooley Identification of an Allatostatin from the Tobacco hornworm *Manduca sexta* Proc Natl Acad Sci U S A **88**, 9458 (1991)
241 Kromer, E, and M Lagueux Studies on *Locusta* Insulin Related Peptide In Perspectives in Comparative Endocrinology (K G Davey, R E Peter, and S S Tobe, eds), pp 220–225 Ottawa National Research Council of Canada 1994
242 Kromer–Metzger, E, and M Lagueux Expression of the Gene Encoding an Insulin-Related Peptide in *Locusta* (Insecta, Orthoptera) Evidence for Alternative Promoter Usage Eur J Biochem **221**, 427 (1994)
243 Kuniyoshi, H, H Nagasawa, T Ando, and A Suzuki N-terminal Modified Analogs of C-terminal Fragments of PBAN with Pheromonotropic Activity Insect Biochem Molec Biol **22**, 399 (1992)
244 Kuniyoshi, H, H Nagasawa, T Ando, A Suzuki, R J Nachman, and G M Holman Cross-activity between Pheromone Biosynthesis Activating Neuropeptide (PBAN) and Myotropic Pyrokinin Insect Peptides Biosci Biotechnol Biochem **56**, 167 (1992)
245 Lagueux, M, E Kromer, and J Girardie Cloning of a *Locusta* cDNA Encoding Neuroparsin A Insect Biochem Molec Biol **22**, 511 (1992)
246 Lagueux, M, L Lwoff, M Meister, F Goltzene, and J A Hoffmann cDNAs from Neurosecretory Cells of Brains of *Locusta migratoria* (Insecta, Orthoptera) Encoding a Novel Member of the Superfamily of Insulins Eur J Biochem **187**, 249 (1990)
247 Lange, A B, I Orchard, and B G Loughton Spontaneous and Neurally Evoked Contractions of Visceral Muscles in the Oviduct of *Locusta migratoria* Arch Insect Biochem Physiol **1**, 179 (1984)
248 Lange, A B, N M Peeff, and I Orchard Isolation, Sequence, and Bioactivity of FMRFamide-related Peptides from the Locust Ventral Nerve Cord Peptides **15**, 1089 (1994)
249 Lea, A O The Medial Neurosecretory Cells and Egg Maturation in Mosquitoes J Insect Physiol **13**, 419 (1967)
250 Lea, A O Regulation of Egg Maturation in the Mosquito by the Neurosecretory System the Role of the Corpus Cardiacum Gen Comp Endocrinol (suppl) **3**, 602 (1972)
251 Lederis, K, A Letter, D McMaster, G Moore, and D Schlesinger Complete Amino Acid Sequence of Urotensin I, a Hypotensive and Corticotropin-Releasing Neuropeptide from *Catostomus* Science **218**, 162 (1982)
252 Lederis, K P, Y Okawara, D Richter, and S D Morley Evolutionary Aspects of Corticotropin Releasing Hormones In Progress in Comparative Endocrinology (A Epple, C G Scanes, and M H Stetson, eds) pp 467–472 New York Wiley–Liss 1990
253 Lee, M J, and G J Goldsworthy Acetate Uptake Test, the Basis of a Rapid Method for Determining Potencies of Adipokinetic Peptides for Structure-Activity Studies J Insect Physiol **41**, 163 (1995)
254 Lee, M J, and G J Goldsworthy The Preparation and Use of Dispersed Cells from Fat Body of *Locusta migratoria* in a Filtration Plate Assay for Adipokinetic Peptides Anal Biochem **228**, 155 (1995)
255 Lehman, H K, C M Margiuc, T A Miller, T D Lee, and J G Hildebrand Crustacean Cardioactive Peptide in the Sphinx Moth, *Manduca sexta* Peptides **14**, 735 (1993)
256 Lehmberg, E, R B Ota, K Furuya, D S King, S W Applebaum, H -J Ferenz, and

D A Schooley Identification of a Diuretic Hormone of *Locusta migratoria* Biochem Biophys Res Comm **179**, 1036 (1991)

257 Liebrich, W , R Kellner, and G Gade Isolation and Primary Structures of Neuropeptides of the AKH/RPCH Family from Various Termite Species Peptides **16**, 559 (1995)

258 Linck, B , J M Klein, S Mangerich, R Keller, and W M Weidemann Molecular Cloning of Crustacean Red Pigment Concentrating Hormone Precursor Biochem Biophys Res Comm **195**, 807 (1993)

259 Liu, T P , and K G Davey Partial Characterization of a Proposed Antigonadotropin from the Ovaries of the Insect, *Rhodnius prolixus*, Stal Gen Comp Endocrinol **24**, 405 (1974)

260 Lorenz, M W , R Kellner, and K H Hoffmann A Family of Neuropeptides that Inhibit Juvenile Hormone Biosynthesis in the Cricket, *Gryllus bimaculatus* J Biol Chem **270**, 21103 (1995)

261 Lorenz, M W , R Kellner, and K H Hoffmann Identification of Two Allatostatins from the Cricket, *Gryllus bimaculatus* de Geer (Ensifera, Gryllidae) Additional Members of a Family of Neuropeptides Inhibiting Juvenile Hormone Biosynthesis Regul Pept **57**, 117 (1995)

262 Lottspeich, F , T Houthaeve, and R Kellner The Edman Degradation In Microcharacterization of Proteins (R Kellner, F Lottspeich, and H E Meyer, eds), pp 117–130 Weinheim VCH Verlagsgesellschaft 1994

263 Lundquist, C T , F L Clottens, G M Holman, R Nichols, R J Nachman, and D R Nassel Callitachykinin I and II, Two Novel Myotropic Peptides Isolated from the Blowfly, *Calliphora vomitoria*, that Have Resemblances to Tachykinins Peptides **15**, 761 (1994)

264 Lundquist, T , and D R Nassel Substance P, FMRFamide and Gastrin/Cholecystokinin-Like Neurons in the Thoracico-Abdominal Ganglia of the Flies *Drosophila* and *Calliphora* J comp Neurol **294**, 161 (1990)

265 Ma, M , K -P Sieber, J Ballarino, and S -J Wu ELISA and Monoclonal Antibodies In Immunological Techniques in Insect Biology (L I Gilbert and T A Miller, eds), Chapter 2, pp 43–73 New York Springer 1988

266 Maddrell, S H P Characteristics of Epithelial Transport in Insect Malpighian Tubules In Current Topics in Membranes and Ion Transport (F Bronner and A Kleinzeller, eds), Vol 14, pp 427–463 New York Academic Press 1980

267 Maeda, S Increased Insecticidal Effect by a Recombinant Baculovirus Carrying a Synthetic Diuretic Hormone Gene Biochem Biophys Res Comm **165**, 1177 (1989)

268 Marti, T , K Takio, K A Walsh, G Terzi, and J W Truman Microanalysis of the Amino Acid Sequence of the Eclosion Hormone from the Tobacco Hornworm *Manduca sexta* FEBS Lett **219**, 415 (1987)

269 Maruyama, K , H Hietter, H Nagasawa, A Isogai, S Tamura, A Suzuki, and H Ishizaki Isolation and Primary Structure of Bombyxin-IV, a Novel Molecular Species of Bombyxin from the Silkworm, *Bombyx mori* Agric Biol Chem **52**, 3035 (1988)

270 Maruyama K , H Nagasawa, A Isogai, S Tamura, H Ishizaki, and A Suzuki Synthesis of Bombyxin-IV, an Insulin-Like Heterodimeric Peptide from the Silkworm, *Bombyx mori* Peptides **11**, 169 (1990)

271 Maruyama K , K Nagata, M Tanaka, H Nagasawa, A Isogai, H Ishizaki, and A Suzuki Synthesis of Bombyxin-IV, an Insulin Superfamily Peptide from the Silkworm, *Bombyxin mori*, by Stepwise and Selective Formation of Three Disulfide Bridges J Protein Chem **11**, 1 (1992)

272 Masler, E P, A K Raina, R M Wagner, and J P Kochansky Isolation and Identification of a Pheromonotropic Neuropeptide from the Brain-Suboesophageal Ganglion Complex of *Lymantria dispar* a New Member of the PBAN Family Insect Biochem Molec Biol **24**, 829 (1994)

273 Matsumoto, S, M R Brown, J W Crim, S R Vigna, and A O Lea Isolation and Primary Structure of Neuropeptides from the Mosquito, *Aedes aegypti*, Immunoreactive to FMRFamide Antiserum Insect Biochem **19**, 277 (1989)

274 Matsumoto, S, M R Brown, A Suzuki, and A O Lea Isolation and Characterization of Ovarian Ecdysteroidogenic Hormones from the Mosquito, *Aedes aegypti* Insect Biochem **19**, 651 (1989)

275 Matsumoto, S, A Fonagy, M Kurihara, K Uchiumi, T Nagamine, M Chijimatsu, and T Mitsui Isolation and Primary Structure of a Novel Pheromonotropic Neuropeptide Structurally Related to Leucopyrokinin from the Armyworm Larvae, *Pseudaletia separata* Biochem Biophys Res Comm **182**, 534 (1992)

276 Matsumoto, S, A Fonagy, L Schoofs, A de Loof, M Kurihara, T Nagamine, and T Mitsui Induction of Cuticular Melanization in the Army Worm Larvae, *Pseudaletia separata* by Insect Myotropin Neuropeptides, Possessing FXPRLamide at the C-Terminus J Pestic Sci **18**, 127 (1993)

277 Matsumoto, S, A Kitamura, H Nagasawa, H Kataoka, C Orikasa, T Mitsui, and A Suzuki Functional Diversity of a Neurohormone Produced by the Suboesophageal Ganglion Molecular Identity of Melanization and Reddish Colouration Hormone and Pheromone Biosynthesis Activating Neuropeptide J Insect Physiol **36**, 427 (1990)

278 Matsumoto, S, O Yamashita, A Fonagy, M Kurihara, K Uchiumi, T Nagamine, and T Mitsui Functional Diversity of a Pheromonotropic Neuropeptide Induction of Cuticular Melanization and Embryonic Diapause in Lepidopteran Insects by *Pseudaletia* pheromonotropin J Insect Physiol **38**, 847 (1992)

279 Mayer, R J, and D J Candy Control of Haemolymph Lipid Concentration During Locust Flight an Adipokinetic Hormone from the Corpora Cardiaca J Insect Physiol **15**, 611 (1969)

280 Metzger, J W, and C Eckerskorn Electrospray Mass Spectrometry In Micro-Characterization of Proteins (R Kellner, F Lottspeich, and H E Meyer, eds), pp 167–188 Weinheim VCH Verlagsgesellschaft 1994

281 Meyer, H E Analyzing Post-Translational Protein Modifications In Microcharacterization of Proteins (R Kellner, F Lottspeich, and H E Meyer, eds), pp 131–146 Weinheim VCH Verlagsgesellschaft 1994

282 Milde, J J, R Ziegler, and M Wallstein Adipokinetic Hormone Stimulates Neurones in the Insect Central Nervous System J exp Biol **198**, 1307 (1995)

283 Mizoguchi, A Distribution and Function of Bombyxin In Perspectives in Comparative Endocrinology (K G Davey, R E Peter, and S S Tobe, eds), pp 215–219 Ottawa National Research Council of Canada 1994

284 Mizoguchi, A, H Ishizaki, H Nagasawa, H Kataoka, A Isogai, S Tamura, A Suzuki, M Fujino, and C Kitada A Monoclonal Antibody Against a Synthetic Fragment of Bombyxin (4K-Prothoracicotropic Hormone) from the Silkmoth *Bombyx mori* Characterization and Immunohistochemistry Mol Cell Endocrinol **51**, 227 (1987)

285 Mizoguchi, A, T Oka, H Kataoka, H Nagasawa, A Suzuki, and H Ishizaki Immunohistochemical Localization of Prothoracicotropic Hormone-producing Neurosecretory Cells in the Brain of *Bombyx mori* Dev Growth Diff **32**, 591 (1990)

286 Mohrherr, C J, K Maruska, M Raabe, J P Riehm, and K R Rao Primary

Structure of a Pigment-Dispersing Factor from the Stick Insect, *Carausius morosus* Soc Neurosci Abstr **20**, 914 (1994)

287 MOHRHERR, C J, K R RAO and J P RIEHM Characterization of a Pigment-Dispersing Factor from the American Cockroach Soc Neurosci Abstr **17**, 276 (1991)

288 MONTECUCCHI, P C, and A HENSCHEN Amino Acid Composition and Sequence Analysis of Sauvagine, a New Active Peptide from the Skin of *Phyllomedusa sauvagei* Int J Pept Protein Res **18**, 113 (1981)

289 MOREAU, R, L GOURDOUX, and J GIRARDIE Neuroparsin A New Energetic Neurohormone in the African Locust Arch Insect Biochem Physiol **8**, 135 (1988)

290 MOSHITZKY, P, D F YAMASHIRO, L STUVE, J RAMACHANDRAN, and S W APPLEBAUM Determination of Locust AKH I by Radioimmunoassay and the Identification of an AKH I-Like Factor in the Locust Brain Insect Biochem **17**, 765 (1987)

291 MUEHLEISEN, D P, R S GRAY, E J KATAHIRA, M K THOMAS, and W E BOLLENBACHER Immunoaffinity Purification of the Neuropeptide Prothoracicotropic Hormone from *Manduca sexta* Peptides **14**, 531 (1993)

292 MUREN, J E, C T LUNDQUIST, and D R NASSEL Quantitative Determination of Myotropic Neuropeptide in the Nervous System of the Cockroach *Leucophaea maderae* Distribution and Release of Leucokinins J exp Biol **179**, 289 (1993)

293 NACHMAN, R J, G M HOLMAN, and B J COOK Active Fragments and Analogs of the Insect Neuropeptide Leucopyrokinin Structure-Function Studies Biochem Biophys Res Comm **137**, 936 (1986)

294 NACHMAN, R J, and G M HOLMAN Myotropic Insect Neuropeptide Families from the Cockroach *Leucophaea maderae* Structure-Activity Relationships In Insect Neuropeptides Chemistry, Biology, and Action ACS Symposium Series No 453 (J J MENN, T J KELLY, and E P MASLER, eds), pp 194–214 Washington, D C American Chemical Society Books 1991

295 NACHMAN, R J, G M HOLMAN, B J COOK, W F HADDON, and N LING Leucosulfakinin-II, a Blocked Sulfated Insect Neuropeptide with Homology to Gastrin and Cholecystokinin Biochem Biophys Res Comm **140**, 357 (1986)

296 NACHMAN, R J, G M HOLMAN, and W F HADDON Structural Aspects of Gastrin/CCK-Like Insect Leucosulfakinins and FMRF-Amide Peptides **9**, 137 (1988)

297 NACHMAN, R J, G M HOLMAN, W F HADDON, and T K HAYES Structure Activity Relationships for Myotropic Activity of the Gastrin/CCK-like Insect Sulfakinins Peptide Res **2**, 171 (1989)

298 NACHMAN, R J, G M HOLMAN, W F HADDON, and N LING Leucosulfakinin, a Sulfated Insect Neuropeptide with Homology to Gastrin and Cholecystokinin Science **234**, 71 (1986)

299 NACHMAN, R J, G M HOLMAN, W F HADDON, and W H VENSEL An Active Pseudopeptide Analog of the Leucokinin Insect Neuropeptide Family Int J Peptide Protein Res **37**, 220 (1991)

300 NACHMAN, R J, G M HOLMAN, L SCHOOFS, and O YAMASHITA Silkworm Diapause Induction Activity of Myotropic Pyrokinin (FXPRLamide) Insect Neuropeptides Peptides **14**, 1043 (1993)

301 NACHMAN, R J, V A ROBERTS, H J DYSON, G M HOLMAN, and J A TAINER Active Conformation of an Insect Neuropeptide Family Proc Natl Acad Sci U S A **88**, 4518 (1991)

302 NACHMAN, R J, V A ROBERTS, G M HOLMAN, and R C BEIER Pseudodipeptide Analogs of the Pyrokinin/PBAN (FXPRLa) Insect Neuropeptide Family Containing Carbocyclic Pro-Mimetic Conformation Components Regul Pept **57**, 359 (1995)

303 NASSEL, D R Insect Myotropic Peptides Differential Distribution of Lo-

custatachykinin- and Leucokinin-like Immunoreactive Neurons in the Locust Brain Cell Tissue Res **274**, 27 (1993)

304 NASSEL, D R Neuropeptides in the Insect Brain a Review Cell Tissue Res **273**, 1 (1993)

305 NASSEL, D R Neuropeptides, Multifunctional Messengers in the Nervous System of Insects Verh Dtsch Zool Ges **87**, 59 (1994)

306 NASSEL, D R, E BAYRAKTAROGLU, and H DIRCKSEN Neuropeptides in Neurosecretory and Efferent Neural Systems of Insect Thoracic and Abdominal Ganglia Zool Sci **11**, 15 (1994)

307 NASSEL, D R, R CANTERA, and A KARLSSON Neurons in the Cockroach Nervous System Reacting with Antisera to the Neuropeptide Leucokinin I J Comp Neurol **322**, 45 (1992)

308 NASSEL, D R, and M O'SHEA Proctolin-Like Immunoreactive Neurons in the Blowfly Control Nervous System J Comp Neurol **265**, 437 (1987)

309 NASSEL, D R, P C C M PASSIER, K ELEKES, H DIRCKSEN, H G B VULLINGS, and R CANTERA Evidence that Locustatachykinin I is Involved in Release of Adipokinetic Hormone from Locust Corpora Cardiaca Regul Pept **57**, 297 (1995)

310 NASSEL, D R, S SHIGA, C J MOHRHERR, and K R RAO Pigment Dispersing Hormone-Like Peptide in the Nervous System of the Flies *Phormia* and *Drosophila* Immunocytochemistry and Partial Characterization J Comp Neurol **331**, 183 (1993)

311 NAGASAWA, H Neuropeptides of the Silkworm, *Bombyx mori* Experientia **48**, 425 (1992)

312 NAGASAWA, H, T KAMITO, S TAKAHASHI, A ISOGAI, H FUGO, and A SUZUKI Eclosion Hormone of the Silkworm, *Bombyx mori* Purification and Determination of the N-Terminal Amino Acid Sequence Insect Biochem **15**, 573 (1985)

313 NAGASAWA, H, H KATAOKA, Y HORI, A ISOGAI, S TAMURA, A SUZUKI, F GUO, X ZHONG, A MIZOGUCHI, M FUJISHITA, S Y TAKAHASHI, E OHNISHI, and H ISHIZAKI Isolation and Some Characterization of the Prothoracicotropic Hormone from *Bombyx mori* Gen Comp Endocrinol **53**, 143 (1984)

314 NAGASAWA, H, H KATAOKA, A ISOGAI, S TAMURA, A SUZUKI, H ISHIZAKI, A MIZOGUCHI, Y FUJIWARA, and A SUZUKI Amino-Terminal Amino Acid Sequence of the Silkworm Prothoracicotropic Hormone Homology with Insulin Science **226**, 1344 (1984)

315 NAGASAWA, H, H KATAOKA, A ISOGAI, S TAMURA, A SUZUKI, A MIZOGUCHI, Y FUJIWARA, A SUZUKI, S Y TAKAHASHI, and H ISHIZAKI Amino Acid Sequence of a Prothoracicotropic Hormone of the Silkworm *Bombyx mori* Proc Natl Acad Sci U S A **83**, 5840 (1986)

316 NAGASAWA, H, A KITAMURA, T INOUE, H KATAOKA, S MATSUMOTO, R ARIMA, T ANDO, M UCHIYAMA, and A SUZUKI Isolation of Pheromone Biosynthesis Activating Neuropeptide of the Silkworm, *Bombyx mori* Agric Biol Chem **52**, 2985 (1988)

317 NAGASAWA, H, H KUNIYOSHI, R ARIMA, T KAWANO, T ANDO, and A SUZUKI Structure and Activity of *Bombyx* PBAN Arch Insect Biochem Physiol **25**, 261 (1994)

318 NAGASAWA, H, K MARUYAMA, B SATO, H HIETTER, H KATAOKA, A ISOGAI, S TAMURA, H ISHIZAKI, T SEMBA, and A SUZUKI Structure and Synthesis of Bombyxin from the Silkworm, *Bombyx mori* In Peptide Chemistry (T SHIBA and S SAKAKIBARA, eds), pp 123–126 Osaka Protein Research Foundation 1988

319 NAGASAWA, H, T MIKOGAMI, Y KONO, H FUGO, and A SUZUKI Molecular Heterogeneity of Eclosion Hormone in Adult Heads of the Silkworm, *Bombyx mori* Agric Biol Chem **51**, 1741 (1987)

320 NAMBU, J R, C MURPHY-ERDOSH, P C ANDREWS, G J FEISTNER, and R H SCHELLER

Isolation and Characterization of a *Drosophila* Neuropeptide Gene Neuron **1**, 55 (1988)

321 Nichols, R , Isolation of a Vertebrate Peptide Homologue Present in *Drosophila melanogaster* In Molecular Neurobiology of *Drosophila* (B Ganetsky and J Hall, eds), p 25 Cold Spring Harbour, New York Cold Spring Harbour Laboratory Press, 1987

322 Nichols, R Isolation and Expression of the *Drosophila* Drosulfakinin Neural Peptide Gene Product, DSK-1 Mol Cell Neurosci **3**, 342 (1992)

323 Nichols, R Isolation and Structural Characterization of *Drosophila* TDVDHVFLRFamide and FMRFamide-Containing Neural Peptides J Molec Neurosci **3**, 213 (1992)

324 Nichols, R , S A Schneuwly, and J E Dixon Identification and Characterization of a *Drosophila* homologue to the Vertebrate Neuropeptide Cholecystokinin J Biol Chem **263**, 12167 (1988)

325 Nicolson, S W Diuresis or Clearance is there a Physiological Role for the "Diuretic Hormone" of the Desert Beetle *Onymacris*? J Insect Physiol **37**, 447 (1991)

326 Nicolson, S W The Ionic Basis of Fluid Secretion in Insect Malpighian Tubules Advances in the Last Ten Years J Insect Physiol **39**, 451 (1993)

327 Nijhout, H F Insect Hormones Princeton Princeton University Press 1994

328 Noyes, B E , F N Katz, and M H Schaffer Identification and Expression of the *Drosophila* Adipokinetic Hormone Gene Mol Cell Endocrin **109**, 133 (1995)

329 Noyes, B E , and M H Schaffer The Structurally Similar Neuropeptides Adipokinetic Hormone I and II are Derived from Similar, Very Small mRNAs J Biol Chem **265**, 483 (1990)

330 O'Brien, M A , E J Katahira, T R Flanagan, L W Arnold, G Haughton, and W E Bollenbacher A Monoclonal Antibody to the Insect Prothoracicotropic Hormone J Neurosci **8**, 3247 (1988)

331 Orchard, I Neurosecretion Morphology and Physiology In Endocrinology of Insects (R G H Downer and H Laufer, eds), pp 13–38 New York A Liss Inc 1983

332 Orchard, I Adipokinetic Hormones–an Update J Insect Physiol **33**, 451 (1987)

333 Orchard, I , J H Belanger, and A B Lange Proctolin a Review with Emphasis on Insects J Neurobiol **20**, 470 (1989)

334 O'Shea, M , and M A Adams Proctolin from "Gut Factor" to Model Neuropeptide In Advances in Insect Physiology Vol 19 (P D Evans and V B Wigglesworth, eds), pp 1–28 London Academic Press 1986

335 O'Shea, M , and R C Rayne Adipokinetic Hormones Cell and Molecular Biology Experientia **48**, 430 (1992)

336 O'Shea, M , J Witten, and M Schaffer Isolation and Characterization of two Myoactive Neuropeptides Further Evidence for an Invertebrate Peptide Family J Neurosci **4**, 521 (1984)

337 Otsuka, M , and K Yoshioka Neurotransmitter Functions of Mammalian Tachykinins Physiol Rev **73**, 229 (1993)

338 Oudejans, R C H M , R M Dijkhuizen, F P Kooiman, and A M T Beenakkers Dose-Response Relationships of Adipokinetic Hormones (Lom–AKH–I, II and III) from the Migratory Locust, *Locusta migratoria* Proc Exper Appl Entomol **3**, 165 (1992)

339 Oudejans, R C H M , F P Kooiman, W Heerma, C Versluis, A J Slotboom, and A M T Beenakkers Isolation and Structure Elucidation of a Novel Adipokinetic Hormone (Lom-AKH-III) from the Glandular Lobes of the Corpus Cardiacum of the Migratory locust, *Locusta migratoria* Eur J Biochem **195**, 351 (1991)

340 PAEMEN, L , L SCHOOFS, and A DE LOOF Localization of Lom-AG-Myotropin-I-Like Substances in the Male Reproductive and Nervous Tissue of the Locust, *Locusta migratoria* Cell Tissue Res **268**, 91 (1992)

341 PAEMEN, L , L SCHOOFS, P PROOST, B DECOCK, and A DE LOOF Isolation, Identification and Synthesis of Lom-AG-Myotropin II, a Novel Peptide in the Male Accessory Glands of *Locusta migratoria* Insect Biochem **21**, 243 (1991)

342 PAEMEN, L , A TIPS, L SCHOOFS, P PROOST, J VAN DAMME, and A DE LOOF Lom-AG-Myotropin, a Novel Myotropic Peptide from the Male Accessory Glands of *Locusta migratoria* Peptides **12**, 7 (1991)

343 PANNABECKER, T , and I ORCHARD Octopamine and Cyclic AMP Mediate release of Adipokinetic Hormone I and II from Isolated Neuroendocrine Tissue Mol Cell Endocrin **48**, 153 (1986)

344 PASSIER, P C C M , H G B VULLINGS, J H B DIEDEREN, and D J VAN DER HORST Modulatory Effects of Biogenic Amines on Adipokinetic Hormone Secretion from Locust Corpora Cardiaca *in vitro* Gen Comp Endocrin **97**, 231 (1995)

345 PEEFF, N M , I ORCHARD, and A B LANGE Isolation, Sequence, and Bioactivity of PDVDHVFLRFamide and ADVGHVFLRFamide Peptides from the Locust Central Nervous System Peptides **15**, 387 (1994)

346 PHILLIPS, J E Comparative Physiology of Insect Renal Function Am J Physiol **241**, R241 (1981)

347 PHILLIPS, J E Endocrine Control of Salt and Water Balance Excretion In Endocrinology of Insects (R G H DOWNER and H LAUFER, eds), pp 411–425 New York A Liss Inc 1983

348 PHILLIPS, J E , J HANRAHAN, M CHAMBERLIN, and B THOMSON Mechanisms and Control of Reabsorption in Insect Hindgut Adv Insect Physiol **19**, 329 (1986)

349 PHILLIPS, J E , B THOMSON, J L PEACH, A P STAGG, and N AUDSLEY Mechanisms of Acid-Base Transport and Control in Locust Excretory System Physiol Zool **67**, 95 (1993)

350 POPE, M M , L K GASTON, and T C BAKER Composition, Quantification and Periodicity of Sex Pheromone Volatiles from Individual *Heliothis zea* Females J Insect Physiol **12**, 943 (1984)

351 PRATT, G E , D E FARNSWORTH, and R FEYEREISEN Changes in the Sensitivity of Adult Cockroach Corpora Allata to the Brain Allatostatin Mol Cell Endocrin **70**, 185 (1990)

352 PRATT, G E , D E FARNSWORTH, K F FOK, N R SIEGEL, A L McCORMACK, J SHABANOWITZ, D F HUNT, and R FEYEREISEN Identity of a Second Type of Allatostatin From Cockroach Brains An Octadecapeptide Amide with a Tyrosine-Rich Address Sequence Proc Natl Acad Sci U S A **88**, 2412 (1991)

353 PRATT, G E , D E FARNSWORTH, N R SIEGEL, K F FOK, and R FEYEREISEN Identification of an Allatostatin from Adult *Diploptera punctata* Biochem Biophys Res Comm **163**, 1243 (1989)

354 PRATT, G E , D E FARNSWORTH, N R SIEGEL, K F FOK, and R FEYEREISEN Two Types of Allatostatic Peptides from Brains of the Cockroach *Diploptera punctata* In Insect Neuropeptides Chemistry, Biology, and Action ACS Symposium Series 453 (J J MENN, T J KELLY, and E P MASLER, eds), pp 177–192 Washington D C American Chemical Society Books 1991

355 PREDEL, R , H AGRICOLA, D LINDE, L WOLLWEBER, J A VEENSTRA, and H PENZLIN The Insect Neuropeptide Corazonin Physiological and Immunocytochemical Studies in Blattariae Zoology **98**, 35 (1994)

356 PREDEL, R , D LINDE, J RAPUS, S VETTERMANN, and H PENZLIN Periviscerokinin

(Pea-PVK) A Novel Myotropic Neuropeptide from the Perisympathetic Organs of the American Cockroach Peptides **16**, 61 (1995)
357 PROUX, J P, and J-R HERAULT Cyclic AMP a Second Messenger of the Newly Characterized AVP-Like Insect Diuretic Hormone, the Migratory Locust Diuretic Hormone Neuropeptides **12**, 7 (1988)
358 PROUX, J P, C A MILLER, J P LI, R L CARNEY, A GIRARDIE, M DELAAGE, and D A SCHOOLEY Identification of an Arginine Vasopressin-Like Diuretic Hormone from *Locusta migratoria* Biochem Biophys Res Comm **149**, 180 (1987)
359 PROUX, J, and G ROUGON RAPUZZI Evidence for Vasopressin-like Molecule in Migratory Locust Radioimmunological Measurements in Different Tissues Correlation with Various States of Hydration Gen Comp Endocrin **42**, 378 (1980)
360 PUIROUX, J, and B G LOUGHTON Degradation of the Neuropeptide Proctolin by Membrane Bound Proteases of the Hindgut and Ovary of *Locusta migratoria* and the Effects of Different Inhibitors Arch Insect Biochem Physiol **19**, 193 (1992)
361 PUIROUX, J, A PEDELABORDE, and B G LOUGHTON Characterization of Proctolin Binding Sites on Locust Hindgut Membranes Insect Biochem Molec Biol **22**, 547 (1992)
362 PUIROUX, J, A PEDELABORDE, and B G LOUGHTON Characterization of Proctolin Binding Sites on Locust Oviduct Membranes Insect Biochem Molec Biol **22**, 859 (1992)
363 PUIROUX, J, A PEDELABORDE, and B G LOUGHTON The Effect of Proctolin Analogues and Other Peptides on Locust Oviduct Muscle Contractions Peptides **14**, 1103 (1993)
364 QUISTAD, G B, M E ADAMS, R M SCARBOROUGH, R L CARNEY, and D A SCHOOLEY Metabolism of Proctolin, a Pentapeptide Neurotransmitter in Insects Life Sci **34**, 569 (1984)
365 RAABE, M Etudes des Phenomenes de Neurosecretion au Niveau de la Chaine Nerveuse Ventrale des Phasmides Bull Soc Zool **90**, 631 (1965)
366 RAABE, M Insect Neurohormones New York Plenum Press 1982
367 RAFAELI, A, J HIRSCH, V SOROKER, B KAMENSKY, and A K RAINA Spatial and Temporal Distribution of Pheromone Biosynthesis-Activating Neuropeptide in *Helicoverpa* (*Heliothis*) *armigera* Using RIA and *in vitro* Bioassay Arch Insect Biochem Physiol **18**, 119 (1991)
368 RAINA, A K, and G GADE Insect Peptide Nomenclature Insect Biochem **18**, 785 (1988)
369 RAINA, A K, H JAFFE, T G KEMPE, P KIEM, R W BLACHER, H M FALES, C T RILEY, J A KLUN, R L RIDGWAY, and D K HAYES Identification of a Neuropeptide Hormone that Regulates Sex Pheromone Production in Female Moths Science **244**, 796 (1989)
370 RAINA, A K, H JAFFE, J A KLUN, R L RIDGWAY, and D K HAYES Characteristics of a Neurohormone that Controls Sex Pheromone Production in *Heliothis zea* J Insect Physiol **33**, 809 (1987)
371 RAINA, A K, and T G KEMPE A Pentapeptide of the C-Terminal Sequence of PBAN with Pheromonotropic Activity Insect Biochem **20**, 849 (1990)
372 RAINA, A K, and T G KEMPE Structure Activity Studies of PBAN of *Helicoverpa zea* (Lepidoptera Noctuidae) Insect Biochem Molec Biol **22**, 221 (1992)
373 RAINA, A K, and J A KLUN Brain Factor Control of Sex Pheromone Production in the Female Corn Earworm Moth Science **225**, 531 (1984)
374 RAINA, A K, J A KLUN, and E A STADELBACHER Diel Periodicity and Effect of Age and Mating on Female Sex Pheromone Titer in *Heliothis zea* (Lepidoptera Noctuidae) Ann Entomol Soc Am **79**, 128 (1986)

375 RAINA, A K , and J J MENN Pheromone Biosynthesis Activating Neuropeptide from Discovery to Current Status Arch Insect Biochem Physiol **22**, 141 (1993)

376 RAINA, A , L PANNELL, J KOCHANSKY, and H JAFFE Primary Structure of a Novel Neuropeptide Isolated from the Corpora Cardiaca of Periodical Cicadas having Adipokinetic and Hypertrehalosaemic Activities Insect Biochem Molec Biol **25**, 929 (1995)

377 RAO, K R , C J MOHRHERR, S L BONOMELLI, J P RIEHM and T G KINGAN Insect Neuropeptides Influence on Color Change in Insects and Chromatophoral Pigment Movements in Crustaceans In Insect Neuropeptides Chemistry, Biology, and Action ACS Symposium Series No 453 (J J MENN, T J KELLY, and E P MASLER, eds), pp 110–122 Washington D C American Chemical Society 1991

378 RAO, K R , C J MOHRHERR, J P RIEHM, C A ZAHNOW, S NORTON, L JOHNSON, G E TARR Primary Structure of an Analog of Crustacean Pigment-Dispersing Hormone from the Lubber Grasshopper *Romalea microptera* J Biol Chem **262**, 2672 (1987)

379 RAO, K R , and J P RIEHM Pigment-Dispersing Hormones In The Melanotropic Peptides Ann N Y Acad Sci , Vol 680, pp 78–88 1993

380 RAO, K R , J P RIEHM, C A ZAHNOW, L H KLEINHOLZ, G E TARR, L JOHNSON, S NORTON, M LANDAU, O J SEMMES, R M SATTELBERG, W H JORENBY, and M F HINTZ Characterization of a Pigment-Dispersing Hormone in Eyestalks of the Fiddler Crab *Uca pugilator* Proc Natl Acad Sci U S A **82**, 5319 (1985)

381 RAYNE, R C , and M O'SHEA Inactivation of Neuropeptide Hormones (AKH I and AKH II) Studied in vivo and in vitro Insect Biochem Molec Biol **22**, 25 (1992)

382 RAYNE, R C , and M O'SHEA Structural Requirements for Processing of Pro-Adipokinetic Hormone I Eur J Biochem **217**, 905 (1993)

383 RAYNE, R C , and M O'SHEA Reconstitution of Adipokinetic Hormone Biosynthesis in vitro Indicates Steps in Prohormone Processing Eur J Biochem **219**, 781 (1994)

384 REAGAN, J D Expression Cloning of an Insect Diuretic Hormone Receptor J Biol Chem **269**, 9 (1994)

385 REMY, C , and J GIRARDIE Anatomical Organization of two Vasopressin-Neurophysin-Like Neurosecretory Cells Throughout the Central Nervous System of the Migratory Locust Gen Comp Endocrin **40**, 27 (1980)

386 REYNOLDS, S E , and J W TRUMAN Eclosion Hormone In Endocrinology of Insects (R G H DOWNER and H LAUFER, eds), pp 217–233 New York A Liss Inc 1983

387 RINEHART, K L , T G HOLT, N L FREGEAU, A L STALEY, A G THOMPSON, K -I HARADA, J M CURTIS, L -S RONG, F SUN, L S SHIELD, G GADE, C J P GRIMMELIKHUIJZEN, C C DOUGHTY and C E GRIMSHAW Applications of High-Resolution Tandem FAB Mass Spectrometry In Biological Mass Sectrometry (A L Burlingame and J A McCloskey, eds), pp 233–257 Amsterdam Elsevier Science Publishers B V 1990

388 RIVIER, J , J SPIESS, and W VALE Characterization of Rat Hypothalamic Corticotropin-Releasing Factor Proc Natl Acad Sci U S A **80**, 4851 (1983)

389 ROBB, S , L C PACKMAN, and P D EVANS Isolation, Primary Structure and Bioactivity of SchistoFLRF-amide, a FMRF-amide-like Neuropeptide from the Locust, *Schistocerca gregaria* Biochem Biophys Res Comm **160**, 850 (1989)

390 SAEGUSA, H , A MIZOGUCHI, H KITAHORA, H NAGASAWA, A SUZUKI, and H ISHIZAKI Change in the Titer of Bombyxin–Immunoreactive Material in Hemolymph during the Postembryonic Development of the Silkmoth *Bombyx mori* Dev Growth Differ **34**, 595 (1992)

391 SATO, Y , M IKEDA, and O YAMASHITA Neurosecretory Cells Expressing the Gene for

Common Precursor for Diapause Hormone and Pheromone Biosynthesis-Activating Neuropeptide in the Suboesophageal Ganglion of the Silkworm, *Bombyx mori* Gen Comp Endocrin **96**, 27 (1994)

392 Sato, Y , Y Nakazawa, N Menjo, K Imai, T Komiya, H Saito, M Shin, M Ikeda, K Sakakibara, M Isobe, and O Yamashita A New Diapause Hormone Molecule of the Silkworm, *Bombyx mori* Proc Japan Acad **68B**, 75 (1992)

393 Sato, Y , M Oguchi, N Menjo, K Imai, H Saito, M Ikeda, M Isobe, and O Yamashita Precursor Polyprotein for Multiple Neuropeptides Secreted from the Suboesophageal Ganglion of the Silkworm *Bombyx mori* Characterization of the cDNA Encoding the Diapause Hormone Precursor and Identification of Additional Peptides Proc Natl Acad Sci U S A **90**, 3251 (1993)

394 Scarborough, R M , G C Jamieson, F Kalish, S J Kramer, G A McEnroe, C A Miller, and D A Schooley Isolation and Primary Structure of Two Peptides with Cardioacceleratory and Hyperglycemic Activity from the Corpora Cardiaca of *Periptlaneta americana* Proc Natl Acad Sci USA **81**, 5575 (1984)

395 Schaffer, M H Isolation and Characterization of Neuropeptides In Neurochemical Techniques in Insect Research (H Breer and T A Miller, eds), pp 47–78 Berlin Springer 1985

396 Schaffer, M H , and B E Noyes Adipokinetic Hormone Neuropeptide Family Applying Recombinant DNA Techniques In Insect Neuropeptides Chemistry, Biology, and Action ACS Symposium Series No 453 (J J Menn, T J Kelly, and E P Masler, eds), Chapter 20, pp 226–233 Washington, D C American Chemical Society Books 1991

397 Schaffer, M H , B E Noyes, C A Slaughter, G C Thorne, and S J Gaskell The Fruitfly *Drosophila melanogaster* Contains a Novel Charged Adipokinetic-Hormone-Family Peptide Biochem J **269**, 315 (1990)

398 Scharrer, B Uber sekretorisch tatige Nervenzellen bei wirbellosen Tieren Naturwissenschaften **9**, 131 (1937)

399 Scharrer, B Neurosecretion II Neurosecretory Cells in the Central Nervous System of Cockroaches J Comp Neurol **74**, 93 (1941)

400 Scharrer, B , and E Scharrer Neurosecretion VI A Comparison Between the Intercerebralis-Cardiacum-Allatum System of the Insects and the Hypothalamo-Hypophyseal System of Vertebrates Biol Bull **87**, 243 (1944)

401 Scharrer, E Die Lichtempfindlichkeit blinder Elritzen I Untersuchungen uber das Zwischenhirn der Fische Z Vergl Physiol **7**, 1 (1928)

402 Schneider, L E , and P H Taghert Isolation and Characterization of a *Drosophila* Gene that Encodes Multiple Neuropeptides Related to Phe–Met–Arg–Phe–NH_2 (FMRFamide) Proc Natl Acad Sci U S A **85**, 1993 (1988)

403 Schoofs, L , G M Holman, T K Hayes, J P Kochansky, R J Nachman, and A de Loof Locustatachykinin III and IV Two Additional Insect Neuropeptides with Homology to Peptides of the Vertebrate Tachykinin Family Regul Pept **31**, 199 (1990)

404 Schoofs, L , G M Holman, T K Hayes, R J Nachman, and A De Loof Locustatachykinin I and II, Two Novel Insect Neuropeptides with Homology to Peptides of the Vertebrate Tachykinin Family FEBS Lett **261**, 397 (1990)

405 Schoofs, L , G M Holman, T K Hayes, R J Nachman, and A De Loof Isolation and Identification of a Sulfakinin-like Peptide with Sequence Homology to Vertebrate Gastrin and Cholecystokinin from the Brain of *Locusta migratoria* In Chromatography and Isolation of Insect Hormones and Pheromones (A R McCaffery, and I D Wilson, eds), pp 231–241 New York Plenum Press 1990

406 SCHOOFS, L, G M HOLMAN, T K HAYES, R J NACHMAN, and A DE LOOF Isolation, Identification and Synthesis of Locustamyotropin II, an Additional Neuropeptide of *Locusta migratoria* Member of the Cephalomyotropic Peptide Family Insect Biochem **20**, 479 (1990)

407 SCHOOFS, L, G M HOLMAN, T K HAYES, R J NACHMAN, and A DE LOOF Isolation, Primary Structure, and Synthesis of Locustapyrokinin a Myotropic Peptide of *Locusta migratoria* Gen Comp Endocrin **81**, 97 (1991)

408 SCHOOFS, L, G M HOLMAN, T K HAYES, R J NACHMAN, and A DE LOOF Isolation, Identification and Synthesis of Locustamyoinhibiting Peptide (Lom-MIP), a Novel Biologically Active Neuropeptide from *Locusta migratoria* Regul Pept **36**, 111 (1991)

409 SCHOOFS, L, G M HOLMAN, T K HAYES, R J NACHMAN, J P KOCHANSKY, and A DE LOOF Isolation, Identification and Synthesis of Locustamyotropin III and IV, Two Additional Neuropeptides of *Locusta migratoria* Members of the Locustamyotropin Peptide Family Insect Biochem Molec Biol **22**, 447 (1992)

410 SCHOOFS, L, G M HOLMAN, T K HAYES, A TIPS, R J NACHMAN, F VANDESANDE, and A DE LOOF Isolation, Identification and Synthesis of Locustamyotropin (Lom-MT), a Novel Biologically Active Insect Peptide Peptides **11**, 427 (1990)

411 SCHOOFS, L, G M HOLMAN, R NACHMAN, P PROOST, J VAN DAMME, and A DE LOOF Isolation, Identification and Synthesis of Locustapyrokinin II from *Locusta migratoria*, Another Member of the FXPRL-amide Peptide Family Comp Biochem Physiol **106C**, 103 (1993)

412 SCHOOFS, L, G M HOLMAN, L PAEMEN, D VEELAERT, M AMELINCKX, and A DE LOOF Isolation, Identification and Synthesis of PDVDHVFLRFamide (SchistoFLRFamide) in *Locusta migratoria* and its Association with the Male Accessory Glands, the Salivary Glands, the Head and the Oviduct Peptides **14**, 409 (1993)

413 SCHOOFS, L, G M HOLMAN, P PROOST, J VAN DAMME, T K HAYES, and A DE LOOF Locustakinin, a Novel Myotropic Peptide from *Locusta migratoria*, Isolation, Primary Structure and Synthesis Regul Pept **37**, 49 (1992)

414 SCHOOFS, L, A TIPS, G M HOLMAN, R J NACHMAN, and A DE LOOF Distribution of Locustamyotropin-like Immunoreactivity in the Nervous System of *Locusta migratoria* Regul Pept **37**, 237 (1992)

415 SCHOOFS, L, J VAN DEN BROECK, and A DE LOOF The Myotropic Peptides of *Locusta migratoria* Structures, Distribution, Functions and Receptors Insect Biochem Molec Biol **23**, 859 (1993)

416 SCHOOFS, L, D VEELAERT, G M HOLMAN, T K HAYES, and A DE LOOF Partial Identification, Synthesis and Immunolocalization of Locustamyoinhibin, the Third Myoinhibiting Neuropeptide Isolated from *Locusta migratoria* Regul Pept **52**, 139 (1994)

417 SCHOOLEY, D A, and F C BAKER Juvenile Hormone Biosynthesis In Comprehensive Insect Physiology, Biochemistry and Pharmacology (G A KERKUT and L I GILBERT, eds) Vol 7, pp 363–389 Oxford Pergamon Press 1985

418 SCHOOLEY, D A, H KATAOKA, S J KRAMER, and A TOSCHI Isolation Techniques for Insect Neuropeptides In Insect Neurochemistry and Neurophysiology 1989 (A B BORKOVEC and E P MASLER, eds), pp 39–62 Clifton, N J The Humana Press Inc 1990

419 SCHOOLEY, D A, C A MILLER, and J P PROUX Isolation of Two Arginine Vasopressin-like Factors from Ganglia of *Locusta migratoria* Arch Insect Biochem Physiol **5**, 157 (1987)

420 SCHOONEVELD, H, H M ROMBERG–PRIVEE, and J A VEENSTRA Adipokinetic Hormone-immunoreactive Peptide in the Endocrine and Central Nervous System of

Several Insect Species a Comparative Immunocytochemical Approach Gen Comp Endocrin **57**, 184 (1985)

421 Schooneveld, H , H M Romberg–Privee, and J A Veenstra Immunocytochemical Differentiation between Adipokinetic Hormone (AKH)-like Peptides in Neurons and Glandular Cells in the Corpus Cardiacum of *Locusta migratoria* and *Periplaneta americana* with C-terminal and N-terminal Specific Antisera to AKH Cell Tissue Res **243**, 9 (1986)

422 Schooneveld, H , G I Tesser, J A Veenstra, and H M Romberg–Privee Adipokinetic Hormone and AKH-like Peptide Demonstrated in Corpora Cardiaca and Nervous System of *Locusta migratoria* by Immunocytochemistry Cell Tissue Res **230**, 67 (1983)

423 Schooneveld, H , and J A Veenstra Immunocytochemistry In Immunological Techniques in Insect Biology (L I Gilbert and T A Miller, eds), Chapter 4, pp 93–133 New York Springer 1988

424 Schulz–Aellen, M F , E Roulet, J Fisher–Lougheed, and M O'Shea Synthesis of a Homodimeric Neurohormone Precursor of Locust Adipokinetic Hormone Studied by *in vitro* Translation and cDNA Cloning Neuron **2**, 1369 (1989)

425 Schwartz, L M , and J W Truman Peptide and Steroid Regulation of Muscle Degeneration in an Insect Science **215**, 1420 (1982)

426 Schwarz, T L , C M H Lee, K K Siwicki, D G Standaert, and E A Kravitz Proctolin in the Lobster the Distribution, Release, and Chemical Characterization of a Likely Neurohormone J Neurosci **4**, 1300 (1984)

427 Serwe, M , and H E Meyer Microseparation Techniques I High Performance Liquid Chromatography In Microcharacterization of Proteins (R Kellner, F Lottspeich, and H E Meyer, eds), pp 29 45 Weinheim VCH Verlagsgesellschaft 1994

428 Shimonishi, Y , and T Takao Methods for Determination of Protein Sequences by Fast Atom Bombardment Mass Spectrometry In Laboratory Methodology in Biochemistry Amino Acid Analysis and Protein Sequencing (C Fini, A Floridi, V N Finelli, and B Wittman–Liebold, eds), pp 239–256 Boca Raton, Florida, U S A CRC Press Inc 1990

429 Siegert, K , P Morgan, and W Mordue Primary Structures of Locust Adipokinetic Hormones II Biol Chem Hoppe-Seyler **366**, 723 (1985)

430 Siegert, K J , and W Mordue Elucidation of the Primary Structures of the Cockroach Hyperglycaemic Hormones I and II Using Enzymatic Techniques and Gas-phase Sequencing Physiol Entomol **11**, 205 (1986)

431 Siegert, K J , and W Mordue Breakdown of Locust Adipokinetic Hormone I by Malpighian Tubules of *Schistocerca gregaria* Insect Biochem **17**, 705 (1987)

432 Siegert, K J , and W Mordue Preliminary Characterization of Enzyme Activities in Malpighian Tubules Involved in the Breakdown of Adipokinetic Hormones Arch Insect Biochem Physiol **19**, 147 (1992)

433 Snyder, S H Brain Peptides as Neurotransmitters Science, **209**, 976 (1980)

434 Spittaels, K , L Schoofs, L Grauwels, H Smet, J van Damme, P Proost, and A De Loof Isolation, Identification and Synthesis of Novel Oviductal Motility Stimulating Head Peptide in the Colorado Potato Beetle, *Leptinotarsa decemlineata* Peptides **12**, 31 (1991)

435 Spring, J H Endocrine Regulation of Diuresis in Insects J Insect Physiol **36**, 13 (1990)

436 Spring, J H , and I Kim Differential Effects of Neuropeptides on the Distal and Mid-Tubules of the House Cricket Arch Insect Biochem Physiol **29**, 11 (1995)

437 Stangier, J , C Hilbich, K Beyreuther, and R Keller Unusual Cardioactive

Peptide (CCAP) from Pericardial Organs of the Shore Crab *Carcinus maenas* Proc Natl Acad Sci U S A **84**, 575 (1987)

438 STANGIER, J , C HILBICH, and R KELLER Occurrence of Crustacean Cardioactive Peptide (CCAP) in the Nervous System of an Insect, *Locusta migratoria* J Comp Physiol **159B**, 5 (1989)

439 STARRATT, A N , and B E BROWN Structure of the Pentapeptide Proctolin, a Proposed Neurotransmitter in Insects Life Sci **17**, 1253 (1975)

440 STARRATT, A N , and B E BROWN Analogs of the Insect Myotropic Peptide Proctolin Synthesis and Structure-activity Studies Biochem Biophys Res Comm **90**, 1125 (1979)

441 STARRATT, A N , and R W STEELE *In vivo* Inactivation of the Insect Neuropeptide Proctolin in *Periplaneta americana* Insect Biochem **14**, 97 (1984)

442 STARRATT, A N , and R W STEELE *In vitro* Inactivation of the Insect Neuropeptide Proctolin in Haemolymph from *Periplaneta americana* Insect Biochem **15**, 511 (1985)

443 STAY, B , S S TOBE, and W G BENDENA Allatostatins Identification, Primary Structures, Functions and Distributions Adv Insect Physiol **25**, 267 (1994)

444 STAY, B , A P WOODHEAD, S JOSHI, and S S TOBE Allatostatins Neuropeptide Inhibitors of Juvenile Hormone Synthesis in Brain and Corpora Allata of the Cockroach *Diploptera punctata* In Insect Neuropeptides Chemistry, Biology, and Action ACS Symposium Series No 453 (J J MENN, T J KELLY, and E P MASLER, eds), pp 164–176 Washington D C American Chemical Society Books 1991

445 STEEL, C G H , and K G DAVEY Integration in the Insect Endocrine System In Comprehensive Insect Physiology, Biochemistry and Pharmacology (G A KERKUT and L I GILBERT, eds), Vol 8, pp 1–35 Oxford Pergamon Press 1985

446 STEELE, J E Occurrence of a Hyperglycaemic Factor in the Corpus Cardiacum of an Insect Nature **192**, 680 (1961)

447 STEELE, J E Hormonal Modulation of Carbohydrate and Lipid Metabolism in Fat Body In Insect Biology in the Future (M LOCKE and D S SMITH, eds), pp 253 271 New York Academic Press 1980

448 STONE, K L , M B LOPRESTI, and K R WILLIAMS Enzymatic Digestion of Proteins and HPLC Peptide Isolation in the Subnanomole Range In Laboratory Methodology in Biochemistry Amino Acid Analysis and Protein Sequencing (C FINI, A FLORIDI, V N FINELLI, and B WITTMAN–LIEBOLD, eds), pp 181–205 Boca Raton, Florida, U S A CRC Press Inc 1990

449 STONE, J V , and W MORDUE Adipokinetic Hormone In Neurohormonal Techniques in Insects (T A Miller, ed), pp 31–80 New York Springer Verlag 1980

450 STONE, J V , W MORDUE, K E BATLEY, and H R MORRIS Structure of Locust Adipokinetic Hormone, a Neurohormone that Regulates Lipid Utilisation During Flight Nature **263**, 207 (1976)

451 STONE, J V , W MORDUE, C E BROOMFIELD, and P M HARDY Structure-Activity Relationships for the Lipid-Mobilising Action of Locust Adipokinetic Hormone Synthesis and Activity of a Series of Hormone Analogues Eur J Biochem **89**, 195 (1978)

452 SULLIVAN, R E , and R W NEWCOMB Structure-Function Analysis of an Arthropod Peptide Hormone Proctolin and Synthetic Analogues Compared on the Cockroach Hindgut Receptor Peptides **3**, 337 (1982)

453 SUZUKI, A , H NAGASAWA, T KONO, B SATO, T KAMITO, H TANAKA, Y SAKAGAMI, A MIZOGUCHI, H ISHIZAKI, and H FUGO *Bombyx* Eclosion Hormone In Insect Neurochemistry and Neurophysiology 1989 (A B BORKOVEC and E P MASLER, eds), pp 211–214 Clifton, N J The Humana Press Inc 1990

454 Taghert, P H, and L E Schneider Interspecific Comparison of a *Drosophila* Gene Encoding FMRFamide-Related Neuropeptides J Neurosci **10**, 1929 (1990)

455 Thorpe, A, and H Duve Insect Neuropeptides In Current Topics in Neuroendocrinology, Vol 9, pp 185–230 Berlin Springer 1988

456 Tips, A, L Schoofs, L Paemen, M Ma, M Blackburn, A Raina, and A De Loof Co-localization of Locustamyotropin- and Pheromone Biosynthesis Activating Neuropeptide-like Immunoreactivity in the Central Nervous System of Five Insect Species Comp Biochem Physiol **106A**, 195 (1993)

457 Tobe, S S, and N Clarke The effect of L-methionine Concentration on Juvenile Hormone Biosynthesis by Corpora Allata of the Cockroach *Diploptera punctata* Insect Biochem **15**, 175 (1985)

458 Troetschler, R G, and S J Kramer Mode of Action Studies on a *Manduca sexta* Diuretic Hormone Archs Insect Biochem Physiol **20**, 35 (1992)

459 Truman, J W Hormonal Control of Ecdysis In Comprehensive Insect Physiology, Biochemistry and Pharmacology (G A Kerkut and L I Gilbert, eds), Vol 8, pp 413–440 Oxford Pergamon Press 1985

460 Truman, J W, P H Taghert, P F Copenhaver, N J Tublitz and L M Schwartz Eclosion Hormone May Control All Ecdyses in Insects Nature **291**, 70 (1981)

461 Tublitz, N J, D L Brink, K S Broadie, P K Loi, and A W Sylwester From Behavior to Molecules an Integrated Approach to the Study of Neuropeptides Trends Neurosci **14**, 254 (1991)

462 Unger, H Untersuchungen zur Neurohormonalen Steuerung der Herztatigkeit bei Schaben Biol Zbl **76**, 204 (1957)

463 Vakharia, V N, A K Raina, T G Kingan, and T G Kempe Synthetic Pheromone Biosynthesis Activating Neuropeptide Gene Expressed in a Baculovirus Expression System Insect Biochem Molec Biol **25**, 583 (1995)

464 Veenstra, J A Isolation and Structure of Corazonin, a Cardioactive Peptide from the American Cockroach FEBS Lett **250**, 231 (1989)

465 Veenstra, J A Isolation and Structure of Two Gastrin/CCK-like Neuropeptides from the American Cockroach Homologous to the Leucosulfakinins Neuropeptides **14**, 145 (1989)

466 Veenstra, J A Presence of Corazonin in Three Insect Species, and Isolation and Identification of [His^7] Corazonin from *Schistocerca americana* Peptides **12**, 1285 (1991)

467 Veenstra, J A Isolation and Identification of Three Leucokinins from the Mosquito *Aedes aegypti* Biochem Biophys Res Comm **202**, 715 (1994)

468 Veenstra, J A, and F Camps Structure of the Hypertrehalosemic Neuropeptide of the German Cockroach, *Blattella germanica* Neuropeptides **15**, 107 (1990)

469 Veenstra, J A, and N T Davis Localization of Corazonin in the Nervous System of the Cockroach *Periplaneta americana* Cell Tissue Res **274**, 57 (1993)

470 Veenstra, J A, and H H Hagedorn Identification of Neuroendocrine Cells Producing a Diuretic Hormone in the Tobacco Hornworm Moth, *Manduca sexta* Cell Tissue Res **266**, 359 (1991)

471 Veenstra, J A, and H H Hagedorn Isolation of Two AKH-Related Peptides from Cicadas Arch Insect Biochem Physiol **29**, 391 (1995)

472 Veenstra, J A, H M Romberg–Privee, and H Schooneveld A Proctolin-like Peptide and its Immunocytochemical Localization in the Colorado Potato Beetle, *Leptinotarsa decemlineata* Cell Tissue Res **240**, 535 (1985)

473 Wagner, R M, C W Woods, J A Hayes, J P Kochansky, J C Hill, and B A Fraser Isolation and Identification of a Novel Peptide from the Accessory Sex Gland of the Female House Fly, *Musca domestica* Biochem Biophys Res Comm **194**, 1336 (1993)

474 WEAVER, R J, Z A FREEMAN, M G PICKERING, and J P EDWARDS Identification of two Allatostatins from the CNS of the Cockroach *Periplaneta americana* Novel Members of a Family of Neuropeptide Inhibitors of Insect Juvenile Hormone Biosynthesis Comp Biochem Physiol **107C**, 119 (1994)

475 WEIGT, C, H E MEYER, and R KELLNER Sequence Analysis of Proteins and Peptides by Mass Spectrometry In Microcharacterization of Proteins (R KELLNER, F LOTTSPEICH, and H E MEYER, eds), pp 189–205 Weinheim VCH Verlagsgesellschaft 1994

476 WHEELER, C H, and G M COAST Assay and Characterisation of Diuretic Factors in Insects J Insect Physiol **36**, 23 (1990)

477 WHEELER, C H, A F DRAKE, C M WILMOT, J M THORNTON, G GADE, and G J GOLDSWORTHY Structures in the AKH family of Neuropeptides In Insect Neurochemistry and Neurophysiology 1989 (A B BORKOVEC and E P MASLER, eds), pp 235–238 Clifton, N J The Humana Press Inc 1990

478 WITTEN, J L, M H SCHAFFER, M O'SHEA, J C COOK, M E HEMLING, and K L RINEHART, JR Structures of Two Cockroach Neuropeptides Assigned by Fast Atom Bombardment Mass Spectrometry Biochem Biophys Res Comm **124**, 350 (1984)

479 WOODHEAD, A P, M A KHAN, B STAY, and S S TOBE Two New Allatostatins from the Brains of *Diploptera punctata* Insect Biochem Molec Biol **24**, 257 (1994)

480 WOODHEAD, A P, B STAY, S L SEIDEL, M A KHAN, and S S TOBE Primary Structure of Four Allatostatins Neuropeptide Inhibitors of Juvenile Hormone Synthesis Proc Natl Acad Sci U S A **86**, 5997 (1989)

481 WOODRING, J P, S DAS, R KELLNER, and G GADE The Sequence of *Acheta* Adipokinetic Hormone and the Variation in Corpus Cardiacum Content and Hyperlipaemic Response with Age Z Naturforsch **45c**, 1176 (1990)

482 XU, W-H, Y SATO, M IKEDA, and O YAMASHITA Molecular Characterization of the Gene Encoding the Precursor Protein of Diapause Hormone and Pheromone Biosynthesis Activating Neuropeptide (DH–PBAN) of the Silkworm, *Bombyx mori* and its Distribution in Some Insects Biochim Biophys Acta **1261**, 83 (1995)

483 XU, W-H, Y SATO, M IKEDA, and O YAMASHITA Stage-Dependent and Temperature-Controlled Expression of the Gene Encoding the Precursor Protein of Diapause Hormone and Pheromone Biosynthesis Activating Neuropeptide in the Silkworm, *Bombyx mori* J Biol Chem **270**, 3804 (1995)

484 YAMASHITA, O Egg Diapause In Endocrinology of Insects (R G H DOWNER and H LAUFER, eds), pp 337–342 New York A Liss Inc 1983

485 YAMASHITA, O, and K SUZUKI Role of Morphogenetic Hormones in Embryonic Diapause In Morphogenetic Hormones in Arthropods (A P GUPTA, ed) Vol 3, pp 81–128 New Brunswick Rutger University Press 1991

486 YI, S Detection, Purification and Identification of Myoactive Peptide in Larval Tissues of *Manduca sexta* (L) (Lepidoptera Sphingidae) Ph D Thesis State University of Gent, Belgium 1993

487 YI, S-X, L TIRRY, C BAI, B DEVREESE, J VAN BEEUMEN, and D DEGHEELE Isolation, Identification, and Synthesis of Mas-MG-MT-I, a Novel Peptide from the Larval Midgut of *Manduca sexta* (Lepidoptera Sphingidae) Arch Insect Biochem Physiol **28**, 159 (1995)

488 YU, C G, T K HAYES, A STREY, W G BENDENA, and S S TOBE Identification and Partial Characterization of Receptors for Allatostatins in Brain and Corpora Allata of

the Cockroach *Diploptera punctata* Using a Binding Assay and Photoaffinity Labeling Regul Pept **57**, 347 (1995)

489 ZIEGLER, R , K ECKART, R D JASENSKY, and J H LAW Structure-activity Studies on Adipokinetic Hormones in *Manduca sexta* Arch Insect Biochem Physiol **18**, 229 (1991)

490 ZIEGLER, R , K ECKART, and J H LAW Adipokinetic Hormone Controls Lipid Metabolism in Adults and Carbohydrate Metabolism in Larvae of *Manduca sexta* Peptides **11**, 1037 (1990)

491 ZIEGLER, R , K ECKART, H SCHWARZ, and R KELLER Amino Acid Sequence of *Manduca sexta* Adipokinetic Hormone Elucidated by Combined Fast Atom Bombardment (FAB)/Tandem Mass Spectrometry Biochem Biophys Res Comm **133**, 337 (1985)

492 ZIEGLER, R , R D JASENSKY, and H MORIMOTO Characterization of the Adipokinetic Hormone Receptor from the Fat Body of *Manduca sexta* Regul Pept **57**, 329 (1995)

493 ZOLLNER, N , and K KIRSCH Über die quantitative Bestimmung von Lipoiden (Mikromethode) mittels der vielen naturlichen Lipoiden (allen bekannten Plasmalipoiden) gemeinsamen Sulfophosphovanillin Reaktion Z Ges Exp Med **135**, 545 (1962)

494 ZUBRZYCKI, I Z , and G GADE Conformational Study on an Insect Neuropeptide of the AKH/RPCH Family by Combined ^{1}H NMR Spectroscopy and Molecular Mechanics Biochem Biophys Res Comm **198**, 228 (1994)

(Received April 1, 1996)

Sesquiterpenoids from *Thapsia* Species and Medicinal Chemistry of the Thapsigargins

S B CHRISTENSEN, A ANDERSEN, and U W SMITT
Department of Medicinal Chemistry, Royal Danish School of Pharmacy, DK-2100 Copenhagen, Denmark

Contents

1. Introduction

For centuries preparations containing resin from the root of *Thapsia garganica* L. (Fig. 1) have been used in Arabian and European medicine for treatment of pulmonary diseases, catarrh and as counterirritants for relief of rheumatic pains (*1*). The properties of the resin were described already by Theophrastos (372–287 B.C.), Dioscorides (approximately A.D. 50), and Plinius (A.D. 24–79) (*2*). *Radix Thapsiae* and *Resina Thapsiae* have been included in several pharmacopoeias, the latest in the French pharmacopoeia from 1937. The two major active principles were about

Fig. 1. *Thapsia garganica*

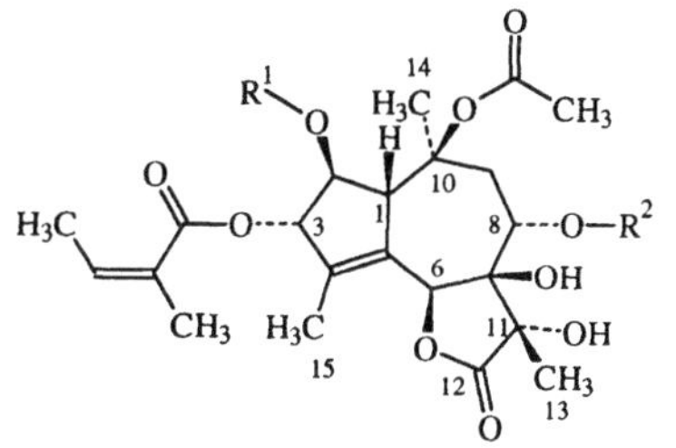

Thapsigargin (**1**), R1= Oct, R2= But

Thapsigargicin (**2**), R1= Hex, R2 = But

Thapsitranstagin (**3**), R1= iVal, R2= 2-MeBut

Thapsivillosin A (**4**), R1= Ang, R2= Sen

Thapsivillosin B (**5**), R1= Ang, R2= 2-MeBut

Thapsivillosin C (**6**), R1= Oct, R2= 2-MeBut

Thapsivillosin D (**7**), R1= 6-MeOct, R2= Sen

Thapsivillosin E (**8**), R1= 6-MeOct, R2= 2-MeBut

Thapsivillosin G (**9**), R1= 6-MeHep, R2= 2-MeBut

Thapsivillosin H (**10**), R1 or R2= Ang or Sen

Thapsivillosin I (**11**), R1= Ang, R2= But

Thapsivillosin J (**12**), R1= iVal, R2= But

Thapsivillosin K (**13**), R1= Sen, R2= 2-MeBut

Chart 1. Hexaoxygenated thapsigargins found in *Thapsia*

two decades ago found to be the sesquiterpene lactones thapsigargin (**1**) and thapsigargicin (**2**) (*3*).

If applied on the skin these compounds induce within 4–5 hours erythema, small vesiculae and intense itching which remains for several days. The present interest in the genus *Thapsia* arose when thapsigargin and thapsigargicin were recognized as highly potent histamine liberators (*3*), general stimulants of the immune system (*4–7*), non-TPA tumour promoters (*8, 9*) and selective inhibitors of the microsomal Ca^{2+}-ATPases (SERCA-ATPases) (*6, 10, 11*). Besides thapsigargin and thapsigargicin a number of related hexaoxygenated guaianolides (**3–13**) only differing in the structure of the acyl groups attached to O(2) and O(8) (*12–14*), and

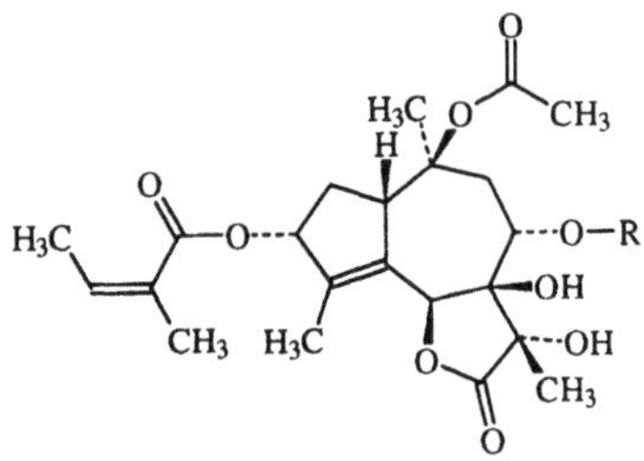

Trilobolide (**14**), R=(*S*)-2-MeBut

Nortrilobolide (**15**), R = But

Thapsivillosin F (**16**), R = Sen

Chart 2. Pentaoxygenated thapsigargins found in *Thapsia* and *Laser trilobum*

three pentaoxygenated guaianolides (**14**–**16**) (*14–16*) have been isolated. Only one of these, trilobolide (**14**) has been isolated from a species not belonging to *Thapsia*, *i.e.* from *Laser trilobum*, Apiaceae (*17*). Without definition the collective term thapsigargins is generally used for the guaianolides (**1**–**16**), which are characterized as 1βH,6αH,3α,7β,8α,10β, 11α-pentaoxygenated-6,12-guaianolides. The 1βH stereochemistry is often found in guaianolides isolated from Apiaceae (18). Hydroxylation of C(7) is only cxceptionally found in guaianolides [*e.g.* 7α-hydroxy-3-deoxyzalazanin C(**17**), isolated from *Podachaenium eminens*, Asteraceae (*19*)], but the 7β-hydroxy group is unique for the thapsigargins. A likely explanation for the unique 7β-hydroxy group is that a precursor possessing a C(7)-C(11) double bond during the biosynthesis is converted into an epoxide, which subsequently is opened into a *trans*-glycol (*18*).

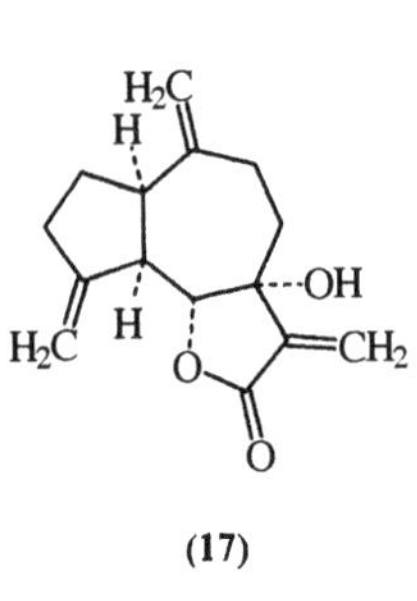

(17)

Chart 3. 7α-Hydroxy-3-deoxyzaluzanin C

2. Taxonomy of *Thapsia*

The genus *Thapsia* belongs to the family Apiaceae, tribe Laserpitiae. In Flora Europaea (*20*) the genus is divided into three species: *T. garganica* L., *T. maxima* Miller and *T. villosa* L. distributed in the Mediterranean area and on the Iberian peninsula. However, recent chemotaxonomic studies based on morphological and anatomical characters, chromosome numbers and secondary metabolites have indicated a need for taxonomic revision of the genus (*14, 21, 22*).

2.1. *Thapsia garganica* and *Thapsia transtagana*

T. garganica L. and *T. transtagana* Brot. are classified as synonymous in *Flora Europaea*. The anatomy of the fruits as well as the profile of the secondary metabolites of the two species, however, are different. Thus, in spite of the same chromosome number $2n = 22(= 2x)$ and the presence of thapsigargins (Table 1) in both there are good reasons for considering *T. garganica* and *T. transtagana* as two different species. Closer studies of *T. garganica* have revealed the presence of at least two chemotypes (*14*).

2.2. *Thapsia maxima*

T. maxima has been shown to include two phytochemically identical phenotypes I and II, having the same chromosome numbers $2n = 22(= 2x)$ (*21*). Neither of the two contains thapsigargins. Based on this finding it is concluded that a specimen previously regarded as *T. maxima* (*23*) should be designated T. villosa type 4 [chromosome number $2n = 44(= 4x)$].

2.3. *Thapsia villosa*

T. villosa, the most heterogeneous species, has been divided into two distinctly different groups, 1 and 2 (*22*). Group 1, which does not contain thapsigargins, is further divided into three types 1–3. Types 1 and 2, both have the chromosome number $2n = 22(= 2x)$ and the names *T. minor* Hoffgg. et Link and *T. laciniata* Rouy, respectively, have been proposed. Type 3 has the chromosome number $2n = 44(= 4x)$. Group 2 includes two types, 4 and 5, both of which contain thapsigargins, with the chromosome numbers $2n = 44(= 4x)$ and $2n = 66(= 6x)$, respectively.

Table 1 *Guaianolides from* Thapsia

Structure Number	Name of Compound	Formula	Plant Source	Plant Organ	Reference(s)
(1)	Thapsigargin	$C_{34}H_{50}O_{12}$	*Thapsia garganica*	Root, fruit	*14, 34, 37*
			T gymnesica	Root, fruit	*14*
(2)	Thapsigargicin	$C_{32}H_{46}O_{12}$	*T garganica*	Root, fruit	*14, 34*
			T gymnesica	Root, fruit	*14*
(3)	Thapsitranstagin	$C_{32}H_{46}O_{12}$	*T transtagana*	Root, fruit	*12, 14, 23*
			T villosa, type 5	Root	*13*
(4)	Thapsivillosin A	$C_{32}H_{42}O_{12}$	*T villosa*, type 4	Root	*12, 23*
			T villosa, type 5	Root	*12, 23*
(5)	Thapsivillosin B	$C_{32}H_{44}O_{12}$	*T villosa*, type 4	Root	*12, 23*
			T villosa, type 5	Root	*12, 23*
			T transtagana	Root, fruit	*14*
(6)	Thapsivillosin C	$C_{35}H_{52}O_{12}$	*T villosa*, type 4	Root	*12*
			T villosa, type 5	Root	*12, 13*
(7)	Thapsivillosin D	$C_{36}H_{52}O_{12}$	*T villosa*, type 4	Root	*12*
			T villosa, type 5	Root	*12*
(8)	Thapsivillosin E	$C_{36}H_{54}O_{12}$	*T villosa*, type 4	Root	*12*
			T villosa, type 5	Root	*12*
(9)	Thapsivillosin G	$C_{35}H_{52}O_{12}$	*T villosa*, type 4	Root	*12*
			T villosa, type 5	Root	*12*
(10)	Thapsivillosin H	$C_{32}H_{42}O_{12}$	*T villosa*, type 4	Root	*12*
			T villosa, type 5	Root	*12*
(11)	Thapsivillosin I	$C_{31}H_{42}O_{12}$	*T garganica*	Root, fruit	*12, 14*
(12)	Thapsivillosin J	$C_{31}H_{44}O_{12}$	*T garganica*	Root, fruit	*12 14*
(13)	Thapsivillosin K	$C_{32}H_{44}O_{12}$	*T transtagana*	Root, fruit	*14*
			T villosa, type 5	Root	*13*
(14)	Trilobolide	$C_{27}H_{38}O_{10}$	*T transtagana*	Root, fruit	*14*
			T villosa, type 5	Root	*23*
			*T garganica**	Root, fruit	*14*
(15)	Nortrilobolide	$C_{26}H_{36}O_{10}$	*T garganica*	Root, fruit	*14, 16*
			T gymnesica	Root, fruit	*14*
(16)	Thapsivillosin F	$C_{27}H_{36}O_{10}$	*T villosa*, type 4	Root	*15, 23*
(18)		$C_{26}H_{38}O_{9}$	*T garganica*	Fruit	*40, 41*
(19)		$C_{28}H_{42}O_{9}$	*T transtagana*	Root, fruit	*41*
(20)		$C_{26}H_{36}O_{10}$	*T villosa*, type 5	Root	*42*
(21)		$C_{26}H_{34}O_{10}$	*T villosa*, type 5	Root	*42*
(22)		$C_{24}H_{34}O_{9}$	*T villosa*, type 5	Root	*42*
(23)		$C_{24}H_{32}O_{9}$	*T villosa*, type 5	Root	*42*
(24)		$C_{20}H_{28}O_{5}$	*T maxima*	Root	*79*
(25)		$C_{20}H_{28}O_{5}$	*T maxima*	Root	*79*

* Only present in some specimens

Table 2. *Thapsane Derivatives from* Thapsia

Structure Number	Name of Compound	Formula	Plant Source	Plant Organ	Reference(s)
(26)	15-*O*-Acetylthapsane-14-al	$C_{17}H_{28}O_3$	*Thapsia villosa* var. minor*	Root	*80*
(27)	6, 14-Thapsene-15-ol	$C_{15}H_{26}O$	*T. villosa*, type 2 (*T. laciniata*)	Root	*81*
(28)	15-*O*-Feruloyl-6, 14-thapsene	$C_{25}H_{34}O_4$	*T. villosa*, type 2 (*T. laciniata*)	Root	*81*
(29)	(1*S*)-1-*O*-Senecioyl-6, 14-thapsene-15-ol	$C_{20}H_{32}O_3$	*T. villosa* var. *minor*	Root	*81*
(30)	(1*S*, 6*R*)-1-*O*-Senecioyl-6, 14-epoxythapsane-15-ol	$C_{20}H_{32}O_4$	*T. villosa* var. *minor*	Root	*80*
(31)	(1*S*, 6*R*)-15-*O*-Acetyl-1-*O*-Senecioyl-6, 14-epoxythapsane	$C_{22}H_{34}O_5$	*T. villosa* var. *minor*	Root	*80*
(32)	14, 15-Epoxythapsane-14-ol	$C_{15}H_{26}O_2$	*T. villosa*, type 2 (*T. laciniata*)	Root	*81*
(33)	(8*R*, 14*S*)-8-*O*-Angeloyl-14,15-epoxythapsane-14-ol	$C_{20}H_{32}O_4$	*T. villosa*, type 2 (*T. laciniata*)	Root	*82*
(34)	(8*R*, 14*S*)-8-*O*-Senecioyl-14, 15-epoxythapsane-14-ol	$C_{20}H_{32}O_4$	*T. villosa* var. *minor*	Root	*24*
(35)	8-*O*-Coumaroyl-14, 15-epoxythapsane-14-ol	$C_{24}H_{32}O_5$	*T. villosa* var. *minor*	Root	*26*
(36)	8-*O*-Feruloyl-14, 15-epoxythapsane-14-ol	$C_{25}H_{34}O_6$	*T. villosa* var. *minor*	Root	*26*
(37)	1-*O*-Senecioyl-14, 15-epoxythapsane-14-ol	$C_{20}H_{32}O_4$	*T. villosa* var. *minor*	Root	*26*
(38)	1-*O*-Angeloyl-14, 15-epoxythapsane-14-ol	$C_{20}H_{32}O_4$	*T. villosa* type 3	Root	*83*
(39)	1-*O*-Tigloyl-14, 15-epoxythapsane-14-ol	$C_{20}H_{32}O_4$	*T. villosa* type 3	Root	*83*
(40)	3-*O*-Angeloyl-14, 15-epoxythapsane-14-ol	$C_{20}H_{32}O_4$	*T. villosa* var. *minor*	Root	*26*
(41)		$C_{40}H_{62}O_7$	*T. villosa* var. *minor*	Root	*26*

* *T. villosa* var. *minor* corresponds phytochemically to *T. villosa*, type 1

Table 3 *Guaiol and Guaiane Esters from* Thapsia

Structure Number	Name of Compound	Formula	Plant Source	Plant Organ	Reference(s)
(42)	Guaiol	$C_{15}H_{26}O$	*Thapsia villosa*, type 2 (*T laciniata*)	Root	*82*
(43)	(4*S*, 5*S*, 7*S*, 8*S*)-8-Senecioyloxy-1(10)-guaien-11-ol	$C_{20}H_{30}O_3$	*T villosa*, type 2 (*T laciniata*)	Root	*84*
(44)	(4*S*, 5*S*, 7*S*, 8*S*)-8-p-Coumaroyloxy-1(10)-guaien-11-ol	$C_{24}H_{32}O_4$	*T villosa*, type 2 (*T laciniata*)	Root	*84*
(45)	(4*S*, 5*S*, 7*S*, 8*S*)-8-Feruloyloxy-1(10)-guaien-11-ol	$C_{25}H_{34}O_5$	*T villosa*, type 2 (*T laciniata*)	Root	*84*

Table 4 *Germacrane Esters from* Thapsia

Structure Number	Name of Compound	Formula	Plant Source	Plant Organ	Reference(s)
(46)	8-O-Angeloyltovarol	$C_{20}H_{32}O_3$	*Thapsia villosa* var *minor**	Root	*28*
			T villosa var *villosa***	Root	*28*
(47)	8-O-Senecioyltovarol	$C_{20}H_{32}O_3$	*T villosa* var *minor*	Root	*28*
(48)	8-O-Coumaroyltovarol	$C_{24}H_{32}O_4$	*T villosa* var *minor*	Root	*28*
(49)	8-O-Feruloyltovarol	$C_{25}H_{34}O_5$	*T villosa* var *minor*	Root	*28*
(50)	12-Hydroxy-8-O-angeloyltovarol	$C_{20}H_{32}O_4$	*T villosa* var *minor*	Root	*27*
(51)	12-O-Angeloyl-8-O-angeloyltovarol	$C_{25}H_{38}O_5$	*T villosa* var *minor*	Root	*27*
(52)	8-O-Angeloylshiromodiol	$C_{20}H_{32}O_4$	*T villosa* var *minor*	Root	*28*
			T villosa var *villosa*	Root	*28*
(53)	6-O-Acetyl-8-O-angeloylshiromodiol	$C_{22}H_{34}O_5$	*T villosa* var *minor*	Root	*28*
			T villosa var *villosa*	Root	*28*
(54)	12-O-Angeloyl-8-O-angeloylshiromodiol	$C_{25}H_{38}O_8$	*T villosa* var *minor*	Umbellas	*27*
(55)	6-O-Acetyl-8-O-Angeloyl-1(10),4(5)-diepoxygermacrane	$C_{22}H_{34}O_6$	*T villosa* var *minor*	Root	*28*
			T villosa var *villosa*	Root	*28*

* *T villosa* var *minor* corresponds phytochemically to *T villosa*, type 1

** *T villosa* var *villosa* corresponds phytochemically to *T villosa*, type 5

Table 5. *Other Sesquiterpenoids from* Thapsia

Structure Number	Name of Compound	Formula	Plant Source	Plant Organ	References
(56)	δ-Cadinene	$C_{15}H_{24}$	*Thapsia villosa* var. *minor*	Umbellas	*27*
(57)	γ-Cadinene	$C_{15}H_{24}$	*Thapsia villosa* var. *minor*	Umbellas	*27*
(58)	γ-Muurolene	$C_{15}H_{24}$	*Thapsia villosa* var. *minor*	Umbellas	*27*
(59)	β-Caryophyllene	$C_{15}H_{24}$	*Thapsia villosa* var. *minor*	Umbellas	*27*
(60)	β-Caryophyllene oxide	$C_{15}H_{24}O$	*Thapsia villosa* var. *minor*	Umbellas	*27*

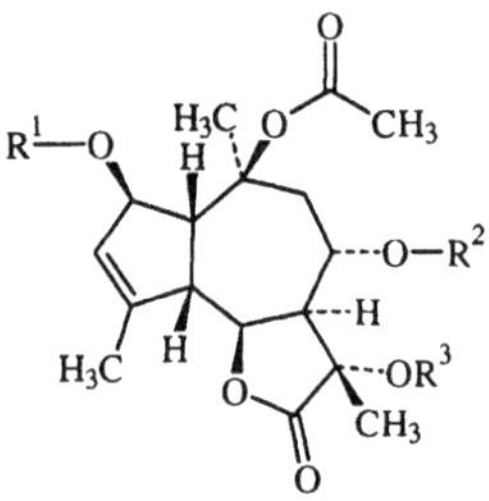

(18), R1= 2-MeBut, R2= But, R3= H

(19), R1= iVal, R2= 2-MeBut, R3= H

(20), R1= Ac, R2= 2-MeBut, R3= Ac

(21), R1= Ac, R2= Sen, R3= Ac

(22), R1= H, R2= 2-MeBut, R3= Ac

(23), R1= H, R2= Sen, R3= Ac

Chart 4. Slovanolides found in *Thapsia*

The heterogeneity of *T. villosa* has caused some confusion in the naming of investigated plant specimens. The name *T. villosa* var. *minor* was used by a Spanish group for plant material, from which they isolated a number of secondary metabolites (*24–28*). The structures of the isolated secondary metabolites make it most likely, that their collection should be designated *T. villosa* type I. Likewise *T. villosa* var. *villosa* (*28*) is assumed to belong to type 5.

Common for all the three types within group 1 is the presence of derivatives of thapsane, tovarol and shiromodiol (Tables 2 and 4) in the

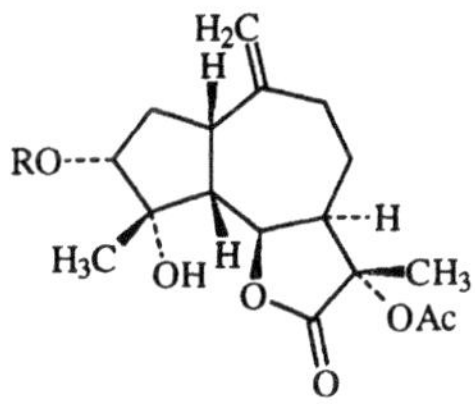

(**24**) R = Ang

(**25**) R = Tig

Chart 5. 10(14)Unsaturated guaianolides found in *Thapsia*

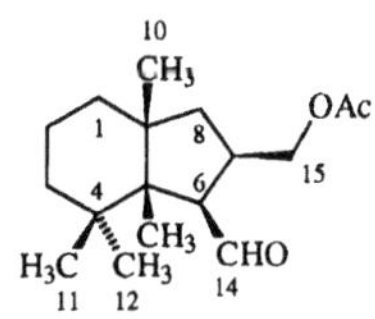

(**26**)

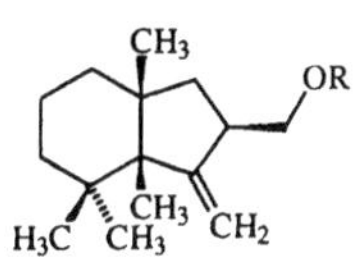

(**27**) R = H

(**28**) R = Fer

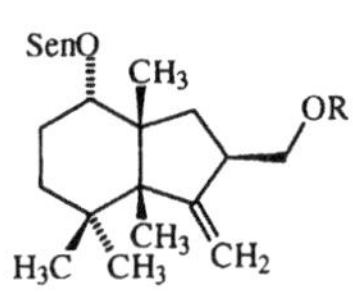

(**29**), R = H

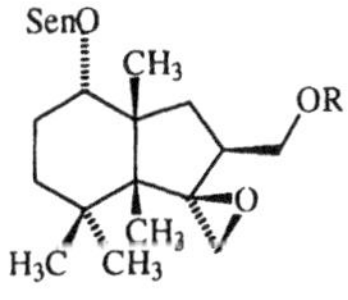

(**30**), R = H

(**31**), R = Ac

Chart 6. Thapsanes found in *Thapsia*

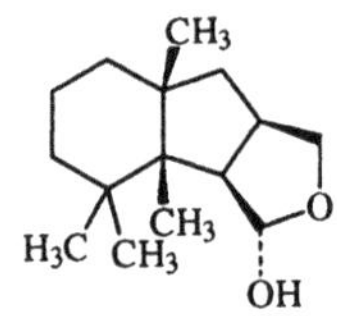

(**32**)

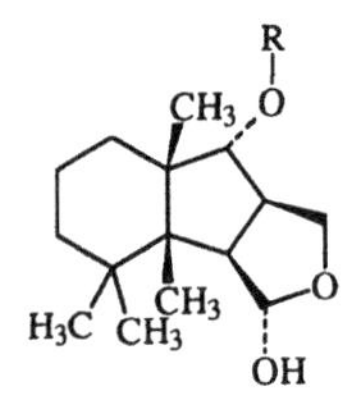

(**33**), R = Ang

(**34**), R = Sen

(**35**), R = *p*-Coum

(**36**), R = Fer

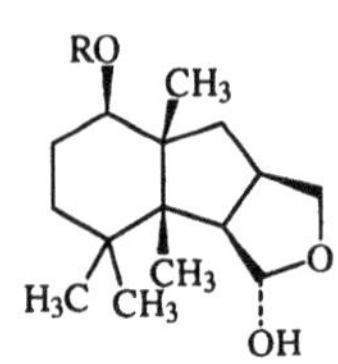

(**37**), R = Sen

(**38**), R = Ang

(**39**), R = Tig

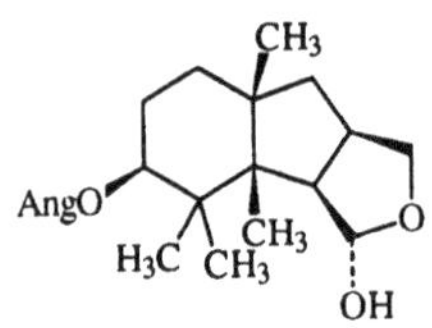

(**40**)

Chart 7. Epoxythapsanes found in *Thapsia*

(**41**)

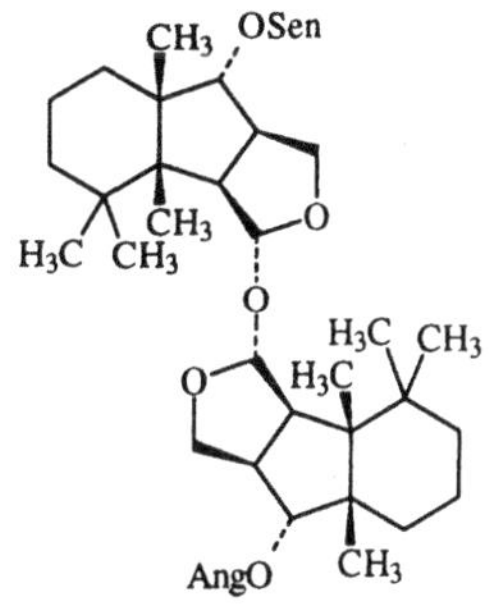

Chart 8. Dimeric epoxythapsanes found in *Thapsia*

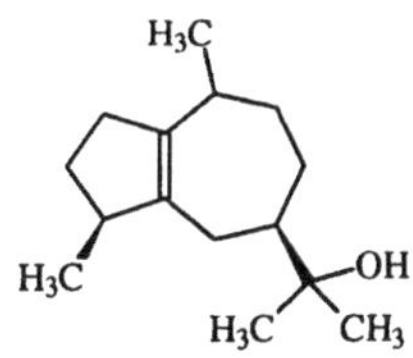

Guaiol (**42**)

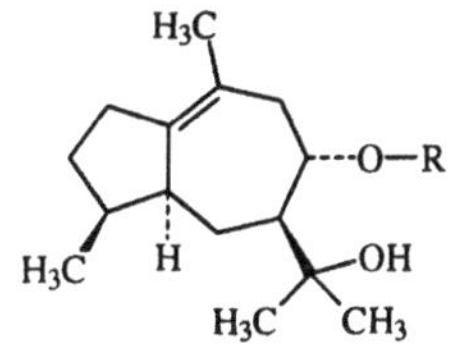

(**43**), R = Sen

(**44**), R = *p*-Coum

(**45**), R = Fer

Chart 9. Guaianes found in *Thapsia*

roots, whereas only type 2 contains guaiol and guaiane esters (Table 3). The major constituent of the essential oil, accounting for 79–89%, from the fruits of all three types is geranyl acetate (*29*).

In contrast, the characteristic constituents of the roots of the two types 4 and 5, within group 2, are thapsigargins and slovanolides (Table 1), phenylpropanoids (*13*) and 6-methoxy-7-geranyloxycoumarin (*23*). Only a few tovarol derivatives (**46, 52, 53** and **55**) have been detected in plants from both groups 1 and 2 (*28*). The essential oils from types 4 and 5 are

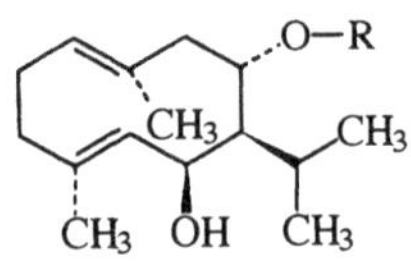

(**46**) R = Ang

(**47**) R = Sen

(**48**) R = p-Coum

(**49**) R = Fer

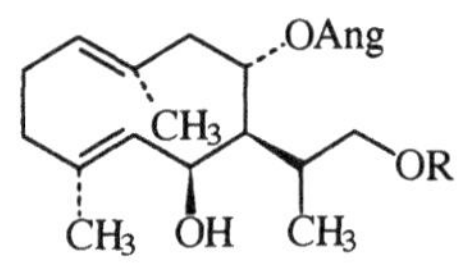

(**50**) R = H

(**51**) R = Ang

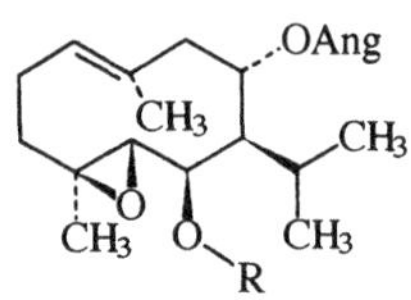

(**52**) R = H

(**53**) R = Ac

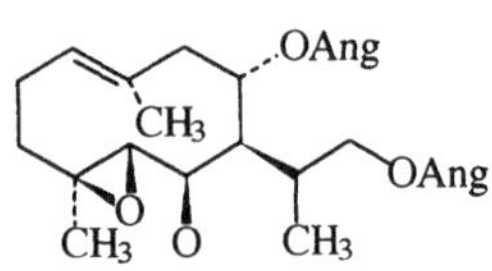

(**54**)

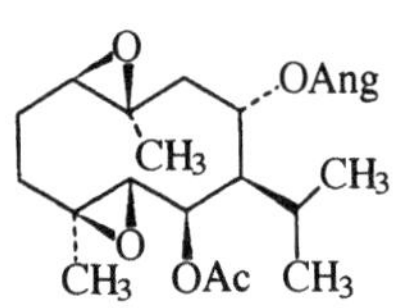

(**55**)

Chart 10. Germacranes found in *Thapsia*

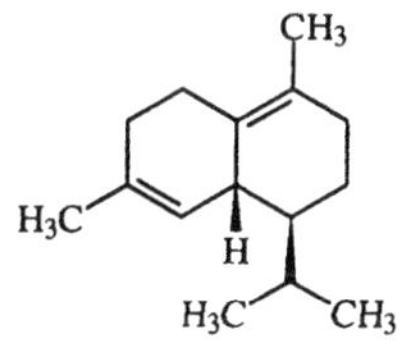

δ-Cadinene (**56**)

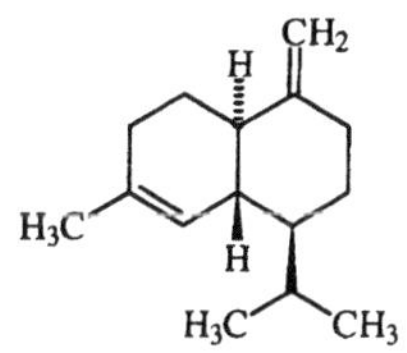

γ-Cadinene (**57**)

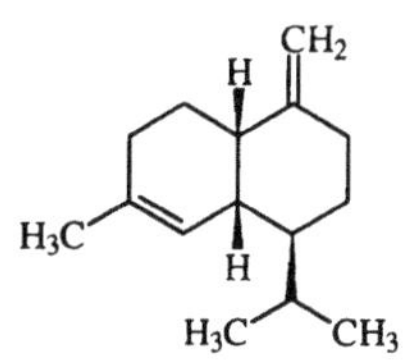

γ-Muurolene (**58**)

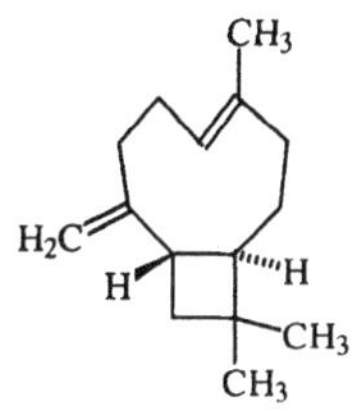

β-Caryophyllene (**59**)

Chart 11. Sesquiterpenes found in *Thapsia*

similar to the essential oil from *T. maxima* in having limonene and methyl eugenol as the two major components which together constitute 80–90% of the oil (*21, 30, 31*).

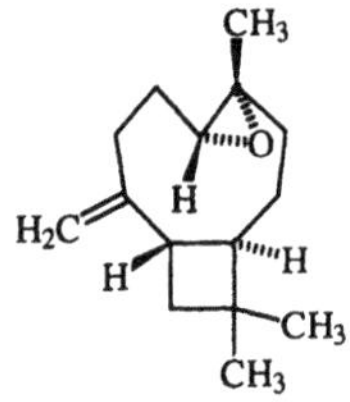

Chart 12. β-Caryophyllene oxide (**60**)

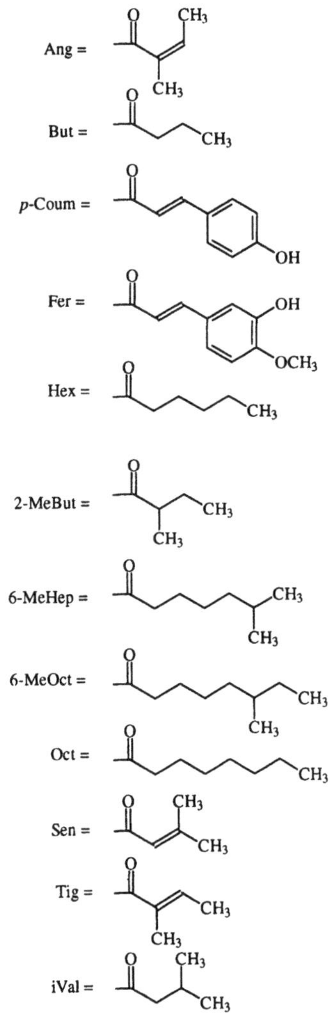

Chart 13. Structure and abbreviations for acyl residues found in *Thapsia*

2.4. *Thapsia gymnesica*

Thapsia gymnesica Rosselló & Pujadas, found only on Mallorca and Minorca, has been described as a new species in 1991 (*32*). Like *T. garganica* the chromosome number is 2n = 22(= 2x) and it contains thapsigargin (**1**), thapsigargicin (**2**) and nortrilobolide (**15**), which previously have been found only in *T. garganica*. The characteristic difference between *T. garganica* and *T. gymnesica* is the much smaller fruits of *T. gymnesica*, which are of the same size as the fruits of *T. maxima* and *T. villosa*.

3. Elucidation of the Structure of Thapsigargin

Comparison of the spectra of thapsigargin (**1**) and thapsigargicin (**2**) (Fig. 2) with those of trilobolide (**14**) (*17*) showed that **1** and **2** were hexaoxygenated guaianolides (*33*). The non-crystalline state of thapsigargin prevented determination of the relative and absolute configuration by an X-ray crystallographic analysis. However, after treatment of thapsigargin with thionyl chloride a crystalline derivative was obtained, the structure of which was determined by X-ray analysis. This analysis established the location of the four acyl groups and the relative configuration, except at C(7) and C(11) (*34*). The X-ray analysis also showed that in analogy with trilobolide (*17*) treatment of the thapsigargin with thionyl chloride converts the vicinal 7,11-diol into the epoxide (**78**) (Scheme 7, p. 156). Although it is easily rationalized thionyl chloride promoted conversion of 1,2-diols into epoxides apparently only occurs if the geometry of the molecule favours intramolecular dehydration (see *e.g.* *35*). The few known analogous reactions did not allow conclusions concerning the stereochemistry of the starting 7, 11-diol.

The unresolved stereochemical questions were elucidated, when the X-ray structure of trilobolide was published (*36*). 8-O-Deacylthapsigargin (**63**) formed an 1,3-dioxane (**82**) upon reaction with acetone (Scheme 7) as did 8-O-deacyltrilobolide (*37*). This common reaction path indicated that the 7-hydroxy group had to be *trans* to the 8- and 11-hydroxy groups. In addition the absolute configurations of C(3) in thapsigargin (**1**) and trilobolide (**14**) were established by taking advantage of the exciton coupling in the allylic ester of the α,β-unsaturated ester residue (*38, 39*). The found absolute configuration of trilobolide (**14**) was confirmed by determination of the absolute configuration of the 2-methylbutyric acid residue (*38*) and taking advantage of the relative stereochemistry as determined by X-ray crystallography (*36*).

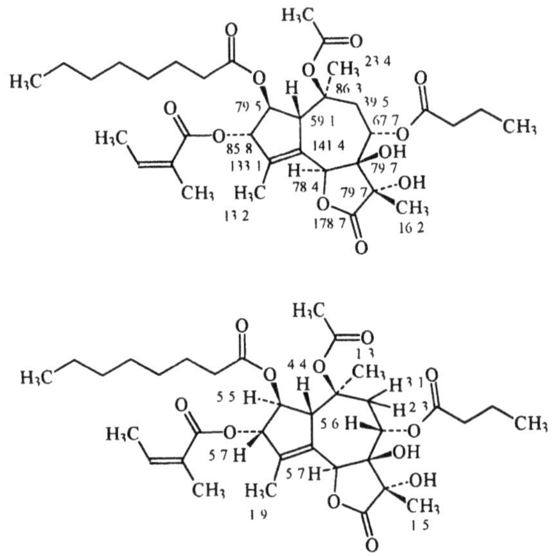

Fig. 2. ^{13}C NMR (CD_3OD) and ^{1}H NMR ($CDCl_3$) Data for Nuclei of the Skeleton of Thapsigargin (*12, 34*)

The acyl groups in the thapsigargins (**2**–**13**) were located by interpretation of the fragmentation pattern of the mass spectra (*12*). This method, however, did not permit locating the isomeric acyl groups in thapsivillosin H (**10**).

4. Proazulenic Slovanolides

In addition to the thapsigargins some 7αH-6,12-guaianolides (**18**–**23**) have been isolated from species belonging to the genus *Thapsia* (*40*–*42*). These guaianolides were originally isolated in order to find the precursor for the azulenes found in the essential oils of the fruits of *T. garganica* (*43*). All of the guaianolides (**18**–**23**), which possess the stereochemistry characteristic of the slovanolides (*44*), are easily converted into azulenes by heating. The mechanism for the degradation of the 11-hydroxy lactones (**18**) and (**19**) to give 1,4-dimethylazulene (**61**) could be a retrograde Prins-like reaction (*45*) and some *cis*-eliminations of carboxylic acids (Scheme 1). This reaction explains the presence of 1,4-dimethylazulene (**61**) in the essential oil of fruits from *T. garganica* and *T. transtagana*.

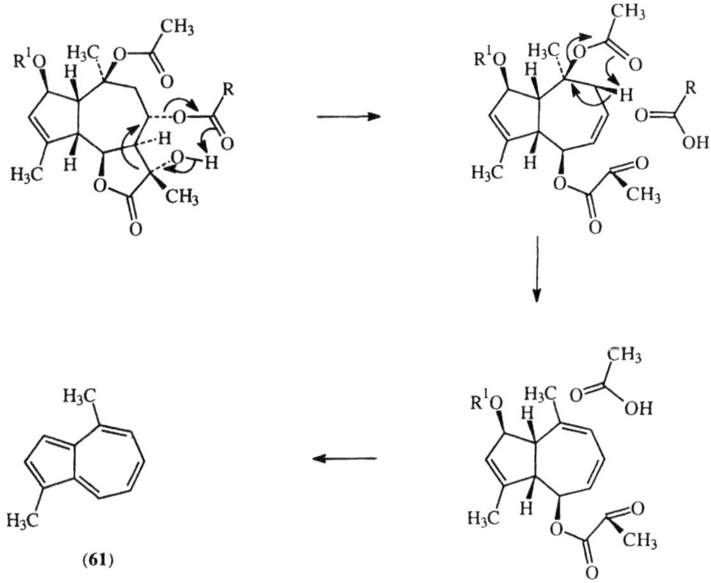

(61)

Scheme 1. Possible mechanism for the formation of 1,4-dimethylzulene

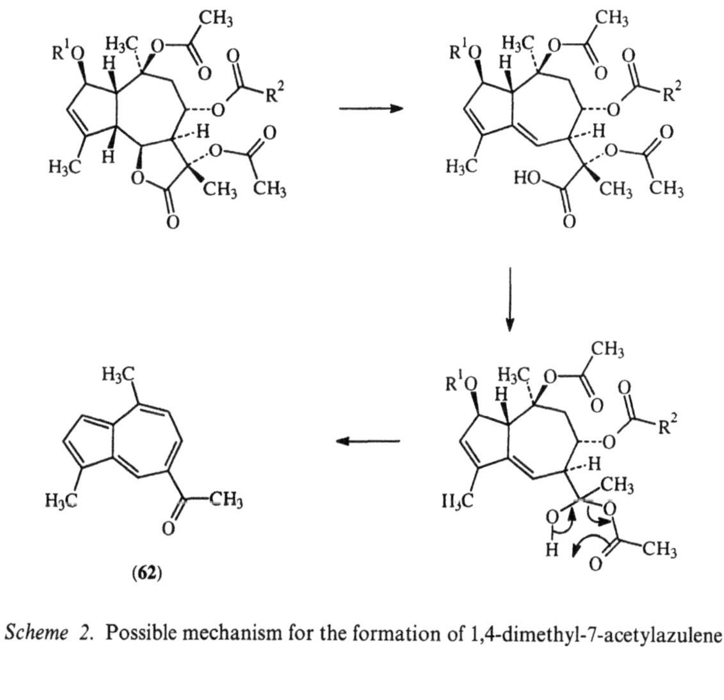

(62)

Scheme 2. Possible mechanism for the formation of 1,4-dimethyl-7-acetylazulene

The proton of the 11-hydroxy group is essential for formation of 1,4-dimethylazulene as depicted. Accordingly only trace amounts of 1,4-dimethylazulene can be found after heating a methanolic solution of the 11α-acetoxyslovanolides (**20**–**23**) whereas the main product is 1,4-dimethyl-7-acetylazulene (**62**). The mechanism for this conversion is obscure, but the decarbonylation of α-oxygenated acids and esters described in the literature (*46*) suggests that the azulene is formed through the reaction path depicted in Scheme 2.

5. Non-lactonic Sesquiterpenoids from *Thapsia*

A number of non-lactonic sesquiterpenoids have been isolated from specimens belonging to *T. villosa* (Tables 2–5). Most interesting from a phytochemical point of view are the 2,3,3a,4,4,7a-hexamethylindan (thapsane) derivatives (**26**–**41**), since natural products possessing this skeleton only have been isolated from plants belonging to *Thapsia*. The unique structure including three contiguous quaternary carbons and five to six chiral centres has made the compounds attractive synthetic targets (*47, 48*).

6. Pharmacological Activity of the Thapsigargins

The mechanism behind the skin irritating effect of the thapsigargins might be related to their ability to release mediators from cells belonging to the immune system. Indeed thapsigargin was demonstrated to activate a broad number of cells including mast cells (*3, 49*), neutrophil and basophil leucocytes, lymphocytes, macrophages and platelets (*4*–*7*). Later studies have verified that thapsigargin activates virtually all kind of cells (*50, 51*), with erythrocytes as exceptions (*4*). Besides causing release of mediators or contraction of muscle cells thapsigargin was shown to be a tumour promoter on mouse skin (*8*). Careful study of the numbers of induced tumours reveals an unusual decrease after 22 weeks. The recently described thapsigargin induced programmed cell death (apoptosis) (*52*) might explain this finding and might indicate a future for thapsigargin in the treatment of cancer.

The broad spectrum of activity indicates that thapsigargin interferes with an ubiquitous target. A clue for the identification of this target was the finding that all effects of thapsigargin were preceded by a dramatic increase in the cytosolic Ca^{2+} concentration (*4, 53*). This effect was rationalized by the observation that thapsigargin was a selective inhibitor

References, pp. 163–167

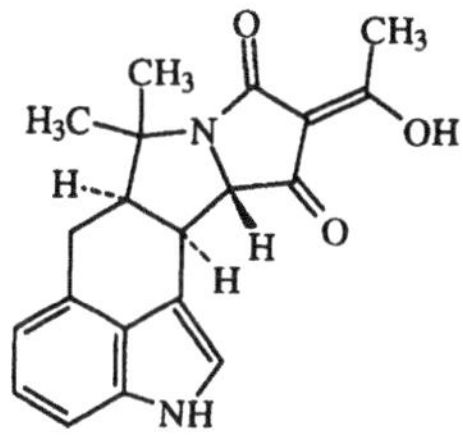

Cyclopiazonic acid

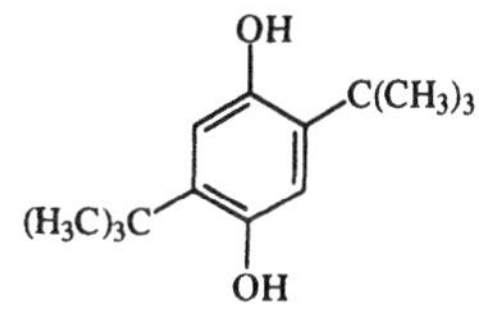

2,5-Di-*tert*-butylhydroquinone

Chart 14. Structure of cyclopiazonic acid and 2,5-di-*tert*-butylhydroquinone

of Ca^{2+} pumps in the sarco- or endoplasmic reticulum (the SERCA family) without affecting either the pumps in the plasma membrane or those in the mitocondrial membrane (*6, 10, 11*). In the resting state of the cells the cytosolic Ca^{2+}-concentration is maintained at a very low level by active transport of Ca^{2+} either into the endo- or sarcoplasmic reticulum or to the extracellular medium. Inhibition of the SERCA pumps is accompanied by a leak in the membranes surrounding the microsomal Ca^{2+}-pools causing an increased cytosolic Ca^{2+} concentration and eventually an opening of Ca^{2+}-channels in the plasma membrane, followed by an influx of extracellular Ca^{2+}. Since Ca^{2+} signal transduction regulates such diverse cellular processes as fertilization, cell growth, muscle contraction, neuronal signal transduction and mediator release, any compound selectively affecting a step in the Ca^{2+} homeostasis is a potential tool for investigating the physiology of the cells.

In addition to the thapsigargins two other compounds, 2,5-di-*tert*-butylhydroquinone and cyclopiazonic acid, have been shown to mobilize Ca^{2+} from the same intracellular pools (*54–57*). However, as it is four order of magnitudes more potent than the latter two compounds, thapsigargin is the preferred tool for investigation of the Ca^{2+} homeostasis (*56, 57*). A still debated question concerning the mobilization of Ca^{2+} during cell activation is whether the depletion of the microsomal Ca^{2+}-pools and the opening of the plasma membrane Ca^{2+}-channels is coupled through an unknown soluble messenger (*58*). Thapsigargin has played a key role in the attempts to elucidate this problem.

7. Molecular Pharmacology

The Ca^{2+}-ATPases belong to the P-type ion pumps. These enzymes are characterised by a transport mechanism which involves occlusion of

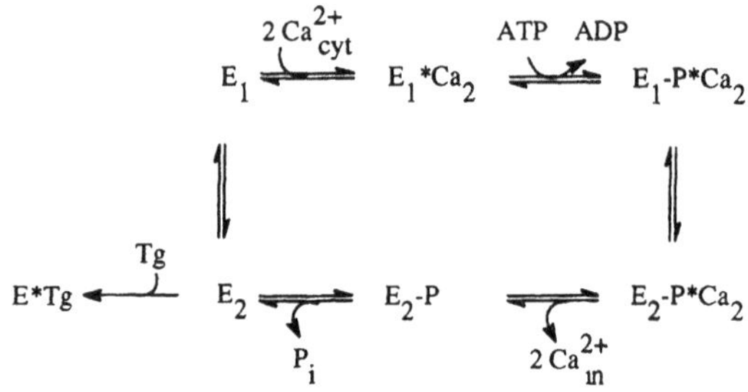

Fig. 3. A model of the transport cycle for SERCA pumps illustrating the dead end complex formed with thapsigargin [modified after (*60*, *61*)]

the cations to be translocated followed by a transfer of the terminal phosphate group of ATP to a β-aspartyl carboxyl. This phosphorylation induces a change of conformation from the E_1 to the E_2 conformation. This conformational change transports the cations through the membrane against the concentration gradient and releases them to the intracellular pool or to the extracellular medium. After release of the cations the pump is dephosphorylated and returns to the E_1 conformation (*59*). Thapsigargin inhibits the SERCA pumps by locking the enzyme into a conformation, which have only a poor if any affinity for Ca^{2+}, ATP and phosphate (*60*, *61*).

In Fig. 3 the complexation between thapsigargin and the ATPase has been drawn as if the reaction were irreversible. In principle, this reaction must be reversible; however, the extremely small dissociation constant [K_d 2.2 pM or less (*62*)] makes this reaction irreversible in practice. Since complexation with thapsigargin locks the enzyme into a dead end complex this binding must inactivate the enzyme by decreasing the flexibility. An improved knowledge of the binding site, thus might contribute to an understanding of the conformation changes involved in the translocation of Ca^{2+}.

At the present the most detailed model for the structure, topology and helix packing of P-type ion pumps has been obtained by electron microscopy (*63*). According to this model the enzyme contains ten transmembrane helices and an ATP binding site and a phosphorylation site on the cytosolic loop combining the fourth and fifth transmembrane segment. The Ca^{2+} binding site is constituted from residues on the fourth, fifth, sixth and eighth transmembrane section (*64*). Studies on chimeric proteins consisting of defined parts of Ca^{2+}-ATPase and Na^+, K^+-ATPase have revealed that the third transmembrane segment is important for the binding of thapsigargin (*65–67*). Studies on the complex between a fluorescent thapsigargin derivative and the pump have revealed that thapsigargin is situated less than 19 Å from tryptophan residue-272 (*68*). An indirect way of characterising the topography of the binding site is to correlate changes of the structure of the molecule with the inhibitory potency of the analogue. This, however, depends on development of methods for selective transformations of thapsigargin.

8. Chemistry of Thapsigargin

Selective modification of the structure of thapsigargin is complicated by the few different functional groups present, although the guaianolide skeleton is heavily substituted.

8.1. Changes at C(8)

Anchimeric assistance by the 11-hydroxy group in the solvolysis of the ester group at C(8) results in selective hydrolysis of the butyrate group to give (**63**) by merely allowing a methanolic solution to stand for some days at room temperature (Schem 3). The reaction is catalysed by addition of a few percent of triethylamine (*69*). In contrast to sodium carbonate catalysed cleavage of the butanoate group (*70*), triethylamine in methanol does not open the lactone ring, a side reaction which after acidification has been shown to afford a mixture of (**63**) and the isomeric 8,12-guaianolide (**64**). Addition of acid to a methanolic solution of thapsigargin decreases the rate of the solvolysis.

An isomer of 8-O-debutanoylthapsigargin has been claimed to be present in a methanolic extract of the roots of *T. garganica* (*71*). The published spectrum of this compound, however, is similar to the spectrum of (**63**) and the time consuming extraction with methanol (7 days) makes it likely, that the compound is (**63**) formed by methanolysis of thapsigargin.

Compound (**63**) has been used as starting material for preparation of radio and fluorescence labelled analogues [*e.g.* (^{3}H-**1**) and (**66**)] (*69, 70*). In spite of the loss in the affinity for the Ca^{2+}-ATPases by insertion of a large fluorescent group, the derivatives have found use as tools for investigation of the Ca^{2+} homeostasis and the topography of the binding sitc.

Esterification of the 8-hydroxy group in (**63**) with vinylacetic acid yields (**65**), which by selective reduction of the terminal double bond by hydridocarbonyltris(triphenylphosphine)-rhodium(I) catalysed hydrogenation using deuterium or tritium gas, gave access to deutero- or tritium

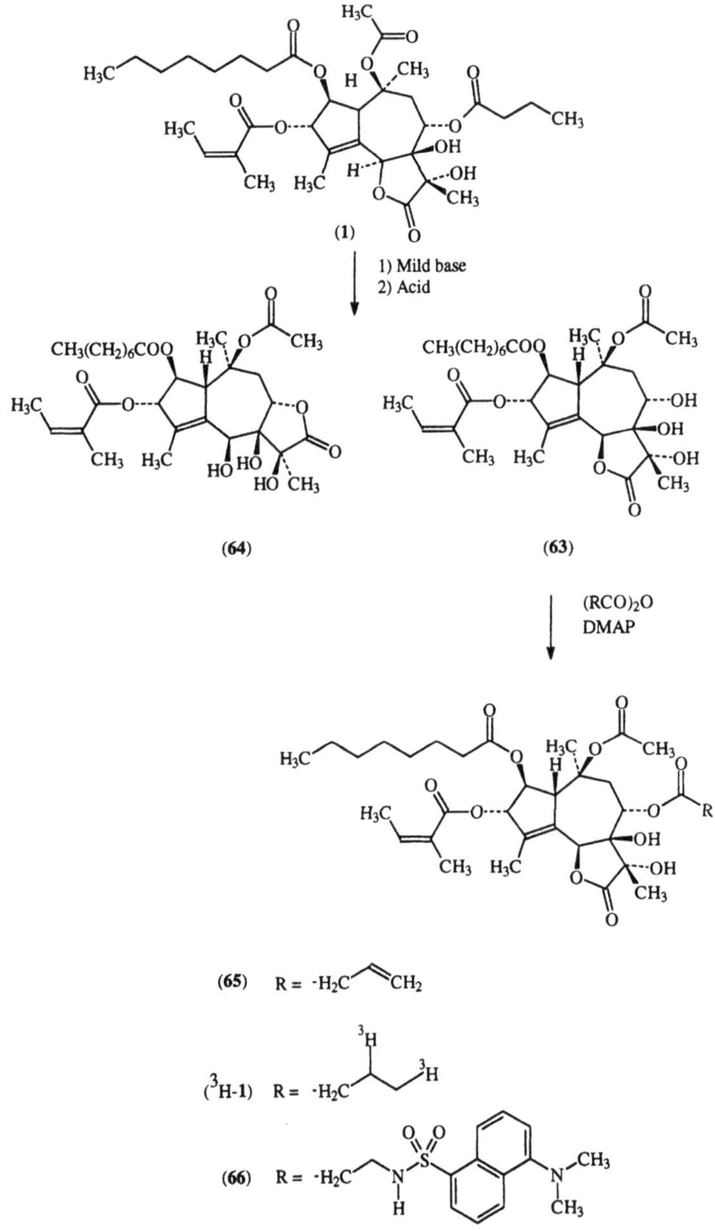

Scheme 3. Replacement of the 8-O-acyl group of thapsigargin

References, pp. 163–167

labelled thapsigargin (^{3}H-**1**) (*70*). Although thapsigargin labelled in the 8-O-acyl group is useful for binding studies (*72*), the derivative is unfit for metabolic studies because of the possible loss of the reporter group. In order to overcome this problem 8-O-debutanoylthapsigargin (**63**) was used for radiolabelling in the guaianolide skeleton. 8-O-Debutanoylthapsigargin (**63**) was oxidized to the ketone (**67**) which by stereoselective reduction with sodium borohydride afforded the starting material (Scheme 4). The use of sodium borotritide permitted tritiation at C(8) in the guaianolide skeleton (*69*). In contrast to the mode of reduction with sodium borohydride, reduction of the 8-ketone with sodium triacetoxyborohydride selectively afforded the 8-hydroxy derivative inverted at C(8) (**68**) (*73*). This might be explained by assuming that the 11-hydroxy group defines the stereochemistry of the product.

8.2. Changes at C(3)

Selective cleavage of the angelate ester at O(3) to give (**71**) was accomplished by permanganate oxidation of the double bond under phase

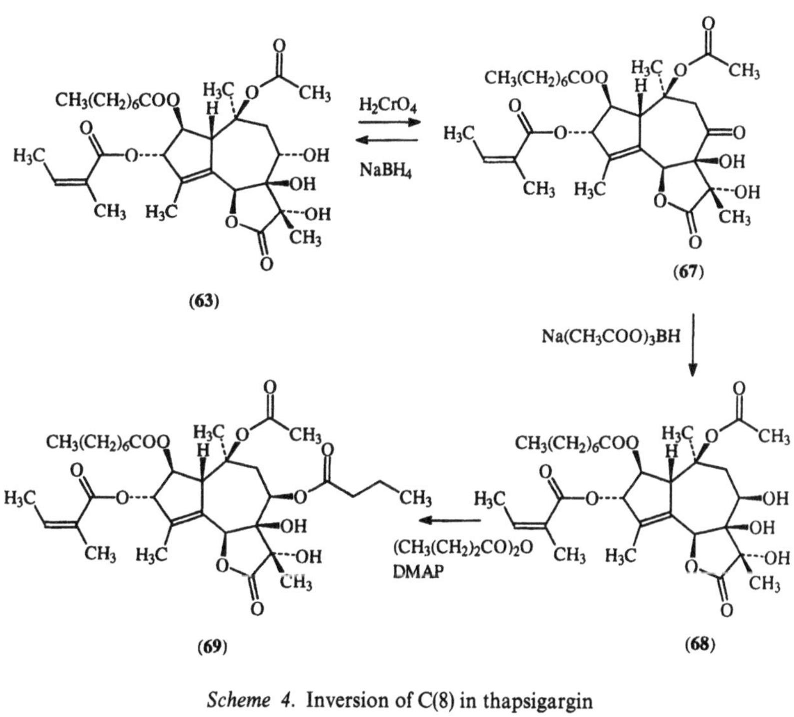

Scheme 4. Inversion of C(8) in thapsigargin

transfer conditions to give the pyruvate (**70**) followed by methanolysis (Scheme 5). Thapsigargin analogues with inverted configuration at C(3) were obtained either by oxidation to the ketone (**72**) followed by

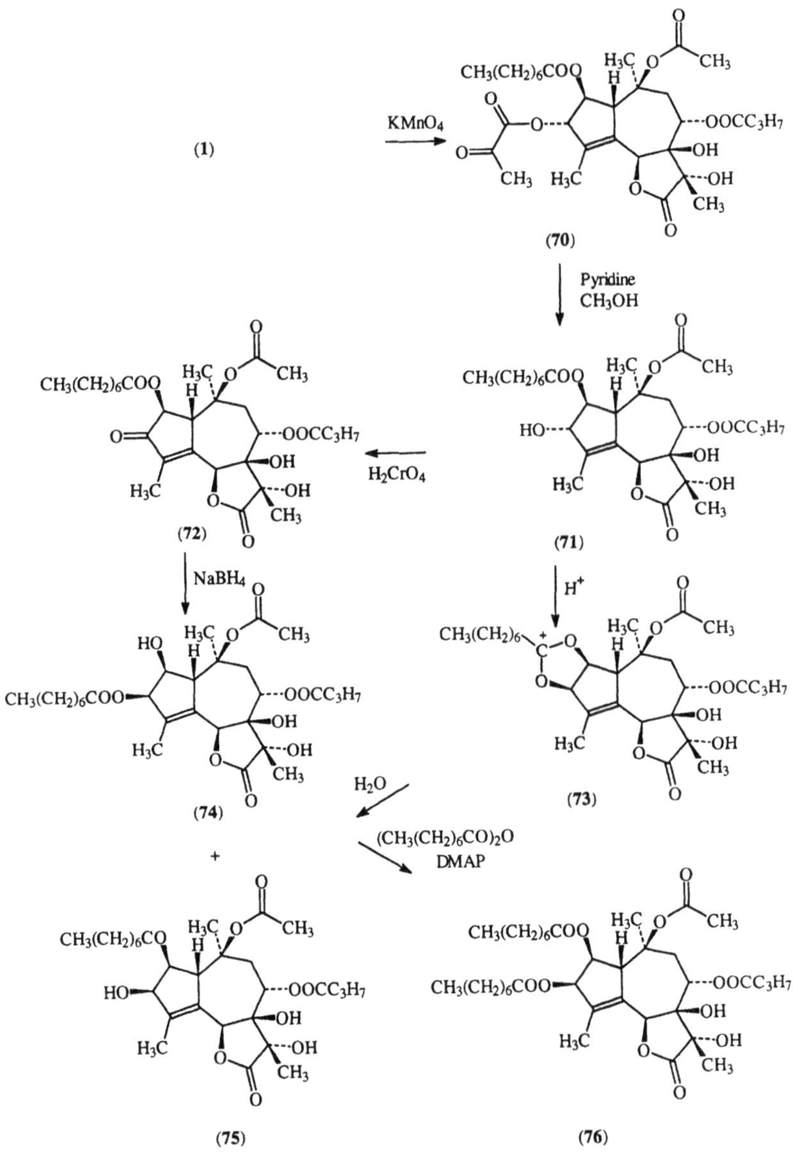

Scheme 5. Inversion of C(3) in thapsigargin

borohydride reduction to give a mixture of (**74**) and (**75**), or by treatment of (**71**) with trifluoromethanesulfonic anhydride to give the same mixture of the two monooctanoates (**74**) and (**75**). The two monooctanoates were found to be easily interconvertible, but treatment of the mixture with octanoic anhydride afforded the stable dioctanoate (**76**) (*73*) (Scheme 6).

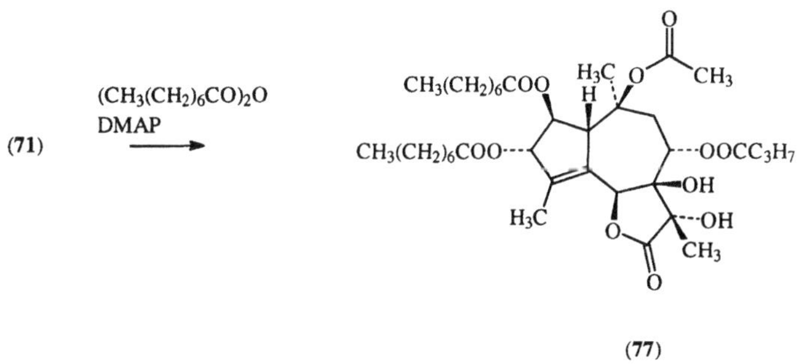

Scheme 6. Replacement of the 3-O-acyl group in thapsigargin

Access to (**71**) has made a number of thapsigargin analogues available, in which the angeloyl group has been replaced with other acyl residues *e.g.* (**77**) (*73*) (Scheme 6). These latter analogues have given important information about the binding site for thapsigargin.

8.3. Changes of the Vicinal Diol

Treatment of thapsigargin with thionyl chloride converts the diol into the β-epoxide (**78**) (*34*) (Scheme 7). Esterification of the two tertiary alcohols affording the diacetate (**81**) only succeeds if 4-dimethylaminopyridine is added as a catalyst. The 11-O-monoacetate (**80**) is formed as the major side product (*73*). Selective esterification of the 7-hydroxy group to give (**79**) is accomplished via the isopropylidene derivative (**82**) (*73*).

8.4. Changes of the Lactone Carbonyl Group

Reduction of thapsigargicin (**2**) with sodium borohydride or preferentially sodium *bis*(2-methoxyethoxy)ethoxy-aluminium hydride (*74*) affords a mixture of the α-and β-lactol (**83**) and (**84**), which has been used as

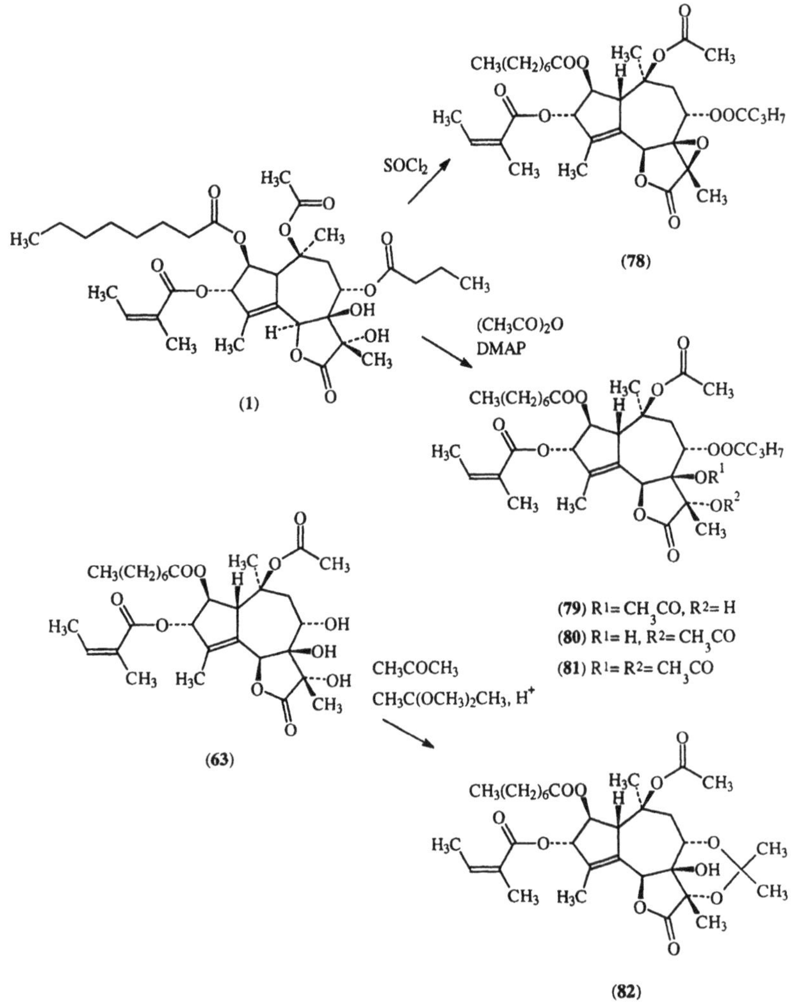

Scheme 7. Derivatives of the glycol residue of thapsigargin

starting material for several analogues of thapsigargicin (**2**) (Scheme 8). Attempts to separate the two epimeric lactols failed, probably because of a phenomenon analogous to mutarotation in carbohydrate chemistry.

Treatment of the lactols (**83**) and (**84**) with trimethyl orthoformate in ethanol affords a mixture of the β-ethyl acetal (**85**) and the two possible ortho formates (**86**) and (**87**) (*74*). In contrast reaction with triethyl

References, pp. 163–167

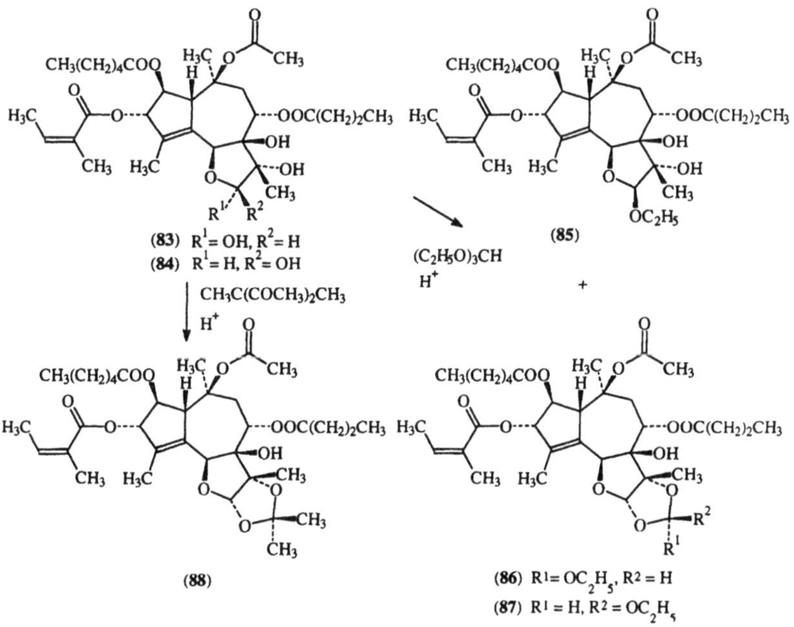

Scheme 8. Derivatives of thapsigargicin lactol

orthoacetate only yields the α-acetate (**91**) (Scheme 9). Reaction between the lactols and 2,2-dimethoxypropane affords the tetracyclic derivative (**88**) (*71*). Treatment of the lactols (**83**) and (**84**) with *N*,*N*-dimethylformamide dimethyl acetal affords the epoxide (**90**).

The α-ethylthioacetal (**94**) obtained by reacting the lactols (**92**) and (**93**) with ethanethiol in the presence of hydrogen chloride was reduced to give the 12-deoxoanalogue of thapsigargin (**95**) (*75*) (Scheme 10), in which the heterocyclic ring cannot be opened under physiological conditions as is the case for thapsigargin (**1**) as well as for the lactols (**92**) and (**93**), (**83**) and (**84**). The reduction, which is catalysed by triphenyltin hydride and α,α′-azoisobutyronitrile follows a radical mechanism and the radicals formed during the reduction also converts the thermodynamically less stable angeloyl residue into a tigloyl residue.

8.5. Changes at O(10)

Selective hydrolysis of the acetate ester can be accomplished indirectly by hydrolysis under more vigorous reaction conditions to give the 2,8,10-

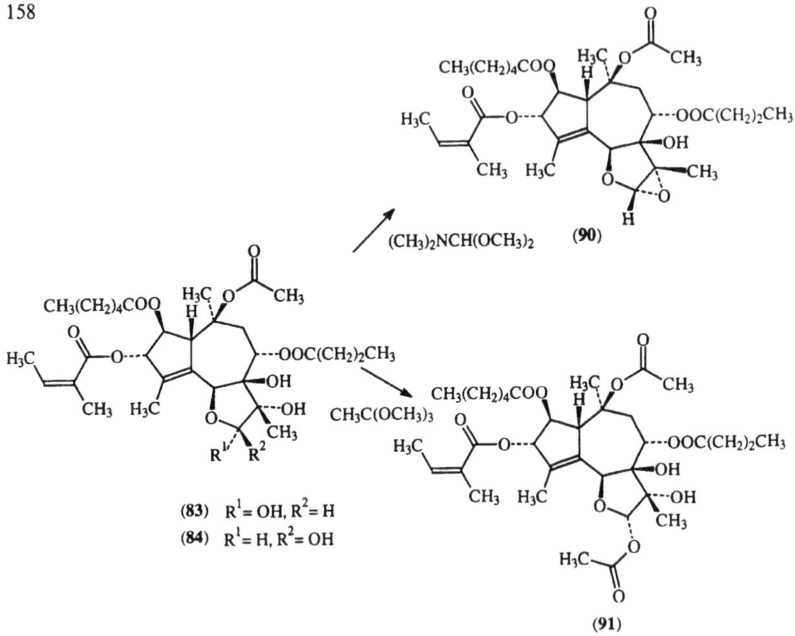

Scheme 9. Derivatives of thapsigargicin lactol

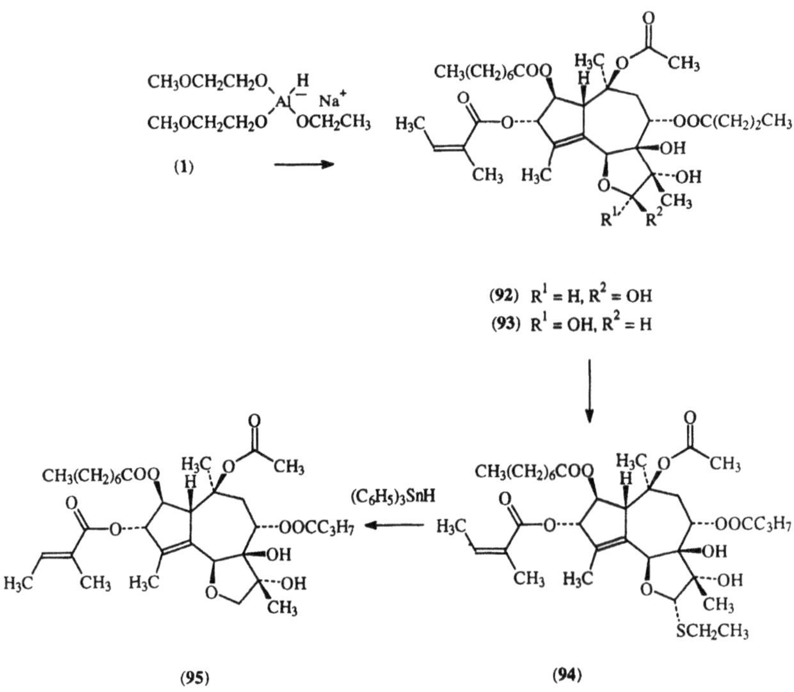

Scheme 10. Synthesis of 12-deoxythapsigargin

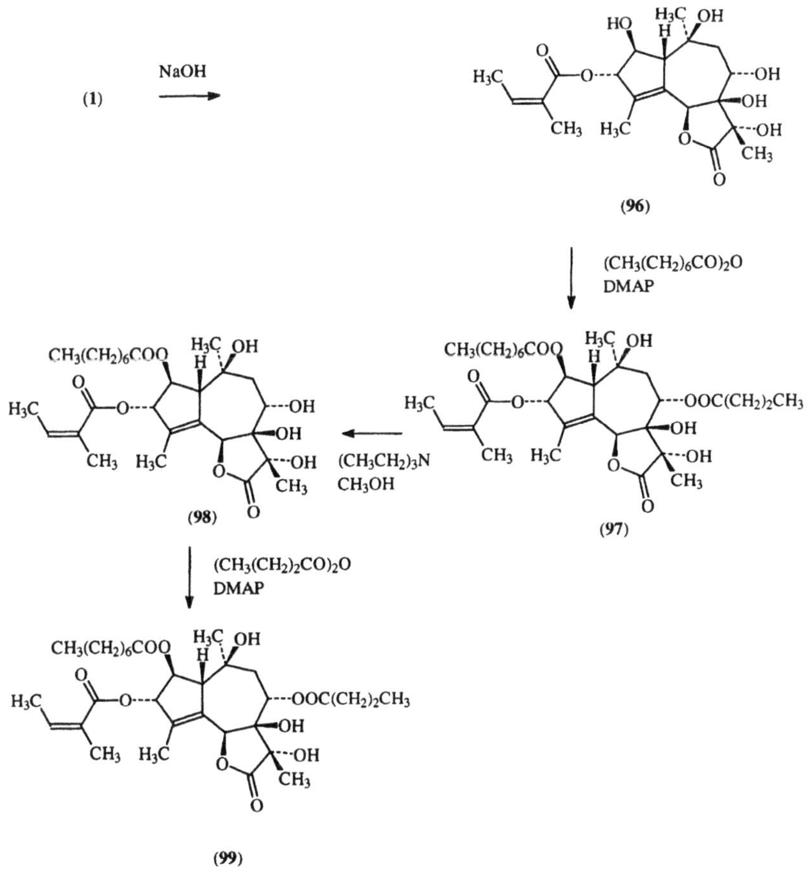

Scheme 11. Selective hydrolysis of the 10-O-acyl group in thapsigargin

O-trideacylderivative (**96**) which by reaction with an excess of octanoic anhydride is converted to the 2,8-dioctanoate (**97**). Selective hydrolysis of the 8-octanoate group to give (**98**) followed by reesterification with butyric anhydride yields 10-O-deacetylthapsigargin (**99**) (*73*) (Scheme 11).

9. Structure Activity Relationships

The very small value of the dissociation constant indicates that thapsigargin (**1**) is very intimately bound to the binding site. This statement is confirmed by the dramatic change in affinity cause by small changes in

Table 6 *Relative potencies of Thapsigargin-Derived Microsomal Ca^{2+}-ATPase Inhibitors**

Compound	Relative Activity (R)
(1)	1
(2)	1 0
(95)	1 1
(85)	1 2
(92 + 93)	1 6
(91)	1 9
(86)	2 5
(80)	2 5
(79)	2 8
(77)	11
(81)	15
(88)	16
(86)	40
(99)	42
(72)	66
(76)	44 × 10
(69)	31 × 10^2

* The R value designates the number obtained by dividing the IC_{50} value of the analogue with the IC_{50} value of thapsigargin (**1**) [IC_{50}(analogue)/IC_{50}(thapsigargin)] The analogues are arranged according to decreasing potencies The numbers are obtained from the IC_{50} values reported in Refs (*72–74*) Notice that the inhibition of the ATPase has been measured in two different ways in the references and that different enzyme preparations have been used

structure (Table 6). Thus, epimerization of C(8) causes the IC_{50} value to increase more than 3000 times [compare (**1**) with (**69**)]. Similar epimerization at C(3) induces a fortyfold decrease in affinity [compare (**77**) with (**76**)]. The carboxylic acid residue at O(3) also has some importance for the affinity, since replacement of angelic acid (**1**) with the larger octanoic acid (**77**) causes an elevenfold decrease in inhibitory activity. The acyl residue at O(10), however, appears to be of major importance for activity since hydrolysis of this ester causes a fortyfold decrease in activity [compare (**1**) with (**99**)]. In contrast, the hydroxyl groups at C(7) and C(11) appear to be of lesser importance since monoacetylation (**79**) and (**80**) only yields a two to threefold decrease in activity. Acetylation of both of these hydroxy groups (**81**), however, produces a somewhat weaker analogue, which might be explained by the bulkiness of the two acetyl groups. Similarly, the lactone carbonyl is not essential for activity, since reduction of this

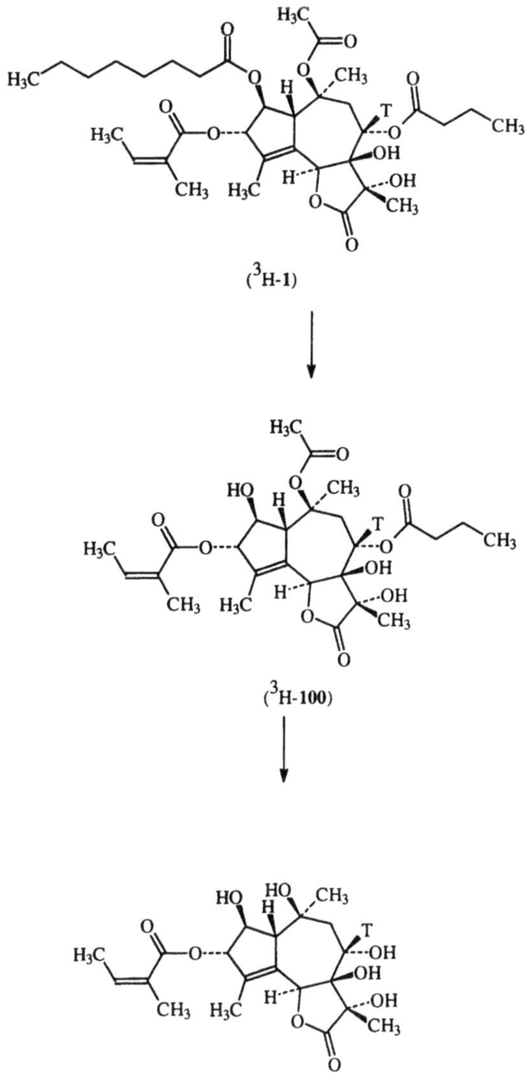

(^{3}H-1)

(^{3}H-100)

(^{3}H-96)

Scheme 12. Metabolic catabolism of thapsigargin

group to a methylene group (**95**) has only a marginal effect. This is confirmed by reduction of the lactone to a mixture of the two lactols (**92**) and (**93**), which has a somewhat smaller activity than thapsigargin. It is tempting to speculate that this remaining activity mainly originates in the β-form, since the β-ethyl acetal (**85**) is only marginally less potent than **1**, whereas the α-acetate (**91**) is only half as potent. Replacement of the butanoic acid residue with larger acid functions causes a decrease in activity, thus limiting the possibilities for introduction of fluorescent groups which in general contain a large aromatic system.

Replacement of the octanoic acid residue at O(2) with hexanoic acid only has a marginal effect on the activity [compare (**1**) with (**2**)]. Unfortunately no chemical method has been developed for selectively replacing this acid. However, Nature produces trilobolide (**14**), which appears to be four times less potent than thapsigargin indicating that the nature of the ester group has some bearing on the activity. It is important to point out that the potencies observed in studying enzyme preparations do not in a simple way correspond to functional assays performed on *e.g.* whole cells. Thus both of the acetates (**79**) and (**80**) have a considerable effect on the isolated enzyme, but they are thirty times less potent as histamine secretagogues (77).

The above structure activity relationships are based on measurements performed on purified enzyme preparations.

10. Metabolic Catabolism of Thapsigargin

No *in vivo* study has been performed on the metabolism of thapsigargin. Incubation of hepatocytes with thapsigargin tritiated at C(8) reveals a quick catabolism of the compound which affords first 2-O-deoctanoylthapsigargin (^{3}H-**100**) and thereafter the trideacylated derivative (^{3}H-**96**) (78) (Scheme 12). Further degradation products could not be detected because of loss of the reporter tritium, probably occurring by an oxidation of the secondary alcohol at C(8). Addition of diethyl *p*-nitrophenyl phosphate strongly protected thapsigargin from metabolic degradation indicating that carboxylesterases catalyze the transformation.

Acknowledgment

This work was supported by the Danish Technical Research Council, the Alfred Benzon Foundation and the Carlsberg Foundation.

References

1 PERROT, R Matieres Premieres Usuelle du Regne Vegetal, p 1630 Paris Mason 1943–44

2 TSCHIRCH, A , and E STOCK Die Harze, Vol 2, 2nd half, 2nd part, p 1540 Berlin Verlag von Gebruder Borntraeger 1936

3 RASMUSSEN, U , S B CHRISTENSEN, and F SANDBERG Thapsigargin and Thapsigargicin, two new Histamine Liberators from *Thapsia garganica* L Acta Pharm Suee **15**, 133 (1978)

4 ALI, H , S B CHRISTENSEN, J C FOREMAN, F L PEARCE, W PIOTROWSKI, and O THASTRUP The Ability of Thapsigargin and Thapsigargicin to Activate Cells Involved in the Inflammatory Response Br J Pharmacol **85**, 705 (1985)

5 SCHARFF, O , B FODER, O THASTRUP, B HOFMAN, J MØLLER, L P RYDER, K D JACOBSEN, E LANGHOFF, E DICKMEISS, S B CHRISTENSEN, P SKINHØJ, and A SVEJGAARD Effect of Thapsigargin on Cytoplasmic Ca^{2+} and Proliferation of Human Lymphocytes in Relation to AIDS Biochim Biophys Acta **972**, 257 (1988)

6 THASTRUP, O , A P DAWSON, O SCHARFF, B FODER, P J BJERRUM, S B CHRISTENSEN, and M R HANLEY Thapsigargin, a Novel Molecular Probe for Studying Intracellular Calcium Release and Storage Agents Actions **27**, 17 (1989)

7 OHUCHI, K , T SUGAWARA, M WATANABE, N HIRASAWA, S TSURUFUJI, H FUJIKI, S B CHRISTENSEN, and T SUGIMURA Analyses of the Stimulative Effect of Thapsigargin, a non-TPA-Type Tumour Promoter, on Arachidonic acid metabolism in Rat Peritoneal macrophages Br J Pharmacol **94**, 917 (1988)

8 HAKII, H , H FUJIKI, M SUGANUMA M NAKAYASU, T TAHIRA, T SUGIMURA, P J SCHEUER, and S B CHRISTENSEN Thapsigargin, a Histamine Secretagogue, is a non-12-O-Tetradecanoylphorbol-13-acetate (TPA) Type Tumour Promoter in Two Stage Mouse Skin Carcinogenesis J Cancer Res Clin Oncol **111**, 177 (1986)

9 THASTRUP, O Role of Ca^{2+}-ATPases in Regulation of Cellular Ca^{2+} Signalling, as Studied with the Selective Microsomal Ca^{2+}-ATPase Inhibitor, Thapsigargin Agents Actions **29**, 8 (1990)

10 THASTRUP, O , P J CULLEN, B K DRØBAK, M R HANLEY, and A P DAWSON Thapsigargin, a Tumor Promoter, Discharges Intracellular Ca^{2+} stores by Specific Inhibition of the endoplasmic reticulum Ca^{2+}-ATPase Proc Acad Sci (USA) **87**, 2466 (1990)

11 LYTTON, J , M WESTLIN, and M HANLEY Thapsigargin Inhibits the Sarcoplasmic or Endoplasmic Reticulum Ca-ATPase Family of Calcium Pumps J Biol Chem **266**, 17067 (1991)

12 CHRISTENSEN, S B , E NORUP, U RASMUSSEN, and J ØGAARD MADSEN Structure of Histamine Releasing Guaianolides from *Thapsia garganica* Phytochemistry **23**, 1659 (1984)

13 De PASCUAL, TERESA, J , J R MORAN, J M HERNANDEZ, and M GRANDE Phenylpropanoids and other Derivatives from *Thapsia villosa* Phytochemistry **24**, 2071 (1985)

14 SMITT, U W , A K JAGER, A ANDERSEN, and L GUDIKSEN Comparative Studies in Phytochemistry and Fruit Anatomy of *Thapsia garganica* and *T transtagana*, Apiaceae (Umbelliferae) Bot J Lin Soc **117**, 281 (1995)

15 NORUP, E , U W SMITT, and S B CHRISTENSEN The Potencies of Thapsigargin and Analogues as Activators of Rat Peritoneal Mast Cells Planta Med 251 (1986)

16 NORUP, E , U W SMITT, and S B CHRISTENSEN Nortrilobolide, a New Potent Guaianolide Secretagogue from *Thapsia garganica* Planta Med **57**, 196 (1991)

17 HOLUB, M , Z SAMEK, R DEGROOTE, V HEROUT, and F SORM The Structure of the

Sesquıterpenıc Trıester Lactone Trılobolıde Collect Czech Chem Comm **38**, 1551 (1973)

18 HOLUB, M, and M BUDESINSKY Sesquıterpene Lactones of the Umbellıferae Phytochemıstry **25**, 2015 (1986)

19 FRONCZEK, F K, D VARGAS, N H FISCHER, and K HOSTETTMANN The Molecular Structure of 7α-Hydroxy-3-desoxy-zaluzanın C, a Molluscıcıdal Sesquıterpene Lactone J Nat Prod **47**, 1036 (1984)

20 TUTIN, T G Thapsıa L In TUTIN, T G, V H HEYWOOD, N A BURGES, D M MOORE, D H VALENTINE, S M WALTHERS, and D A WEBB, eds Flora Europaea, Vol 2, p 370 Cambrıdge Cambrıdge Unıversıty Press 1968

21 AVATO, P, N JACOBSEN, and U W SMITT Chemotaxonomy of *Thapsıa maxıma* Mıller Constıtuents of the Essentıal Oıl of the Fruıts J Ess Oıl Res 4, 467 (1992)

22 SMITT, U W A Chemotaxonomıc Investıgatıon of *Thapsıa vıllosa* L Apıaceae (Umbellıferae) Bot J Lın Soc **118**, 367 (1995)

23 RASMUSSEN, U, S B CHRISTENSEN, and F SANDBERG Phytochemıstry of the Genus *Thapsıa* Planta Med **43**, 336 (1981)

24 DE PASCUAL, M, J TERESA, J R MORAN, and M GRANDE (8*R*, 14*S*)-8-Senecıoyloxy-14, 15-epoxythapsan-14-ol, a New Sesquıterpenoıd from *Thapsıa vıllosa var mınor* A 2D NMR Study Chem Letters 865 (1985)

25 DE PASCUAL, M, J TERESA, M DE PASCUAL, A ARIAS, J M HERNANDEZ, J R MORAN, and M GRANDE Helmantıcın, a Phenylpropanoıd from *T vıllosa* Phytochemıstry **24**, 1773 (1985)

26 DE PASCUAL, M, J TERESA, J R MORAN, A FERNANDEZ, and M GRANDE Hemıacetalıc Thapsane Derıvatıves from *Thapsıa vıllosa var mınor* Phytochemıstry **25**, 703 (1986)

27 DE PASCUAL, M, J TERESA, J R MORAN J M HERNANDEZ, and M GRANDE 12-Hydroxytovarol and other Derıvatıves from *Thapsıa vıllosa* var *mınor* Phytochemıstry **25**, 1165 (1986)

28 DE PASCUAL, M, J TERESA, J R MORAN, J M HERNANDEZ, and M GRANDE Tovarol and Other Germacrance Derıvatıves from *Thapsıa vıllosa* Phytochemıstry **24**, 1779 (1985)

29 AVATO, P, G TRABACE, and U W SMITT Composıtıon of the Essentıal Oıls from Three Types of *Thapsıa vıllosa* L, Apıaceae (Umbellıferae) Phytochemıstry **43**, 609 (1996)

30 AVATO, P, G TRABACE, and U W SMITT Composıtıon of the Essentıal Oıls from Fruıts from Polyploıde Types of *Thapsıa vıllosa* L J Ess Oıl Res **8**, 123 (1996)

31 VELASCO-NEGUERUCLA, A, and M J PEREZ-ALONSO The volatıle Oıl of *Thapsıa vıllosa* L A Medıcınal Plant of the Medıterranean Basın Phytochem (Lıfe Scı Adv) **11**, 125 (1992)

32 PUJADAS, A, J A ROSSELLO, and P BARCELO De Flora balearıca Adnotatıones (10) *Thapsıa gymnesıca* Spec Nov Candollea **46**, 65 (1991)

33 CHRISTENSEN, S B, U RASMUSSEN, and C CHRISTOPHERSEN Thapsıgargın, Constıtutıon of a Sesquıterpene Lactone Hıstamıne Lıberator from *Thapsıa garganıca* Tetrahedron Letters **21**, 3829 (1980)

34 CHRISTENSEN, S B, I KJØLLER LARSEN, U RASMUSSEN, and C CHRISTOPHERSEN Thapsıgargın and Thapsıgargıcın, two Hıstamınc Lıberatıng Sesquıterpene Lactones from *Thapsıa garganıca* X-ray Analysıs of the 7,11-Epoxıde of Thapsıgargın J Organ Chem **47**, 649 (1982)

35 COXON, I M, M P HARTSHORN, and D N KIRK Acıd-Catalysed Reactıons of 13, 17a-Eopxy- and 17a,18-Epoxy-C-nor-D-homo-spırostans Tetrahedron **21**, 2489 (1965)

36 KUTSHABSKY I, G RECK, D PFEIFFER, and H RIPPERGER Dıe Struktur des Sesquıterpens Sılerın Z Chem **24**, 24 (1984)

37 CHRISTENSEN, S B Intepretation of the NMR and Circular Dichroic Data of the Sesquiterpene Lactone Thapsigargin Acta Chem Scand Ser B **42**, 623 (1988)

38 CHRISTENSEN, S B, and E NORUP Absolute Configuration of the Histamine Liberating Sesquiterpene Lactones Thapsigargin and Trilobolide Tetrahedron Letters **26**, 107 (1985)

39 LAURIDSEN, A, C CORNETT, and S B CHRISTENSEN Exciton Coupling in Circular Dichroic Spectroscopy as a Tool for Establishing the Absolute Configuration of α,β-Unsaturated Esters of Allylic Alcohols Acta Chem Scand **45**, 56 (1991)

40 SMITT, U W, P MOLDT, and S B CHRISTENSEN Structure of a pro 1,4-Dimethylazulene Guaianolide from *Thapsia garganica* L Acta Chem Scand Ser B **40**, 711 (1986)

41 AVATO, P, A ANDERSEN, U W SMITT, and S B CHRISTENSEN Localization of the Acyl Groups in Proazulene Guaianolides from *Thapsia transtagana* Brot J Nat Prod **56**, 411 (1993)

42 SMITT, U W, C CORNETT, A ANDERSEN, S B CHRISTENSEN, and P AVATO New Proazulene Guaianloides from *Thapsia villosa* L J Nat Prod **53**, 1479 (1990)

43 AVATO, P Essential Oil of *Thapsia garganica* Planta Med **57**, 585 (1991)

44 SMITALOVA, Z, M BUDESINSKY, D SAMAN, S VASICKOVA, and M HOLUB Components of the Extract from the Underground Parts of *Laserpitium siler* L of Slovenian origin Collect Czech Chem Com **49**, 852 (1984)

45 MARCH, J Advanced Organic Chemistry, 3rd edn, p 856 New York John Wiley and Sons 1985

46 MARCH, J Advanced Organic Chemistry, 3rd edn, p 341 New York John Wiley and Sons 1985

47 SRIKRISHNA, A, and K KRISHNAN Sterospecific Synthesis of Thaps-7(15)-ene and Thaps-6-ene, Probable Biogenetic Precursors of Thapsanes Tetrahedron Letters **30**, 6577 (1989)

48 SRIKRISHNA, A, and K KRISHNAN Stereospecific Construction of Multiple Contiguous Quaternary Carbons Total Synthesis of (±)-*cis,anti,cis*-1,8,12,12-Tetramethyl-4-oxa-tricyclo[6 4 0 0] dodecan-3-ol, a Thapsane Isolated from *Thapsia villosa* var *minor* J Organ Chem **58**, 7751 (1993)

49 PATKAR, S A, U RASMUSSEN, and B DIAMANT On the Mechanism of Histamine Release Induced by Thapsigargin from *Thapsia garganica* L Agents Actions **9**, 53 (1979)

50 MIKKELSEN, E O, O THASTRUP, and S B CHRISTENSEN Effects of Thapsigargin in Isolated Rat Thoracic Aorta Pharmacol Toxicol, **62**, 7 (1988)

51 CHRISTENSEN, S B, A ANDERSEN, A LAURIDSEN, P MOLDT, U W SMITT, and O THASTRUP Thapsigargin, a Lead to Design of Drugs with the Calcium Pump as Target In New Leads and Targets in Drug Research (KROGSGAARD-LARSEN, P, S B CHRISTENSEN, and H KOFOD, eds), p 243 Copenhagen Munksgaard 1992

52 FURUYA, Y, P LUNDMO, A D SHORT, D G GILL, and J T ISAACS Endoplasmic Reticulum Calcium-ATPase as a Therapeutic Target for Activating Programmed Death of Nonproliferating Androgen-Independent Prostatic Cancer Cells Cancer Res **54**, 6167 (1994)

53 JACKSON, T R, S I PATTERSON, O THASTRUP, and M R HANLEY A Novel Tumour Promoter, Thapsigargin, Transiently Increases Cytoplasmic Free Ca^{2+} Without Generation of Inositol Phosphates in NG115-401L Neuronal Cells Biochem J **253**, 81 (1988)

54 KASS, G E N, S K DUDDY, G A MOORE, and S ORRENIUS 2,5-Di(*tert*-butyl)-1,4-benzohydroquinone Rapidly Elevates Cytosolic Ca^{2+} Concentration by mobilizing the inositol 1,4,5-trisphosphate-sensitive Ca^{2+} Pool Biochem J **264**, 15192 (1989)

55 DEMAUREX, N, D P LEW, and K KRAUSE Cyclopiazonic Acid Depletes Intracellular

Ca^{2+} Stores and Activates and Influx Pathway for Divalent Cations in HL-60 Cells. J. Biol. Chem. **267**, 2318 (1992).

56. INESI, G., and Y. SAGARA: Specific Inhibitors of Intracellular Ca^{2+} Transport ATPases J. Membrane Biol. **141**, 1 (1994).
57. FOSKETT, J.K., and A.D. WONG: Calcium Oscillations in Parotid Acinar Cells Induced by Microsomal Ca^{2+}-ATPase Inhibition. Am. J. Physiol. **262**, C656 (1992).
58. PUTNEY, J.W., and G.S. BIRD: The Signal for Capacative Calcium Entry. Cell **75**, 199 (1993).
59. GLYNN, I.M., and S.J.D. KARLISH: Occluded Cations in Active Transport. Ann. Rev. Biochem. **59**, 171 (1990).
60. SAGARA, Y., J.B. WADES, and G. INESI: A Conformational Mechanism for Formation of a Dead-End Complex by the Sarcoplasmic Reticulum ATPase with Thapsigargin. J. Biol. Chem. **267**, 1286 (1992).
61. SAGARA, Y., F. FERNANDEZ-BELDA, L. DE MEIS, and G. INESI: Characterization of the Inhibition of Intracellular Ca^{2+} Transport ATPases by Thapsigargin. J. Biol. Chem. **267**, 12606 (1992).
62. DAVIDSON, G.A., and R.J. VARHOL: Kinetics of Thapsigargin-Ca^{2+}-ATPase (Sarcoplasmic Reticulum) Interaction Reveals a Two Step Binding Mechanism and Picomolar Inhibition. J. Biol. Chem, **270**, 11731 (1995).
63. STOKES, D.L., W.R. TAYLOR, and N.M. GREEN: Structure, Transmembrane Topology and Helix Packing of P-Type Ion Pumps. FEBS Letters **346**, 32 (1994).
64. MACLENNAN, D.H., D.M. CLARKE, T.W. LOO, and I.S. SKERJANC: Sitedirected Mutagenesis of the Ca^{2+} ATPase of Sarcoplasmic Reticulum. Acta Physiol. Scand. **146**, 141 (1992).
65. NØRREGÅRD, A., B. VILSEN, and J.P. ANDERSEN: Transmembrane Segment M3 is Essential to Thapsigargin Sensitivity of the Sarcoplasmic Reticulum Ca^{2+}-ATPase. J. Biol. Chem. **269**, 26598 (1994).
66. NØRREGÅRD, A., B. VILSEN, and J.P. ANDERSEN: Chimeric Ca^{2+}ATPase/Na^+,ATPase Molecules. FEBS Letters **336**, 248 (1993).
67. ANDERSEN, J.P., and B. VILSEN: Structure-Function Relationships of Cation Translocation by Ca^{2+}-and Na^+,K^+-ATPases Studied by Site Directed Mutagenesis. FEBS Letters **359**, 101 (1995).
68. HUA, S., H. MALAK, J.R. LAKOWICZ, and G. INESI: Synthesis and Interaction of Fluorescent Thapsigargin Derivatives with the Sarcoplasmic Reticulum ATPase Membrane-Bound Region. Biochemistry **34**, 5137 (1995).
69. ANDERSEN, A., A. LAURIDSEN, and S.B. CHRISTENSEN: Radio- and Fluorscence Labelling of Thapsigargin, a Selective Inhibitor of Microsomal Calcium-ATPase. J. Label. Comp. Radiopharm. **31** 199 (1992).
70. CHRISTENSEN, S.B.: Radiolabelling of the Histamine Liberating Sesquiterpene Lactone, Thapsigargin. J. Label. Comp. Radiopharm. **22**, 71 (1985).
71. FALSONE, G., H. HADDAD, and D. WENDISCH: Sesquiterpenlactontriester ungewöhnlicher Struktur aus *Thapsia garganica* L. Arch. Pharmaz. (Weinheim) **319**, 372 (1986).
72. CHRISTENSEN, S.B., M. HERGEBNHAHN, H. ROESER, and E. HECKER: Toxicodynamies of Tumour Promoters of Mouse Skin III. Specific Binding of the Tumour Promoter Thapsigargin as Measured by the "Cold Acetone-Filter Assay". J. Cancer Res. Clin. Oncol. **118**, 344 (1992).
73. ANDERSEN, A., C. CORNETT, A. LAURIDSEN, C.E. OLSEN, and S.B. CHRISTENSEN: Selective Transformations of the Ca^{2+} Pump Inhibitor Thapsigargin. Acta Chem. Scand. **48**, 340 (1994).
74. NIELSEN, S.F., O. THASTRUP, R. PEDERSEN, C.E. OLSEN, and S.B. CHRISTENSEN: Structure

Activity Relationships of Analogues of Thapsigargin Modified at O-11 and O-12 J Med Chem **38**, 272 (1995)

75 ANDERSEN, A , M TREIMAN, J J POULSEN, C CORNETT, P MOLDT, C E OLSEN, and S B CHRISTENSEN Ca^{2+}-ATPase Inhibitory Activity of a Locked Analogue of Thapsigargin Bioorg Med Chem Letters **4**, 657 (1994)

76 CHRISTENSEN, S B , A ANDERSEN, J J POULSEN, and M TREIMAN Derivatives of Thapsigargin as Probes of its Binding Site on Endoplasmatic reticulum Ca^{2+} ATPase Steroselectivity and Important Functional Groups FEBS Letters **335**, 345 (1993)

77 ANDERSEN, A Thapsigargin-Analoger, Semisyntese og Struktur-Aktivitets Studier PhD Thesis Royal Danish School of Pharmacy 1994

78 NIELSEN, M S , C E OLSEN, J DICH, S B CHRISTENSEN, and N GRUNNET Metabolism of Thapsigargin in Rat Hepatocytes Drug Metabolism Disposition **22**, 433 (1994)

79 SMITT, U , and S B CHRISTENSEN Personal Communication

80 DE PASCUAL, M , J TERESA, J R MORAN, A FERNANDEZ, and M GRANDE Non Acetalic Thapsane Sesquiterpenoids from *Thapsia villosa var minor* Phytochemistry **25**, 1171 (1986)

81 SMITT, U W , C CORNETT, E NORUP, and S B CHRISTENSEN Novel Hydroindene Sesquiterpenes from *Thapsia villosa* Phytochemistry **29**, 873 (1990)

82 LEMMICH, E , B JENSEN, and U W RASMUSSEN (8*R*,14*S*)-8-Angeloylthapsan-14-ol, Sesquiterpene with a Novel Carbon Skeleton, from *Thapsia villosa* Phytochemistry **23**, 809 (1984)

83 SMITT, U W , P AVATO, and S B CHRISTENSEN Personal Communication

84 LEMMICH, E , U W SMITT, J S JENSEN, and S B CHRISTENSEN Guaiane Esters from *Thapsia villosa* Phytochemistry **30**, 2987 (1991)

(*Received May 6, 1996*)

Pregnane Glycosides

D. DEEPAK, S. SRIVASTAV, and A. KHARE*,
Department of Chemistry, Lucknow University,
Lucknow 226007, India

Contents

---

* In memory of Prof. M. P. KHARE.

1. Introduction

Pregnanes (*1, 2*) are C_{21} steroidal compounds found in nature either in the free state or as glycosides. In pregnane glycosides the sugar moiety is linked to an alcoholic hydroxyl group of the pregnane aglycon, most frequently at C-3 (*3*), C-20 (*4*) or both (bisdesmosidic glycosides) (*5*), through an acetal linkage. However, in some cases, the sugar moiety is linked to hydroxyl functions at C-2 (*6*), C-4 (*7*) or C-21 (*8*). Pregnane glycosides containing one (*9*) to six (*10*) sugar units have been isolated from the extracts of different plant parts, *i.e.* roots, stems, seeds etc.

The last comprehensive review of pregnane glycosides by REICHSTEIN (*1*) covered the literature up to 1967. Although four review articles dealing with certain aspects of pregnanes and their glycosides (*11–14*) have since been published, no comprehensive review has appeared since then. A review article by DEEPAK and co-workers (*2*) dealt in depth with the structural features of plant pregnanes; the present review article is thus a continuation of this earlier review. Besides the structures of isolated pregnanes and their glycosides, new techniques of isolation, recent physicochemical methods of structure elucidation and the biological significance of glycosides reported during the period 1968–1995 have been incorporated.

2. Isolation and Identification

The advent of new chromatographic techniques has made it possible to isolate these compounds in high purity which was not possible earlier. Examples of the use of classical and more recent techniques for isolation of pregnane glycosides are given below.

2.1. Thin Layer and Column Chromatography

Use of thin layer chromatography (*15*) still prevails for preliminary identification and for comparison with authentic samples. The use of reversed phase TLC (RP-8-R_{254} S and RP-18_{254}) for the study of pregnane glycosides has been reported by JIN *et al.* (*16*) and YUAN *et al.* (*17*). Use of high performance TLC (Si 50,000 F-254S) (*6*) and high performance reversed phase TLC (Merck HPTLC RP-18) (*18*) has also been reported.

The most common and successfully employed method for preparative isolation of pregnane glycosides is column chromatography. Normal and reverse phase silica gel columns (Li chromprep RP-8) are being used for such isolations (*17,19*). With reverse-phase packing material, there is increased back pressure which requires a shortening of the column in order to maintain adequate flow rates (*20*). $AgNO_3$ impregnated silica gel has been used for separation of Δ^5- and 5α-H types of pregnane derivatives (*19*). Several bisdesmosidic pregnane glycosides have been isolated by ABE *et al.* (*21*) who used a combination of polystyrene (MCI gel), reverse-phase octadecyl silica (ODS) and silica gel columns.

2.2. Sephadex LH-20 Chromatography

Sephadex LH-20 has been used successfully for the separation of pregnane glycosides. A typical isolation procedure involving silica gel column chromatography and Sephadex LH-20 for the separation of cynanformoside A (**81**) and B (**82**) from *Cynanchum formosanum* has been described by CHEN *et al.* (*22*). Sephadex LH-20 chromatography has been combined with silica gel and ODS chromatography by IDAKA *et al.* (*23*) for the isolation of causiaroside II (**237**).

2.3. Flash Chromatography

Preparative air-pressure (compressed air or nitrogen) driven liquid chromatography (flash chromatography) (*24*) is relatively fast, thus reducing the risk of decomposition and sample loss. Thus, dry column flash chromatography using CH_2Cl_2–MeOH and hexane-Me_2CO has been used by CABRERA *et al.* (*25*) for preliminary separation of the crude glycoside mixture obtained from *Mandevilla pentlandiana.*

2.4. Low Pressure Liquid Chromatography (LPLC)

LPLC (*26*) is a very versatile and simple means of isolating substances on a milligram to gram scale, generally in combination with a prepurification step. In order to increase the effective column length and thus augment loading capacity and separating power, several Lobar columns are connected in series (*20*). The technique makes use of columns containing packing with a particle size of ca 40–60 μm. Thus, YUAN *et al.* (*17*) have isolated marsdekoiside A (**183**), a pregnane triglycoside from *Marsdenia koi*, using Lobar chromatography on a LPLC system with a RP-8 column in combination with Si gel chromatography.

2.5. High Performance Liquid Chromatography (HPLC)

HPLC (*27–29*) is a very efficient technique used for the detection and isolation of pregnane glycosides and is commonly applied as a last step in the purification process. ABE *et al.* (*21, 30*) have purified the bisdesmosidic pregnane glycoside constituents of *Apocynum venetum* and *Trachelospermum asiaticum* using CH_3CN and water as eluent. HPLC using reverse phase packing material is also being successfully employed for isolation purposes. Thus, ITOKAWA *et al.* (*31*) effected the separation of pregnane glycoside constituents of *Periploca sepium* by HPLC on RP-18 column using methanol-water as eluant while toosendanoside (**235**) was isolated from *Melia toosendan* by NAKANISHI *et al.* (*6*) by HPLC on a reverse phase Kusano ODS column (MeOH–H_2O; flow rate 3 ml/min) and Kusano Si–10 silica column ($CHCl_3$–MeOH; flow rate 3 ml/min). Preparative HPLC was used for the isolation of four pregnane glycosides from *Boucerosia aucheriana* by HAYASHI *et al.* (*18*), while AHMAD *et al.* (*32*) used preparative HPLC on a reverse phase column for separation of two pregnane glycosides from *Caralluma tuberculata.* Chiral HPLC columns are being used for confirming the absolute stereochemistry of the sugar moieties obtained by acidic hydrolysis of pregnane glycosides (*19*).

The detection of pregnane glycosides is usually difficult as no diagnostic test or specific reaction for their identity is so far known. Colours observed with non-specific reagents such as chloroformic $SbCl_3$ (*33, 34*) and 50% H_2SO_4 (*35, 36*), although widely used for their detection, are never reliable and conclusive. Still, there are some diagnostic reagents and reactions which are used for characterization, such as the LIEBERMANN–BURCHARDT (*37*) and CARR–PRICE tests (*34*) for steroids. The presence of sugar(s) in these glycosides is established by the MOLISCH test (*38, 39*). 2-Deoxy- and 2,6-dideoxyhexoses are characterized using the xanthydrol test (*3, 40*), WEBB's test (*41*), vanillin-perchloric acid reagent (*42, 43*) and KELLER–KILIANI test (*3, 44*) while the presence of normal (2-hydroxy) sugars is detected by PARTRIDGE (*45*) and FEIGL tests (*9, 46*).

3. Structure Elucidation

The conventional method for structure elucidation of pregnane glycosides involved acid hydrolysis followed by identification of the aglycon and sugar residues separately (*47*), whereas the site of glycosidation was usually determined by comparing the UV absorption of the glycosides with that of the aglycon in the presence or absence of various shift reagents (*48, 49*). In recent years, in addition to mass spectroscopy (EI, CI, FD and

FAB), other physico-chemical techniques of a non-destructive nature such as NMR (^1H, ^{13}C and 2D) etc. are increasingly being used for structure elucidation of pregnane glycosides.

3.1. One-Dimensional NMR Spectroscopy

^1H NMR Spectroscopy

The high frequency (400–500 MHz) ^1H NMR spectra of pregnane glycosides are well resolved; thus the information from the range which contains the signals of anomeric protons is considerable. The anomeric protons of the monosaccharides present in pregnane glycosides appear between 4.3–5.5 ppm (*50, 51*). The anomeric protons of α-glycosides usually resonate 0.3–0.5 ppm downfield from those of the corresponding β-glycosides (*52*). In the case of normal (2-hydroxy) sugars, the anomeric proton usually appears as a doublet (*16*) in the region δ4.4–5.4 (*16, 52*), the magnitude of the splitting depending on the stereochemistry of H-1′ as well as that of H-2′. For example, if the H-2′ is axial (as in the case of gluco and galacto stereochemistry), $J_{1',2'}$ is relatively small (2–4 Hz) for an α-glycosidic linkage, whose H-1′ is equatorial (*53*). In β-anomers of sugars with gluco and galacto configuration H-1′ and H-2′ are trans-diaxial which results in a larger (8–10 Hz) coupling constant (*50*). In sugars having the manno-configuration, such as rhamnose, where H-2′ is equatorial the small dihedral angle gives rise to small values of $J_{1',2'}$ for both α- and β-anomers (*52*). In the case of 2-deoxy sugars, the signals of the anomeric proton appears as a *dd* in the region δ4.2–5.3 (*51, 54*) and sometimes as a triplet (*55*) if $J_{1',2'a} = J_{1',2'b}$ depending on the nature of the glycosidic linkage. Coupling constants of 7–10 and 1–2 Hz are indicative of a β-glycosidic linkage (*3*) with the sugars in the $^4C_1$ conformation (*56*) and H-1′ axial, whereas smaller coupling constants of 3–4 and 1 Hz (*57*) indicate a α-glycoside with the sugar in the $^1C_4$ conformation (*56*) and H-1′ is equatorial. In the higher field region, the signals of the equatorial and axial H-2′ protons of 2-deoxy hexoses appear as two sets of multiplets in the region δ2.0–2.5 and 1.5–2.0 (*53*), respectively, while the characteristic signals of the secondary methyl groups (6′-CH_3) of 6-deoxy sugars appear as doublets (J = 6Hz) between δ1.0–1.5 (*3*).

The ^1H NMR spectra of pregnane glycosides also provide important information about the aglycon. Thus, $-CHOHCH_3$ or $-COCH_3$ side chains at C-17 can be recognized (*15*) by the presence of a three proton doublet in the region δ1.0–1.5 or a three proton singlet at δ2.1, respectively. Two three proton singlets appear in the region δ0.7–1.2 (*15, 53*) due to the angular methyl groups at C-10 and C-13; however, in 18-nor pregnane glycosides the signal of the C-13 angular methyl group is absent (*58*). Signals of the methylene and methine protons occur in the region δ1.5–2.5 (*59*) and δ3–4 (*2*), respectively. The C-11 methine proton under a hydroxyl appears in the region δ3.2–4.6 (*15*, *60*) as a triplet (*15*) or double doublet if a hydroxyl is present at C-12 (**61**) while the C-12 and C-20 methine protons generally are doublets (*62*) and quartets (*63*), respectively, in the same region depending on whether substituents are present on neighbouring carbons. Esterification of the hydroxy functions shifts the signal of corresponding methine proton downfield by 0.6–1 ppm (*2*) compared with its precursor. Most commonly, pregnanes are found as esters of benzoic (*19*), cinnamic (*10*), isovaleric (*64*), tiglic (*65*), nicotinic (*66*), 2-methylbutanoic (*65*), β, β-dimethyl acrylic (ikemic) (*10*, *67*) or acetic acids (*65*).

The number of primary and secondary hydroxyl groups present can be established by counting the acetate peaks at δ2.1–2.3 of acetylated pregnane glycosides (*53*) while the number of tertiary hydroxyl groups can be deduced by D_2O exchange (*16*) and the trichloroacetyl isocyanate reagent (*68*). Decoupling experiments (*7*, *52*, *58*) which are very helpful in confirming the assignments of the anomeric protons and other functional groups are now routine. These experiments can be used for confirming the assignments of the signals due to H-1′, H-2′ and H-5′ of the 2,6-dideoxy sugars besides the C-20 methine and secondary methyl protons present in the side chain of the pregnane aglycon (*63*). Proton spin decoupling and correlated spin–spin coupling experiments (*69*) have been used for establishing the structures of constituent hexoses of pregnane glycosides.

Nuclear Overhauser Effect (NOE) measurements can also be used to prove the point of attachment of the sugar moiety to the aglycon. Irradiation of H-3α of the aglycon (when the sugar is linked to 3-OH) results in an NOE at the anomeric proton of the sugar (S_1) directly attached to the glycon (*5*). Similarly, irradiation of the anomeric proton of the second sugar (S_2) causes enhancement of H-4 proton (in case of a 1 → 4 linkage) of the first sugar (S_1) and vice versa (*70*), thus providing information regarding the sugar sequence and site of glycosidation in pregnane oligoglycosides. The technique is also helpful in determining the structure of the constituent sugars (*7*, *32*) and the stereochemistry at C-17 and C-20 of the pregnane aglycon (*19*).

The point of attachment of the sugar moiety to the pregnane genin can also be ascertained by comparison with the O-acetyl derivative of the pregnane glycoside (*53*). A downfield shift of 0.5–1.0 ppm is observed in the signal of the acetylated methine proton as compared with the parent precursor while the chemical shift of the methine signal involved in the glycosidic linkage remains unaffected.

^{13}C *NMR Spectroscopy*

In recent years ^{13}C NMR spectroscopy which is complementary to ^{1}H NMR spectroscopy has become much more useful due to the greater chemical shift dispersion and the lack of complexities arising from spin–spin coupling and overlap of resonances. It is instrumental in assigning the number, sequence and linkage of sugars (*52*) within the molecule. In the case of oligoglycosides (*52, 71–75*), the identity of the sugar(s) may be established (*18, 19*) on the basis of the chemical shift of the anomeric carbon(s). Moreover, it supplements ^{1}H NMR spectrometry in helping to establish the point of attachment of ester functions present (*52*).

In the ^{13}C NMR spectra of pregnane glycosides the resonances of the anomeric carbons are found in a well-separated chemical shift range of δ96–112 (*52, 76*) and not only greatly aid in determining the number of monosaccharide units but also provide information on the nature of the glycosidic linkages. The signals due to β-linkages usually appear 2–6 ppm downfield from their α-counterparts (*52*). The other resonances due to the carbohydrate part of the glycoside appear in the region δ16–19 (*31, 77*); δ55–62 (*19, 76*); δ60–63.5 (*23, 53*) and δ65–85 (*76*) for the secondary methyl of 6-deoxy sugars, methoxy functions, CH_2OH of normal hexoses and the ring carbons, respectively.

As for the pregnane part of the glycosides the signals of the C-18 angular methyl group appears in the region δ7–15.8 (*22, 51*) while the position of the angular methyl at C-10 varies between δ12–24.5 (*18, 78*). Any variation in the structure of the aglycon, affects (*78*) the chemical shifts of these two angular methyls. If H-5 is α or if a 5,6-double bond is present the signal of C-19 angular methyl group appears between δ10.8–17.0 (*54, 64*) and δ15.5–20.0 (*23, 53*) respectively, while if the double bond is between C6 and C7 it is found between δ14.4–14.7 (*30, 55*). C-21 appears in the region δ15.0–24 unless next to a carboxyl (*6, 22*). Methylene and methine carbons to which no oxygen function is attached absorb between δ35–54 (*79, 80*) while carbons carrying an –OH group have signals in the region δ60–90 (*53, 77*).

Esterification of a hydroxyl deshields the corresponding carbon by 0.6–3.5 ppm (*52, 81*) compared with unacylated precursor. These acylation shifts are important in deducing the position of esterification as the downfield shift of the esterified carbon is accompanied by an upfield shift of the adjacent carbon resonances (the β-carbons) by 1.2–4.0 ppm (*82, 83*). The carbonyl carbon of the ester appears in the region δ165–171 (*51, 65*) depending on the presence or absence of unsaturation in the esters while the other carbons of acid part exhibit their customary shifts (viz. δ20–22 for CH_3 of acetate (*19, 22*), δ128–135 for the aromatic carbons of benzoyl and cinnamoyl residues (*51, 65*), δ117–145 for vinylic carbons of tigloyl and cinnamoyl (*65, 84*) and δ160–164 for the sp^2 hybridized carbon carrying the methyl group of ikemoyl (*10*). The vinylic C-5 and C-6 carbons of the aglycon appear between δ140–144 and δ117–123 (*84*, 85). A $CH_3C{=}O$ side chain attached to C-17 can easily be identified as the carbonyl C-20 resonates between 208–217 (*65, 86*).

The glycosidation shifts are analogous to the acetylation shifts and are instrumental in determining the point of attachment of the sugar chain to the aglycon. The carbon involved in glycosidation shifts to lower field by 3–6 ppm (*87*) while the upfield shift of the adjacent carbons ranges between 0.5–4 ppm as compared with the native genin (*87*). These glycosidation shifts (*88–94*) are being used to ascertain the glycosidation site in the pregnane glycosides (*18, 69*) and in all the reported cases, where sugar is glycosidically linked to C-3 of the genin, the shielding experienced by C-4 is about twice that suffered by C-2 (*19, 58*).

The sugar sequence in the glycoside can be ascertained (*25, 95–98*) by spin lattice relaxation time (T_1) measurements, as the average NT_1 values for the sugar carbons in each unit increase with increasing distance from the aglycon moiety (*99*). This is due to segmental motion in the oligosaccharide chain with the aglycon part exhibiting an anchoring effect (*99*). Differences in the peak intensities of the inner and terminal sugar observed in partially relaxed Fourier transform (PRFT) measurements (*100–103*) in the ^{13}C NMR spectrum also provide information for identification of the terminal sugar and the sugar sequence in pregnane glycosides (*19, 70*). In diglycosides, the anomeric carbon of the terminal sugar resonates 2–4 ppm downfield from that of the inner sugar (*104*).

Long-range selective proton decoupling (LSPD) (*105–108*) has also been used to establish the location of ester functions within the aglycon of pregnane glycosides. This technique has made it possible to correlate protons under ester groups with the corresponding carbonyl carbons, particularly in cases when esters are attached to C-11 and C-12 of a pregnane genin (*19, 107*). This technique also served to identify the chemical shifts of the angular methyl carbons at C-10 and C-13 and the site of the glycosidic linkage. Thus irradiation of the signals due to H-9 and H-12 results in an increase in the intensity of the C-19 and C-18 signals (*107*), respectively, while irradiation of an anomeric proton changes the splitting of that carbon to which it is glycosidically linked, hence permitting identification of the site of glycosidation (*109*).

Primary, secondary and tertiary carbons can be identified by single frequency off resonance decoupling (SFORD) (87, *110*) which reduces CH couplings to such an extent that only the largest coupling constants [*J*(CH)] give rise to residual splittings, thus allowing determination of the number of attached hydrogens (*22*). Thus a quarternary carbon gives rise to a singlet, a methine carbon to a doublet, a methylene to a triplet and a methyl group to a quartet. Information regarding the multiplicity of carbons can also be obtained by newer techniques such as the attached proton test (APT) (*62, 111–113*), distortionless enhancement by polarization transfer (DEPT) (*22, 32, 114–116*) and insensitive nuclei enhanced by polarization transfer (INEPT) (*117–120*). Selective INEPT (*120, 121*) has been used to establish connectivity (*122–124*) between the anomeric proton and carbon atom of the aglycon. Irradiation of the anomeric proton selectively enhances the carbon signal of the aglycon to which it is linked; similarly, irradiation of the aglycon proton leads to the appearance of the anomeric carbon of the glycon residue (*52*).

Berger *et al.* have used the technique of selective proton-decoupling in gated decoupled ^{13}C NMR for the structure revision (*54*) of condurangogenins A, B, C, D and E and their glycosides. The results indicated that the acetoxy group was attached to C-11 at 11α-OH and the cinnamate to C-12 which was the reverse of the originally proposed structures (*125–130*).

3.2. Two-Dimensional NMR Spectroscopy

Although one-dimensional NMR methods (^{1}H and ^{13}C) provide useful information for determining the basic structure of pregnane glycosides, the severe problems encountered due to substantial overlap of multiplets does not generally allow unambiguous assignments of all signals leading to a complete structure of the molecule. These difficulties may be overcome by the use of various two-dimensional techniques developed in recent years (*52, 131–133*). The application of such techniques to solve problems in the field of pregnane glycosides will be discussed briefly.

2D ^{1}H–^{1}H COSY (Homocorrelated Spectroscopy)

This is also referred to as homonuclear shift correlation through *J*-coupling (*134–136*). The information obtained from the spectrum is the scalar coupling connectivity network of the molecule concerned using cross peaks. Assignment of signals requires an initial point for identification of the individual spin systems – in pregnane glycosides, the anomeric proton which is connected to a carbon bearing two oxygen atoms appears downfield and is conveniently taken as a starting point for assignments. Within a typical aldohexopyranosyl ring, the coupling network is unidirectional *i.e.*, H-1 couples to H-2, H-2 couples to H-1 and H-3 and so on

(*52*). In the aglycon portion, the scalar (*J*) coupling pathways leading from H-3α to H-4α, H-4β and to H-2α, H-2β and finally to H-1α, H-1β can be elucidated from a ^{1}H-^{1}H COSY experiment (*113*). The method has also been used to assign the position of an ester function within the aglycon or sugar portion (*17,55*). One fundamental limitation of COSY, however, is that couplings must be at least partially resolved before they can give rise to a cross-peak.

COSY45

COSY-45 (*133*) has two advantages over basic COSY:

(a) By reducing the intensity of transfer between parallel transitions as a result of reducing cross peaks within multiplets and by thus simplifying the appearance of the spectrum around the diagonal in a complex spectrum the technique makes it possible to identify correlations that would otherwise be hidden in the cluster of peaks close to the diagonal.

b) By restricting multiplet transfers largely due to directly connected transitions the method allows determination of the relative sign of coupling constant in a system with three or more spins. AHMAD *et al.* (*32*) have made use of the spin couplings in the COSY-45 (*32*) experiment to identify the sugar of caratuberside A (**58**) from *Caralluma tuberculata*. Sequence information on the sugars of the glycoside could also be deduced from the long-range (^{1}H–^{1}H) COSY-45 experiment (*137*).

Double-Quantum Filtered COSY (DQF–COSY)

Multiple quantum filters (*138*) for elucidating NMR coupling networks have been described; the most widely used filtration method is through double quantum coherence (*139–141*). The great advantage (*86*) of double quantum filtration is that it suppresses the strong signals emanating from singlets, *i.e.* from tertiary methyls and solvents, and that therefore hidden multiplets which are isochronous to tertiary methyls can be assigned unambiguously from the spectrum. It not only provides characteristic multiplicity within the cross-peak, enabling identification of particular sugar units, but also provides semiquantitative information on the coupling constants of protons involved in cross peaks. In the aglycon part of the pregnane glycosides all H–H connectivities except for those next to the angular methyl groups (Me – 18, – 19) can thus be determined by DQF COSY (*7*).

Relayed Coherence Transfer COSY (RCT2D)

In an AMX system where J_{AM} and J_{MX} represent vicinal couplings and $J_{A,X}$ equals zero (for a saturated compound), the corresponding COSY spectrum would show cross peaks between A and M and M and X, but not between A and X. A technique for establishing connectivity between A and X, *i.e.* between two remote nuclei within a given spin system, is known as relayed Coherence Transfer (RCT). RCT COSY (*142–144*) propagates the magnetization transfer from A to M on through further couplings experienced by M. Recently, HUGHES has used RCT 2D NMR spectroscopy for determining proton chemical shifts in steroids (*145*). As the heteronuclear RCT 2D spectrum contains both the direct $^1H-^{13}C$ responses and relayed responses which arise from $^1H-^1H$ vicinal couplings (*146*), it allows the proton–proton and carbon–carbon connectivity network to be deduced irrespective of congestion in the proton spectrum if the carbon spectrum can be resolved. On the basis of RCT2D spectrum, the connectivities in the aglycon portion (C-2 to C-4, the five proton bearing carbon segment from C-6 to C-11, C-14 to C-17 and C-20 to C-21) have been established (*113*).

Nuclear Overhauser Effect Spectroscopy (NOESY)

This experiment offers a means of determining spatial relationships, thus providing the information about the spatial structure of the molecule. Cross peaks are observed in 2D NOESY (*147–148*) spectra between proton pairs that are close in space (*i.e.* typically less than 5A°). In general, 1,3-diaxial and equatorial-axial proton pairs in pyranosyl rings produce intra NOESY cross peaks, *i.e.* for the β-glucopyranosyl residue cross-peaks are observed between H-1 and H-3 (and H-5) whereas a strong cross peak is observed between H-1 and H-2 in the α-glucopyranosyl configuration (*52*). It is also used for sugar sequencing and for determining the sites of glycosidic linkages. In a glycoside (G-O-S_1-O-S_2), where the proton on C-1 of S_2 is close enough to the proton on C-4 of S_1 (in case of a $1 \rightarrow 4$ linkage), a cross peak between H-1 of S_2 and H-4 of S_1 would be observed. Thus, it is possible to demonstrate a linkage between the two sugars from a NOESY experiment (*31, 32*). The experiment is also used for deciding the stereochemistry of substituents (*e.g.* that of the C-17 side chain) in a pregnane aglycon (*39*).

Homonuclear Hartmann–Hahn Spectroscopy (HOHAHA)

The most useful method of relay in coherence along the chain of spins is the isotropic mixing experiment in which the net magnetization is transferred under spin-locking. From a HOHAHA (*149–152*) spectrum, a so-called '*J*-network' can be determined (*39*) where a *J*-network is defined as a group of protons that are serially linked via $^1H-^1H$ J (scalar) couplings. For example, all protons of a single saccharide unit belong to the same *J*-network. A complete spin system can thus be identified (*86*) if there is at least one resonance in the spin system, such as the anomeric proton, which is well isolated and has a resonably large coupling to its neighbouring spin. Therefore, a slice through a HOHAHA spectrum (*39*) at each anomeric proton along the diagonal yields a ^{1}H subspectrum containing all scalar-coupled protons within that sugar residue. However, the distribution of magnetization around the spin system can be impeded by small couplings (*e.g.* H-4 and H-5 in the galactosyl residue) which lead to cross peaks up to H-4 but no further (*52*).

Homonuclear J-resolved Two-dimensional Spectroscopy (HOMO 2DJ)

J-resolved spectroscopy (*153*) is used to resolve overlapping multiplets by producing spectra which have chemical shifts on one axis and scalar coupling on the other. It can provide unprecedented dispersion of the ^{1}H NMR spectra (*154–155*) but leaves unsolved assignment of individual resonances when strongly coupled nuclei are involved and/or multiplets originating from different spin system overlap (*156*). The usefulness of the method declines with increasing number of sugar residues and becomes of limited value in studies of oligoglycoside structure due to overlapping of mutually coupled signals which causes distortions in the multiplet pattern and prevents the use of cross sections for observing individual multiplets and for extraction of the desired ^{1}H–^{1}H couplings (*31, 32*).

Heteronuclear 2D-NMR Spectroscopy

In heterocosy (*157–161*), heteronuclei such as ^{1}H and ^{13}C are correlated in 2D experiments. This, one of the most powerful of 2D experiments, combines the excellent resolving power of decoupled ^{13}C NMR with the ease of interpretation of proton chemical shifts and allows the resolution of single sites in all but the most intractable spin systems. Thus, ^{1}H–^{13}C correlation spectroscopy is useful for identification of protons bonded to individual carbons in pregnane glycosides (*113, 124, 162*).

^{13}C–^{1}H Long Range COSY

Two-dimensional heteronuclear correlation (*163–165*) via long-range coupling has been found to be useful in determining the connectivity of sugar to aglycon and the sequence of the sugars. The technique (*39*) has been employed by ITOKAWA *et al.* (*162*) for determining the sequence of six sugars in the glycosidic chain of periplocoside A (**217**).

Heteronuclear Multiple-Quantum Coherence (HMQC)

Heteronuclear Multiple-Quantum Coherence (HMQC) (*166–168*) is a powerful method for the unambiguous assignment of ^{1}H and ^{13}C NMR (*86*) spectra of pregnane glycosides and the C-H correlation assignment. Kashman *et al.* (*7*) have used HMQC for geminal C-H correlations in deducing the structure of verrucoside (**238**).

3.3. Mass Spectrometry

Mass spectrometry (MS) is obviously of prime importance in structure determination of pregnane glycosides (*2, 169–170*) which are frequently obtained from natural sources only in very small quantities, particularly when it is used in conjunction (*53*) with information obtained from ^{1}H and ^{13}C NMR spectral data. In recent years, better inlet techniques (*171*) have overcome the problem of low volatility.

In electron impact mass spectrometry (EIMS), (*79, 172*) fragments of lower mass value are more evident which often provides valuable structural information (*57*). Sometimes, fragments corresponding to the aglycon and the sugar are obtained (*173*). In addition to producing fragments arising from the common loss of the elements of water, methanol and CH_3CHO in different sequences (*174*), the oligosaccharides of pregnane glycosides also decompose by retro–Diels–Alder fragmentation (*170*) initiated by a double bond created between C2 and C3 by the loss of water or methanol (*175*). Another important mode of fragmentation of oligosaccharides involves the radical ion cleavage of the C1 and C2 bond of the terminal sugar followed by the migration of the methoxyl (or hydroxyl) (*176*) group from C3 to C1 of the same sugar, a process which results in cleavage of the terminal sugar (*50, 53, 170, 175, 177–178*). Further fragmentation of the residual oligosaccharide or glycosides takes place by the characteristic fragmentation patterns reported by Brown *et al.* (*172*). The presence of a methoxy function at C-3 of a normal sugar can be ascertained by the loss of mass fragment $C_3H_6O_2$ from the sugar fragment. Similarly, loss of mass fragment $C_4H_7O_2$ from a 2-deoxy sugar, present at the reducing end, shows the presence of a methoxy function at C-3 (*170*).

EIMS is also very useful in assigning the substituent groups within the aglycon part of polyhydroxy pregnane glycosides (*2*). Studies of the MS of polyhydroxy-pregnanes enabled Fukuoka *et al.* (*179*) and Budzikiewicz *et al.* (*180*) to deduce correlations between structures and fragmentation patterns which have been summarised by Deepak and co-workers (*2*). Mass spectra have been of great utility in establishing the presence of C-8 or/and C-14 hydroxy functions (*181–183*) which being tertiary in nature are not acetylated (*181*) and consequently cannot be easily detected by NMR methods. Mass spectra are also useful in assigning the position of hydroxy functions at C-11, C-12, C-15 and C-16 (*6, 181–182, 184–185*). The loss of ions of m/z 45 or 43 shows the presence of a $-CHOHCH_3$ or $COCH_3$ side chain at C-17 (*3, 61*) and can be used to establish the point of attachment of the sugar chain to either the C-3 or the C-20 hydroxy function of aglycon (*53*). The stereochemistry (α or β orientation) of the C-17 side chain can also be determined by MS (*183*).

Field desorption mass spectra (FDMS) (*186*) of pregnane oligoglycosides (*17, 173*) often contain only the molecular ion $[M^+]$, the protonated molecular ion $[M+H]^+$ or the $[M + \text{cation}]^+$ ion if NH_3 or a metal salt is added and is a reliable method for confirming the molecular formula of pregnane glycosides (*18–19*).

In fast atom bombardment (FAB) (*171*) and secondary ion (SI) MS (*31, 187*) an abundant molecular ion, usually a protonated species $[M+H]^+$ or a cationic species $[M + \text{cation}]^+$, is observed. The MS also contains mass fragments of intermediate and lower mass value which thus provides comprehensive information (*5, 16, 39*) about the oligoglycoside. As evident from FABMS of pregnane glycosides, the individual monosaccharide units become detached from the molecular ion at the glycosidic linkage along with displacement of the hydroxyl group to which it was linked. Starting from the terminal end, the stepwise elimination of monosaccharide units leads to the formation of a fragment corresponding to the genin (*53*). Often, fragments corresponding to $[M^+\text{–genin}]$ and $[M^+\text{–sugar}]$, *i.e.* the oligosaccharide and genin, (*61*) are obtained; these fragment further by repeated H-transfers accompanied by elimination of the terminal sugar less water, thus giving rise to an ion of the same mass as the molecular ion of the corresponding oligosaccharide with one less monosaccharide residue and so on until only the monosaccharide remains. The sequence of sugars and aglycon can be determined from the mass difference of major fragments (*53*). Thus the differences in mass between $G\text{-}S_1\text{-}S_2\text{-}S_3$, $G\text{-}S_1\text{-}S_2$ and $G\text{-}S_1$ in the FAB MS provide information on the sequence of sugars in the glycoside and also indicate which sugar is directly linked to the aglycon (*16, 30*). At what point the sugar residue is attached to the aglycon can also be established (*53*, 61).

While up to a certain point assignment of stereochemistry can also be achieved by mass spectrometry (*183*) a severe limitation of the mass spectrometry approach is the inaccessibility of finer stereochemical details such as the configuration of glycosidic linkage.

3.4. I.R. Spectroscopy

The role of IR spectroscopy in structure elucidation of pregnane glycosides cannot be ignored (*54, 80*), although it has been largely superseded by the techniques discussed earlier. IR spectrometry establishes the presence of carbonyl functions ($\simeq$ 1740–1715 cm^{-1}) (*32, 124*) thus differentiating between hydroxyethyl or acetyl side chains on C-17, and also shows the probable presence of ester functions (*124*). IR spectrometry also establishes the presence of associated and free hydroxyl groups ($\simeq$ 3400 cm^{-1}) (*18, 187*) and unsaturation in pregnane glycosides.

3.5. U.V. Spectroscopy

The absence of a conjugated system in pregnanes has limited the use of U.V. spectroscopy. However, the technique may be useful when pregnane esters containing α, β-unsaturated and/or aromatic acids are encountered (*18–19, 54, 124*).

3.6. Optical Rotatory Dispersion

The C-20 stereochemistry of pregnanes with a CHOH–CH_3 can be established by o.r.d. as was shown by NAGAI (*188*) who reported that the C-20 *o*-nitrobenzoates of pregnane derivatives exhibited a Cotton effect at ca 330 nm due to the $n \rightarrow \pi^*$ transition of the aromatic nitro group whose sign depends on the configuration. Thus 20-*R*-*o*-nitrobenzoates exhibited a negative Cotton effect while 20-*S*-*o*-nitrobenzoates exhibit a positive Cotton effect. It has been reported that polar functional groups present near the nitrobenzoate, such as a 17-OH, strongly influence the Cotton effect. HAYASHI *et al.* (*189*) have used this property to assign absolute configurations to the C-20 carbinol group of sarcostin, utendin and tomentogenin.

The 17-acetyl function of pregnanes may be α- or β-oriented. When no other substituent is present on C-17, compounds with a 17-β-acetyl side chain show a positive Cotton effect whereas those with an α-oriented side chain exhibit a Cotton effect of opposite sign (*190*).

3.7. Hydrolysis of Pregnane Glycosides

Although modern physicochemical techniques link NMR and mass spectrometry play a very important role in structure elucidation of pregnane glycosides, classical degradative methods have not lost their significance. In particular, methods for cleaving sugars from the parent compounds form a vital part of structure determination, especially since they provide confirmation of structural features arrived at by spectrometry. Different conditions of acid hydrolysis, *i.e.* from strong to very mild depending on the nature of the sugar present in the glycoside, are used for identification of sugar and aglycon. Generally, mild acid hydrolysis (0.1N H_2SO_4/Dioxan; RANGASWAMI and REICHSTEIN) (*191–192*) is used for glycosides containing 2-deoxy sugars. Mild acid conditions are required to prevent the destruction of acid sensitive 2-deoxysugars and acid labile tertiary hydroxyl groups in the genin (*193*). Sometimes, hydrolysis is carried out in the presence of methanol (H_2SO_4 or HCl in MeOH) (*17, 31, 39, 70*) or ethanol (AcOH, H_2SO_4 or HCl in EtOH) (*34, 55, 124*). In the case of normal sugar glycosides, strong acidic conditions such as the KILIANI method are required for hydrolysis (*194*). Hydrolysis affords genin and oligosaccharide or monosaccharides. The oligosaccharides are identified either by direct comparison with authentic samples ($[\alpha]_D$, TLC and PC) or by chemical degradation (*23*, *195*). The comparison may involve physical properties (such as PC, $[\alpha]_D$) as well as conversion to lactones, acid phenylhydrazides (*196–197*) and other derivatives such as alditol acetates (*25*), tetramethylsilyl ethers (*16, 113*), partially methylated alditol acetates (*198*) etc., which may be identified by GLC or GC with authentic samples. The absolute configuration of the isolated sugars can be determined by analysing their 3,5-dinitrocarbamate methyl glycoside derivatives on a chiral HPLC column (*18, 199*).

To sequence the sugars in oligoglycosides containing 2-deoxyhexoses controlled partial hydrolysis (*195*) under very mild acid conditions (0.01N H_2SO_4 in dioxane) is used. During hydrolysis, aliquots are taken at different time intervals to obtain intermediate products until only the aglycon is left (*60*). MANNICH and SIEWERT hydrolysis (conc. HCl/acetone) (*200*) is employed for determination of the sugar sequence if the oligoglycoside contains both normal and 2-deoxy sugars (*53, 80*). The sequence can also be deduced by permethylation studies (*23*) using HAKOMORI's method (*201*) followed by acidic hydrolysis.

Enzymatic hydrolysis (*21*) of pregnane oligoglycosides is effective only in eliminating the terminal glucose units (*16*). β-Glucosidase enzyme preparations obtained from snails are used for cleaving terminal β-glucose (*70, 202*). Molsin (protease type XIII from *Aspergillus saitoi*) (*6, 203*) and sulfatase (having β-glucuronidase activity from *Helix pomatia*) (*198*) are also used to cleave terminal β-glucose; the latter effected the cleavage of glycosidic linkages resistant to β-glucosidases (*198*). Specific enzymes cleave specific glycosidic linkages thus providing information on the nature of glycosidic bonds.

References, pp. 309–325

4. Pregnane Aglycons

More than eleven dozen pregnane aglycons have been so far isolated (*1,2*) from natural sources. Basic skeletons are listed in Chart 1. The structural features of plant pregnanes have been discussed in detail in a review article by DEEPAK and co-workers (*2*). The pregnane aglycons isolated since then are listed in Chart 2. Modifications of the pregnane skeleton are also known, for example cyclic ethers closed to C-20 (*1,204–206*). Some 8, 14-seco-(*137*), 14, 15–seco-(*58*) and 13,14; 14,15-disecopregnanes (*79,207–212*) have also been isolated.

5. Sugars of Pregnane Glycosides

5.1. General and Monosaccharides

Most of the sugars obtained from the acid hydrolysate of pregnane glycosides are 6-deoxy- and 2,6-dideoxyhexoses or their oligosaccharides (*1*). Such deoxy sugars have seldom been found in higher plants although they have been reported to occur in microorganisms (*213–214*). The oligoglycosides of pregnane glycosides generally contain a linear (*215*) rather than branched sugar chain although two exceptions have so far been found (*23,216*). A detailed study of the sugar linkages in the glycosides revealed that in the β-D-type, the hexopyranose ring is present in the $^4C_1$ conformation with the aglycon equatorial (*62*) whereas in the α-L-type, the hexopyranose exists in the $^1C_4$ conformation with the aglycon preferentially axial (*217*). The chemistry of naturally occurring deoxysugars has been reviewed by REICHSTEIN (*1*).

New monosaccharides reported as constituents of pregnane glycoside since the last review are L-sarmentose (*55*), 3-O-methyl-D-galactose (*53*), 4-O-acetyl-L-sarmentose (*55*), 2 deoxy L-fucose (*59*) and D-holosamine (4-desoxy, 4-amino-D-cymarose) (*218*).

5.2. Disaccharides from Pregnane Glycosides

The preparative isolation of sugars by hydrolysis of pregnane glycosides has afforded in addition to monosaccharides some novel reducing disaccharides containing 2,6-dideoxyhexose at the reducing end with a normal sugar at the non-reducing end. This was possible because of the very slow rate of hydrolysis of the normal sugar glycosidic linkage compared with 2-deoxysugar glycosidic linkages which being weaker

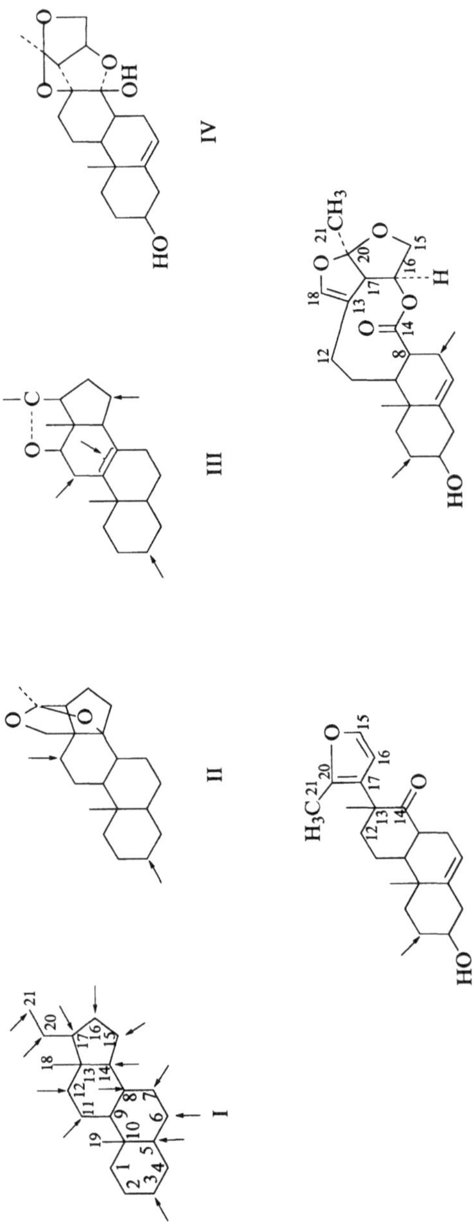

Chart 1. Basic Skeletons of Pregnane Genins
(arrows indicate the reported positions of oxygen functions in pregnane genins)

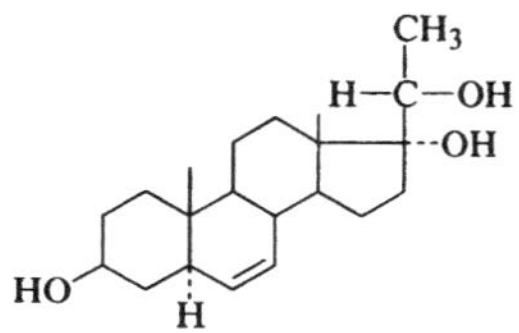

TEIKAGENIN (*244*)
220–23° [−145.2°]

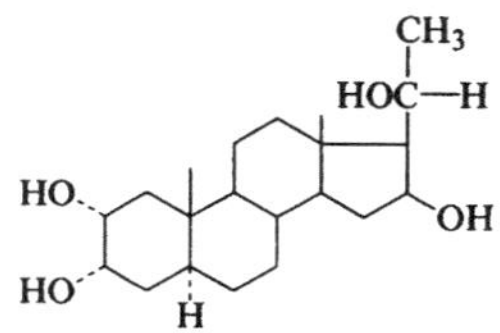

(20 R)-5α-PREGNANE-
2α, 3α, 16β, 20 TETROL (*10*)
281–83° [+22.0°]

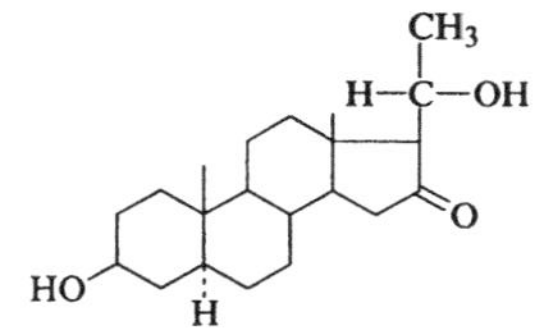

TOOSENDANSTEROL A (*279*)

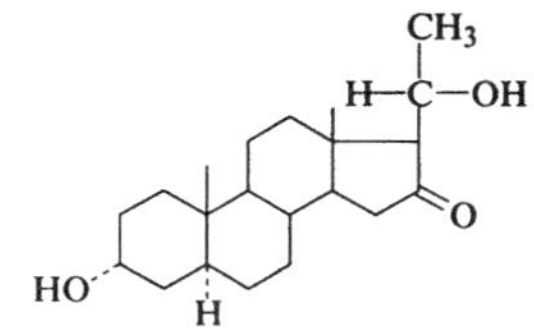

TOOSENDANSTEROL B (*279*)

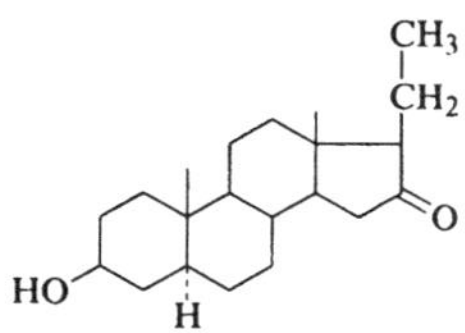

3β-HYDROXY-5α-PREGNAN-
16-ONE (*280*)

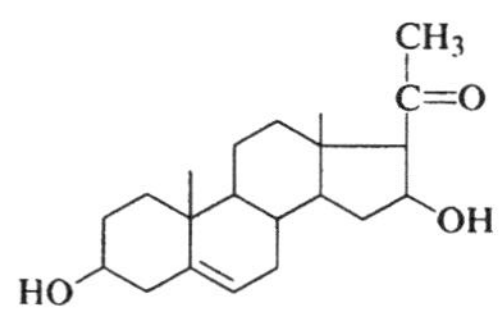

3β, 16β-DIHYDROXY-PREGN-
5-EN-20-ONE (*23*)

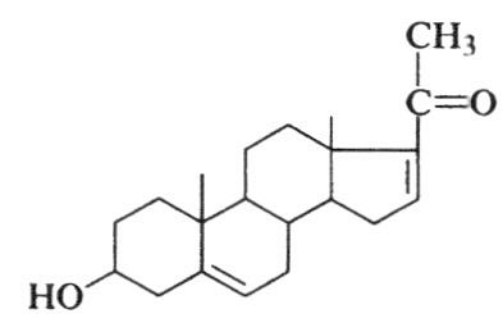

PREGNA-5,16-DIEN-3β-
OL-20-ONE (*216*)
203–204° [−35.9° $CHCl_3$]

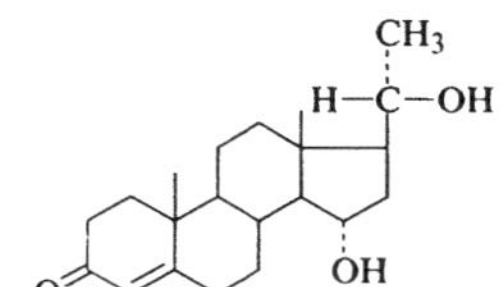

15α, 20β-DIHYDROXY-
Δ^4-PREGNEN-3-ONE (*4*)

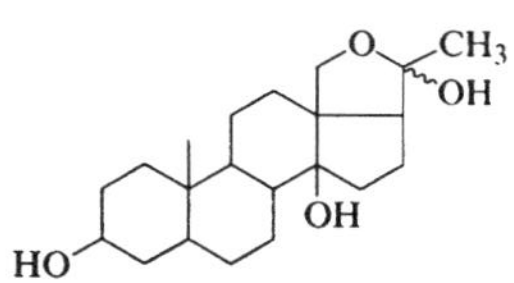

HOLANTOGENIN (*218*)
125° [− 31°]

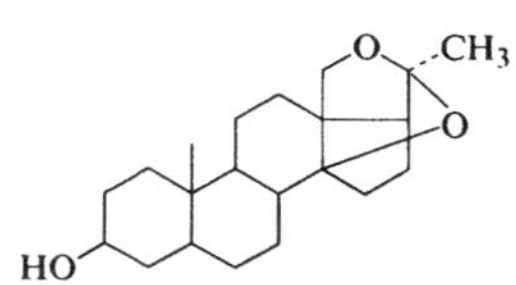

ANHYDROHOLANTOGENIN (*218*)

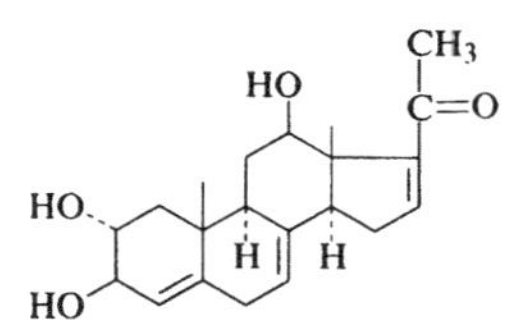

2α, 3β, 12β-TRIHYDROXY-
PREGNA-4,7,16-TRIENE-20-
ONE (*236*)

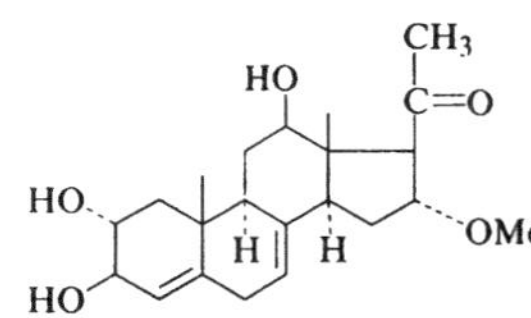

16α- METHOXY-2α, 3β,
12β-TRIHYDROXY-PREGNA-
4,7-DIEN-20-ONE (*236*)

Chart 2

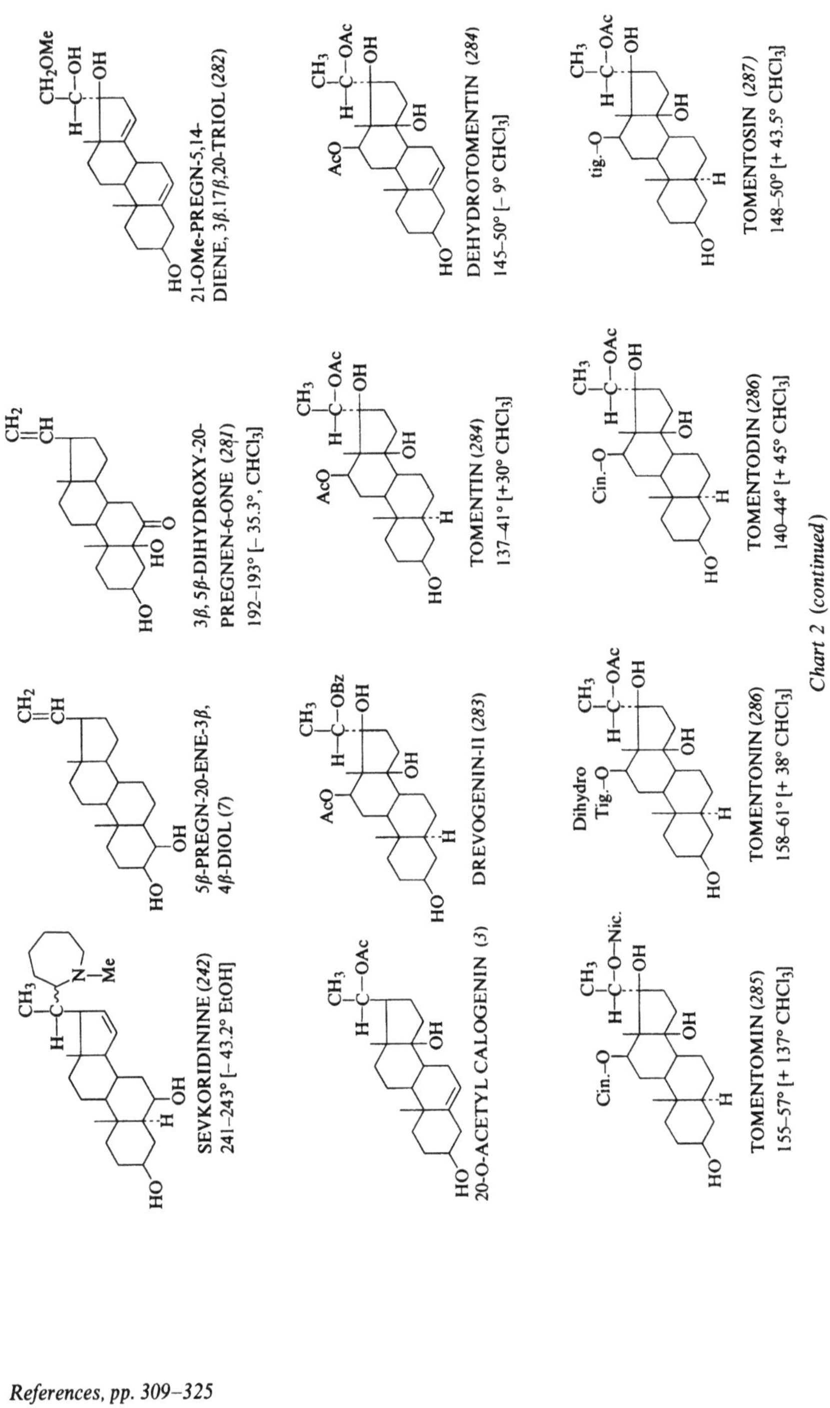

SEVKORIDININE (*242*)
241–243° [– 43.2° EtOH]

5β-PREGN-20-ENE-3β,
4β-DIOL (*7*)

3β, 5β-DIHYDROXY-20-
PREGNEN-6-ONE (*281*)
192–193° [– 35.3°, CHCl$_3$]

21-OMe-PREGN-5,14-
DIENE, 3β.17β,20-TRIOL (*282*)

20-O-ACETYL CALOGENIN (*3*)

DREVOGENIN-II (*283*)

TOMENTIN (*284*)
137–41° [+30° CHCl$_3$]

DEHYDROTOMENTIN (*284*)
145–50° [– 9° CHCl$_3$]

TOMENTOMIN (*285*)
155–57° [+ 137° CHCl$_3$]

TOMENTONIN (*286*)
158–61° [+ 38° CHCl$_3$]

TOMENTODIN (*286*)
140–44° [+ 45° CHCl$_3$]

TOMENTOSIN (*287*)
148–50° [+ 43.5° CHCl$_3$]

Chart 2 (continued)

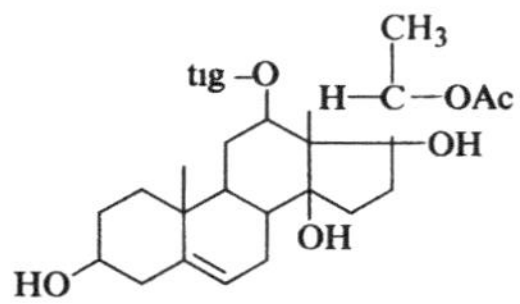
DEHYDROTOMENTOSIN *(286)*
180–83° [+ 42° $CHCl_3$]

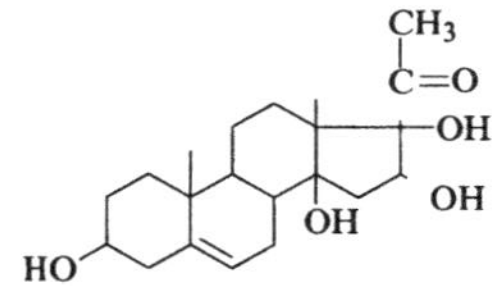
BREGENIN *(182)*
238–40°

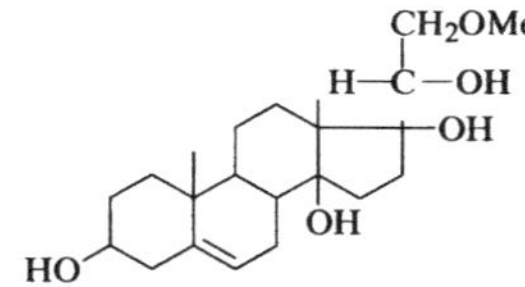
21-OMe-PREGN-5-ENE-
3β, 14β, 17β, 20 TETROL
(282)

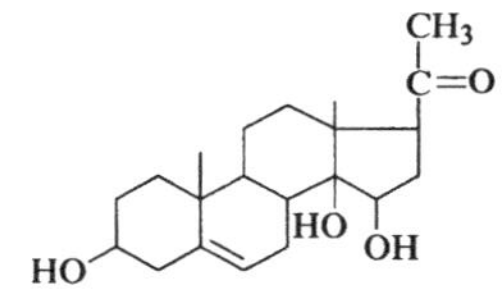
3β, 14β, 15β-TRIHYDROXY-
PREGN-5-ENE-20-ONE *(39)*

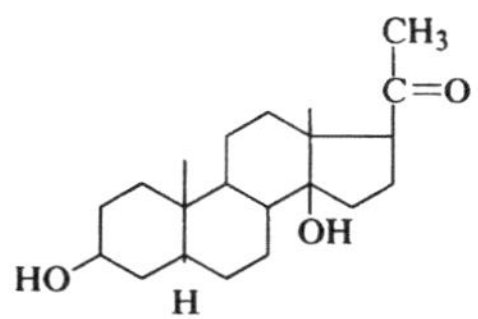
3β,14β-DIHYDROXY-5α-
PREGNAN-20-ONE *(8)*
240–243° [+ 25 3°]

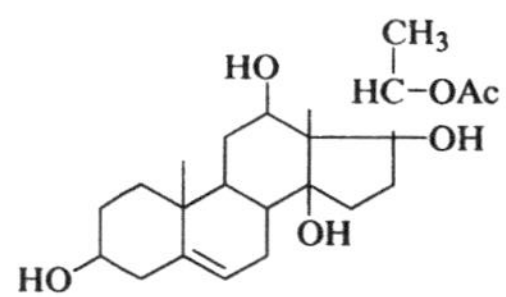
20-O-ACETYL UTENDIN *(22)*

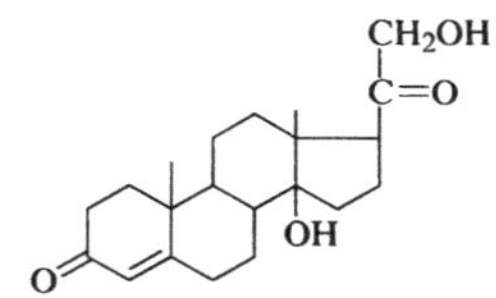
14β, 21-DIHYDROXYPREGN-4-
ENE-3,20-DIONE *(8)*

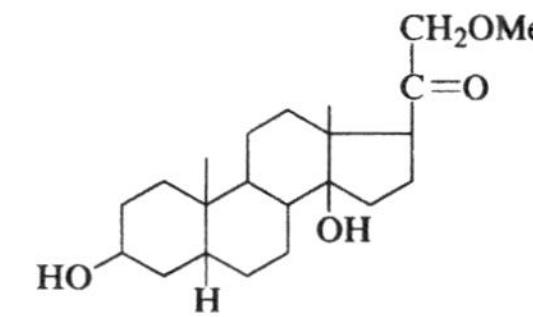
3β,14β-DIHYDROXY-21-
O-METHOXY-5β-PREGNAN-
20-ONE *(25)*
82–87° [70°, $CHCl_3$]

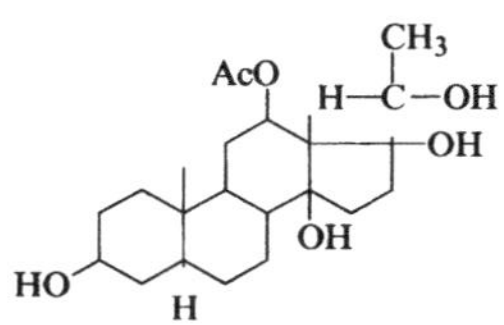
12β-O-ACETYL-
TOMENTOGENIN (288)
165–68° [+ 25° $CHCl_3$]

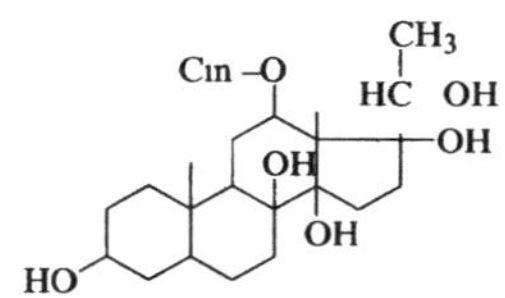
12β-O-CINNAMOYLDIHYDRO-
SARCOSTIN *(17)*

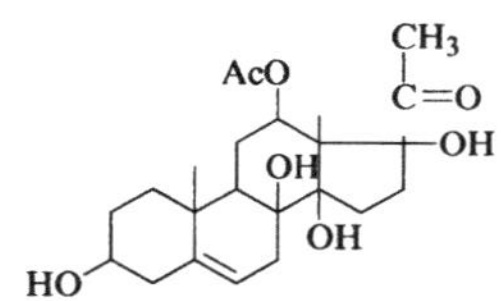
METAPLEXIGENIN *(289)*
264° [– 22 5° MeOH]

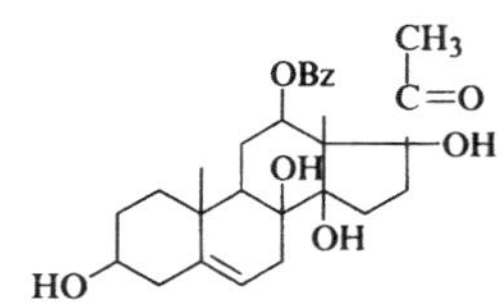
12β-O-BENZOYLDEACYL-
METAPLEXIGENIN *(84)*
142–145°

Chart 2 (continued)

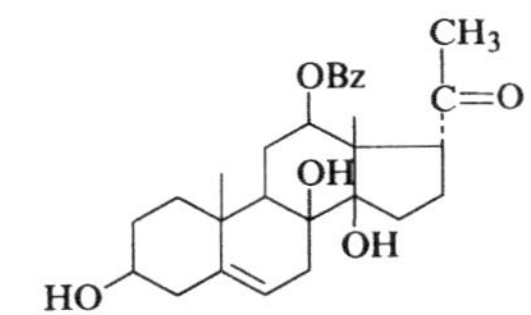
12β-O-BENZOYL LINEOLON (*51*) [– 48°, $CHCl_3$]

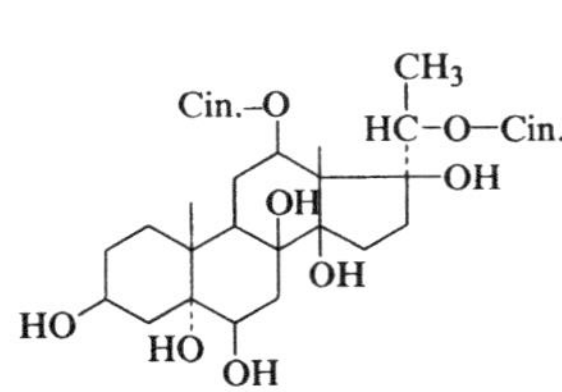
THEROGENIN (*184*) 125–128° [+128.4° MeOH]

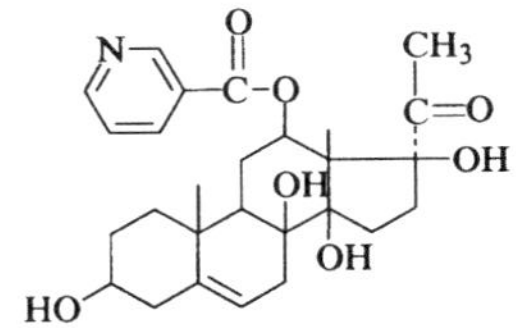
ROSTRATAMINE (*183*)

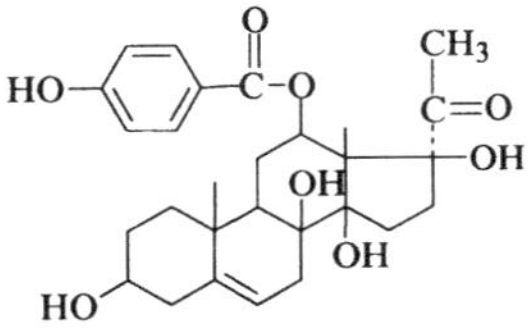
QINGYANGSHENGENIN (*183*)

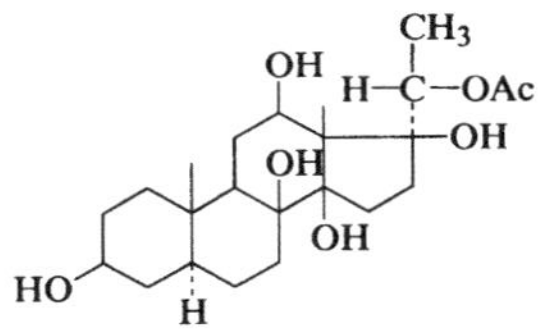
DREGOGENIN (*291*)

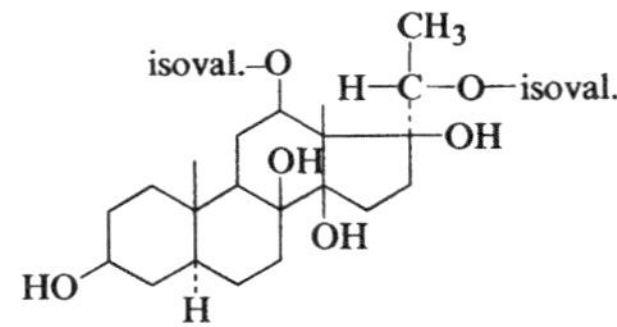
DREVOGENIN-I (*283*)

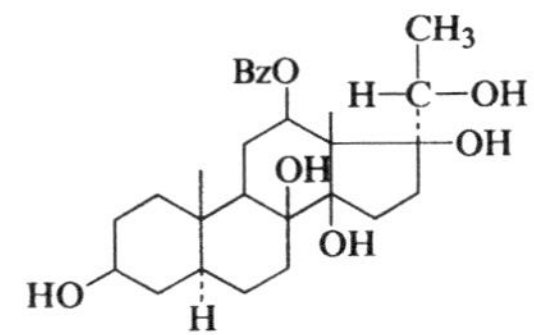
DRESGENIN (*292*)

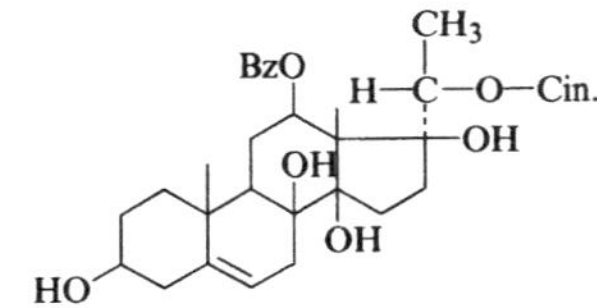
GYMNEMARS GENIN (*293*) 155–57° [+ 145.96° $CHCl_3$]

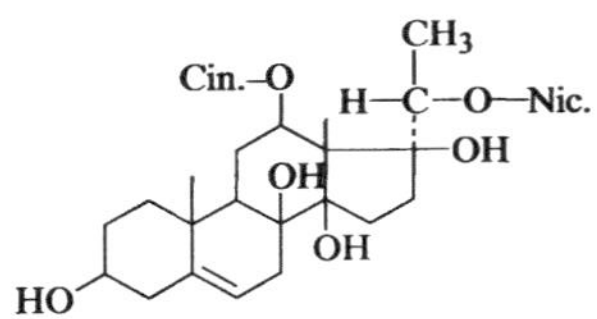
GAGAMININ (*294*) 166–70° [+ 143.8° $CHCl_3$]

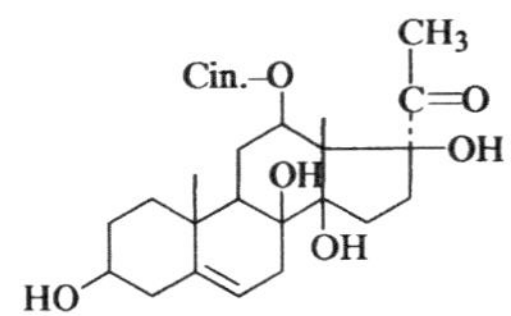
KIDJOLANIN (*295*) 147–49° [+ 63.2° $CHCl_3$]

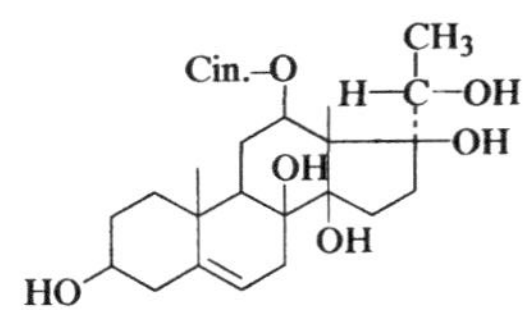
PENUPOGENIN (*296*)

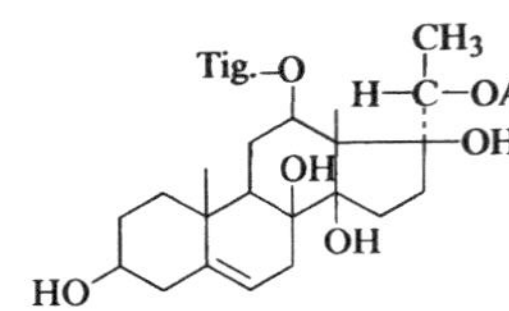
KIDJOLADININ (*297*) 250–52° [+ 43.4° $CHCl_3$]

Chart 2 (continued)

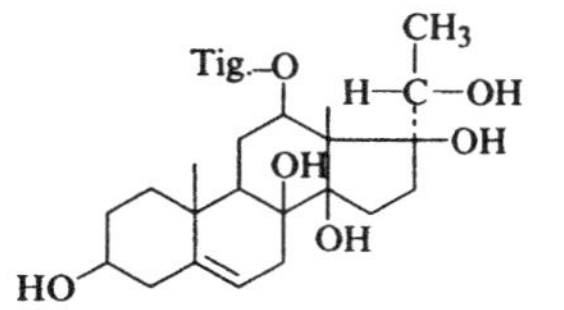

DEACETYLKIDJOLADININ (*297*)
205–09° [+ 9.5° $CHCl_3$]

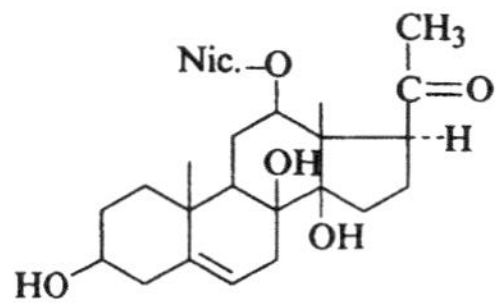

FUKUJUSONE (*298*)
224–27° [+ 100° $CHCl_3$]

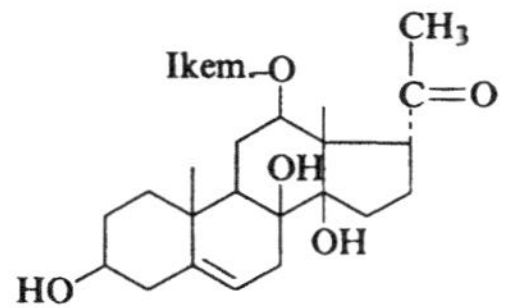

CYNANCHOGENIN (*299*)

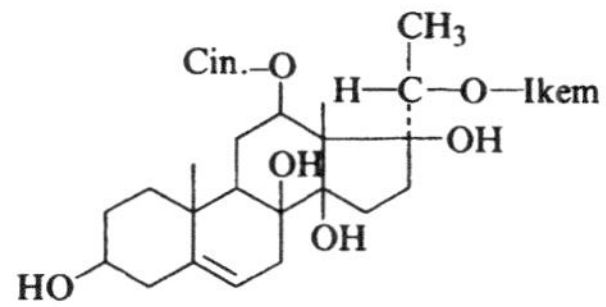

AGLYCONE-D (*300*)

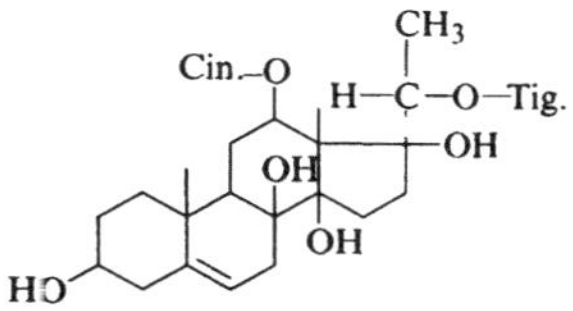

AGLYCONE-E (*300*)

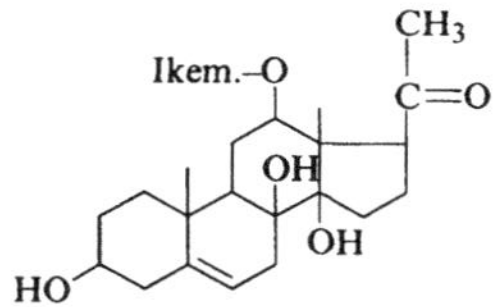

SIBRIGENIN (*301*)

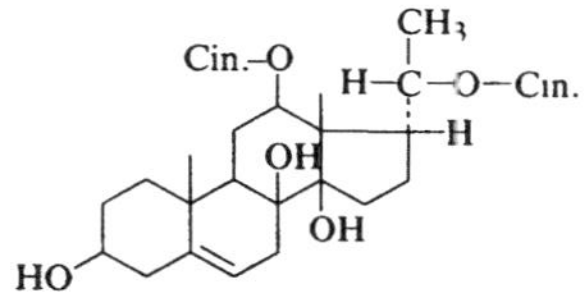

ORNOGENIN (*196*)
75–80°[+111.4°, MeOH]

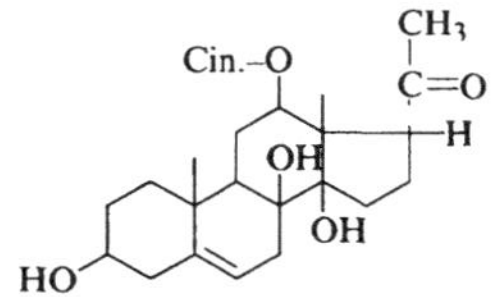

IKEMAGENIN (*302*)

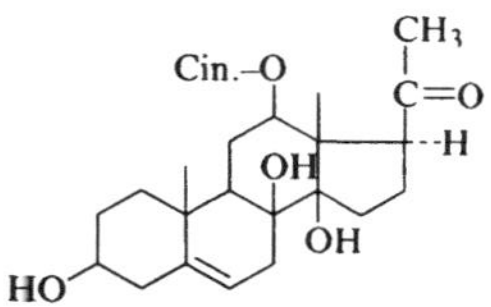

ISOIKEMAGENIN (*302*)

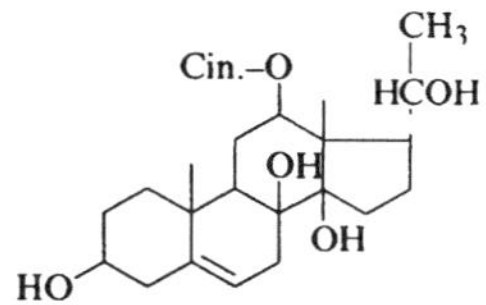

12β-O-CINNAMOYL-
IKEMAGENOL (*303*)

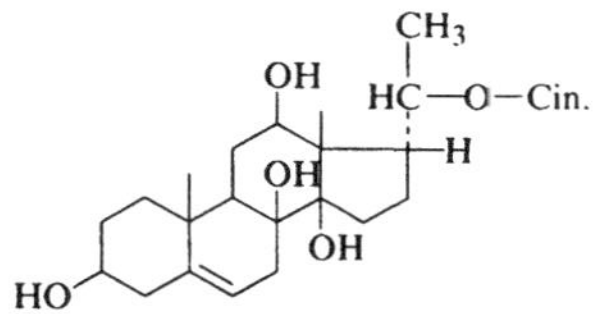

20-O-CINNAMOYL-
IKEMAGENOL (*303*)

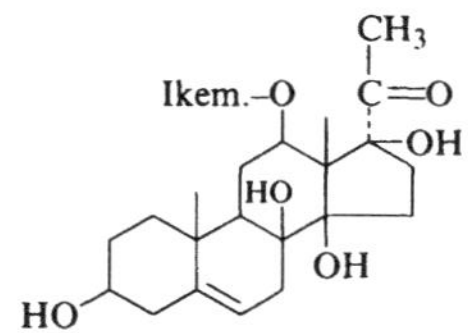

CAUDATIN (*304*)

Chart 2 (continued)

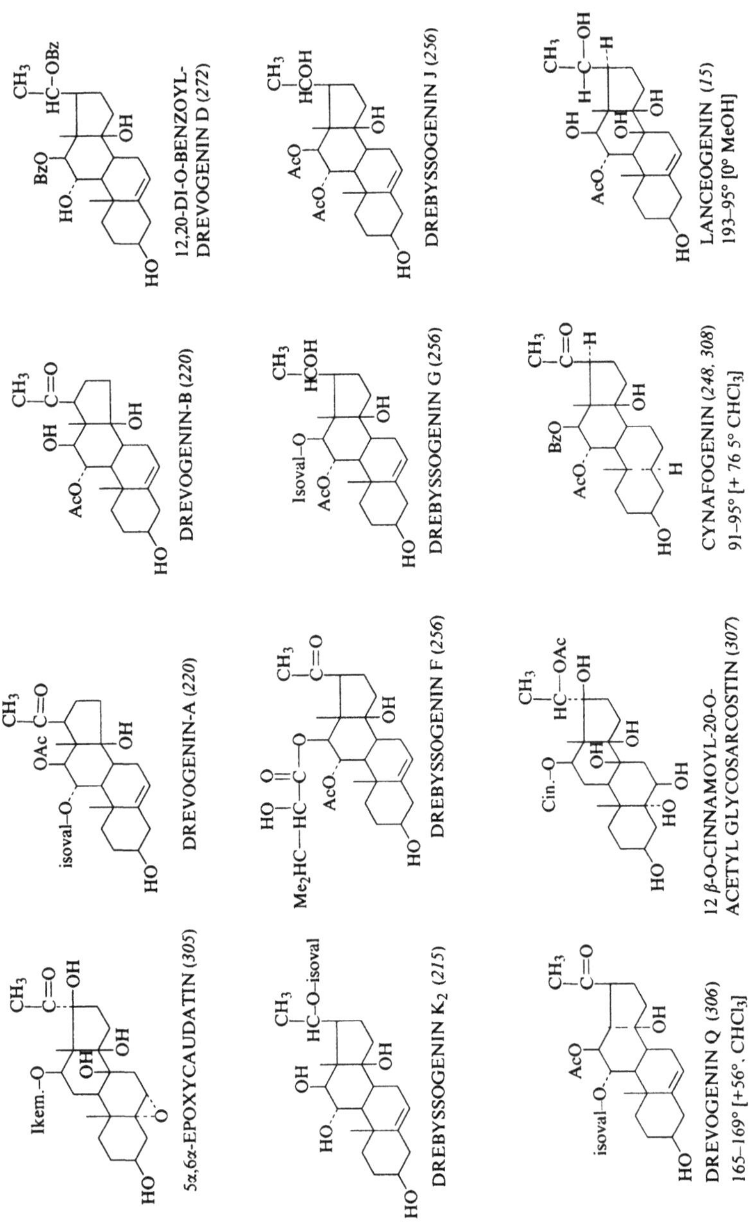

5α,6α-EPOXYCAUDATIN (305)

DREVOGENIN-A (220)

DREVOGENIN-B (220)

12,20-DI-O-BENZOYL-DREVOGENIN D (272)

DREBYSSOGENIN K$_2$ (215)

DREBYSSOGENIN F (256)

DREBYSSOGENIN G (256)

DREBYSSOGENIN J (256)

DREVOGENIN Q (306) 165–169° [+56°, CHCl$_3$]

12 β-O-CINNAMOYL-20-O-ACETYL GLYCOSARCOSTIN (307)

CYNAFOGENIN (248, 308) 91–95° [+ 76 5° CHCl$_3$]

LANCEOGENIN (15) 193–95° [0° MeOH]

Chart 2 (continued)

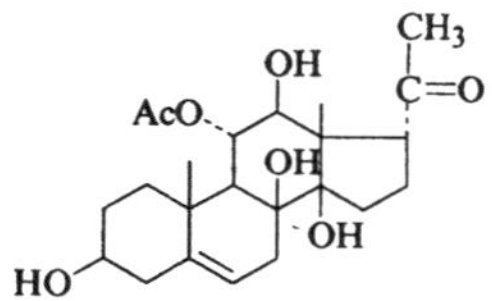

DREGENIN (*43*)
162° [+ 20° MeOH]

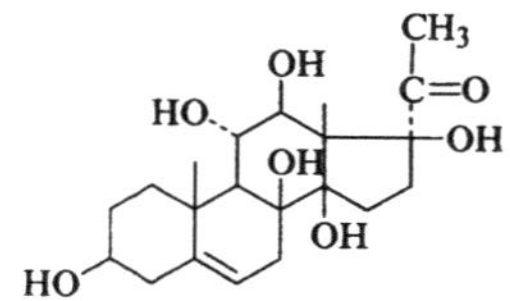

SARCOGENIN (*181*)
105–108° [+ 49°]

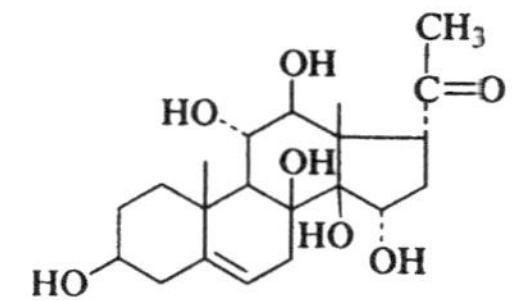

ORGOGENIN (*185*)
180–83° [+ 27°]

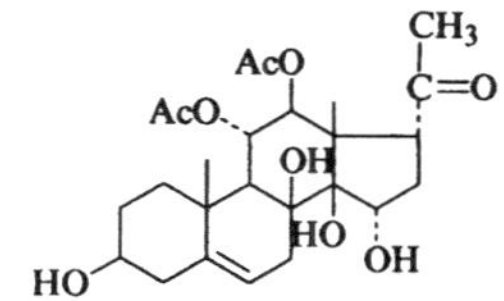

11α, 12β-DI-O-ACETYL ORGOGENIN (*61*)

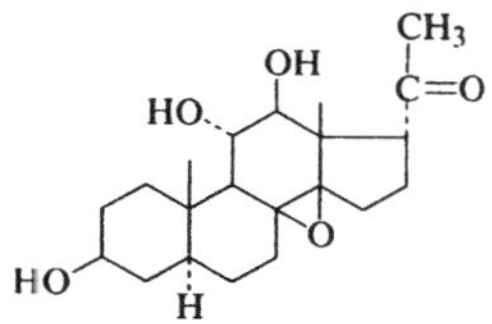

TENACIGENIN-B (*309*)

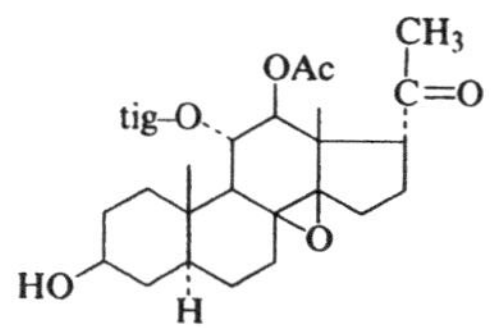

11α-O-TIGLOYL-12β-O-ACETYLTENACIGENIN B (*65*) [+ 0.5°, $CHCl_3$]

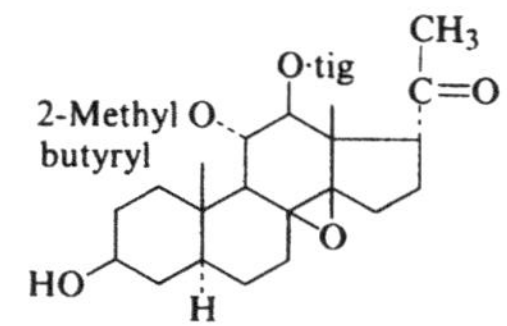

11α-O-2 METHYLBUTYRYL-12-β-O-TIGLOYLTENACIGENIN B (*65*)

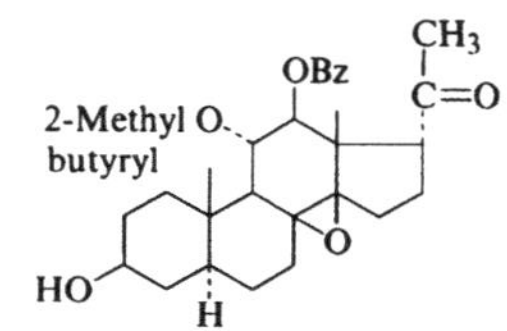

11α-O-2 METHYLBUTYRYL-12β-O-BENZOYL TENACIGENIN B (*65*)

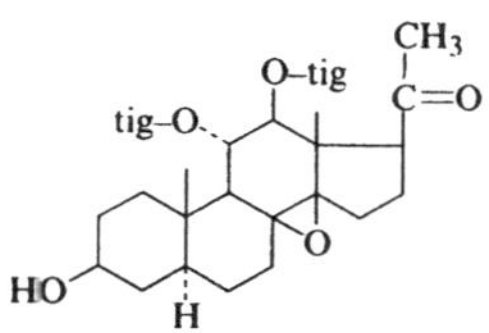

11α, 12β-O, O-DITIGLOYL-17β-TENACIGENIN B (*65*)

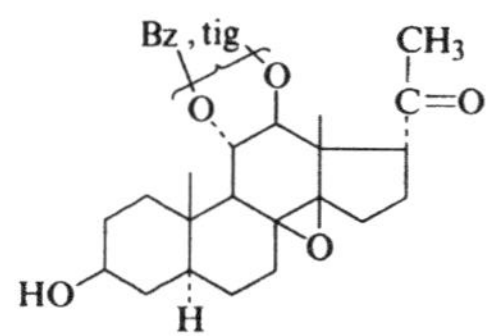

TENACIGENIN B-III (*266*)

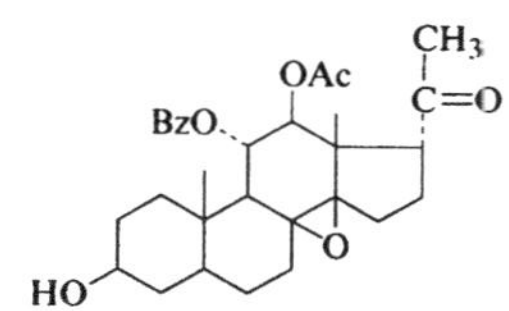

11α-O-BENZOYL-12β-O-ACETYL TENACIGENIN B (*65*) [– 5°, $CHCl_3$]

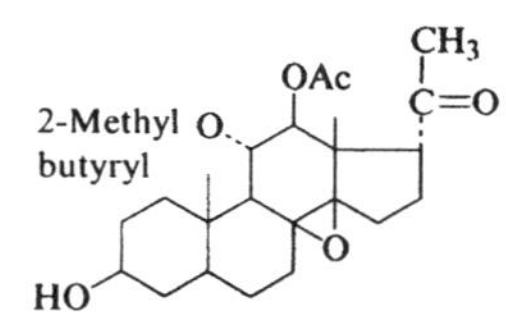

11α-O-2-METHYLBUTYRYL-12β-O-ACETYLTENACIGENIN B (*65*) [+ 10.5°, $CHCl_3$]

Chart 2 (continued)

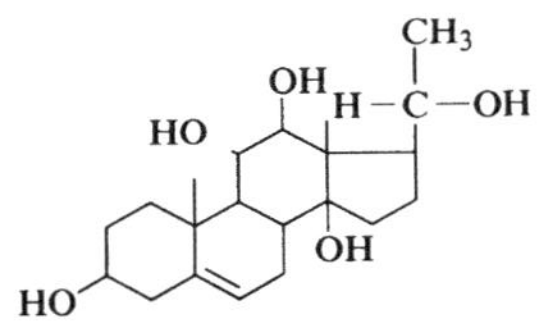

CISSOGENIN (*310*)
210–214° [– 2 34° MeOH]

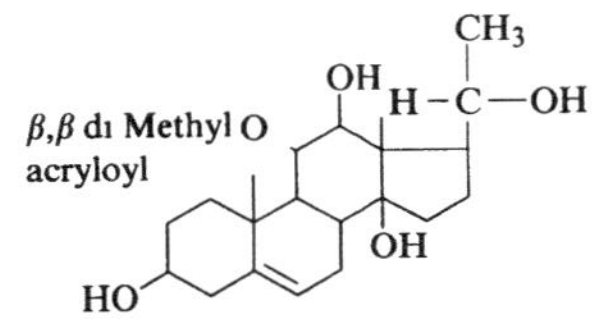

TENASOGENIN (*67*)
180–83° [+ 9 63° MeOH]

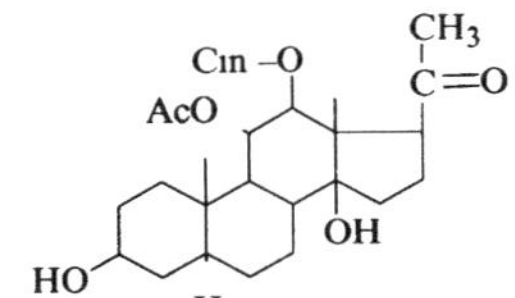

CONDURANGOGENIN A (*125–128*)

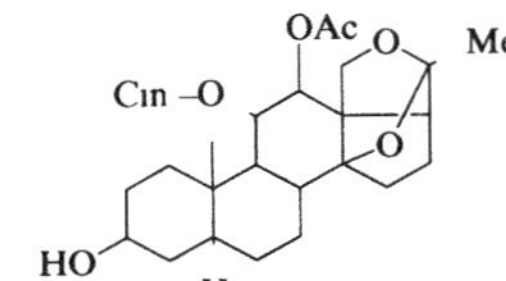

CONDURANGOGENIN B (*130*)

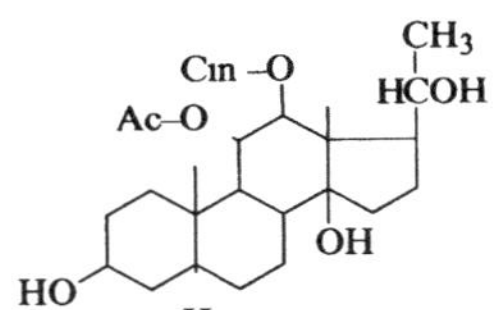

CONDURANGOGENIN C (*128*)

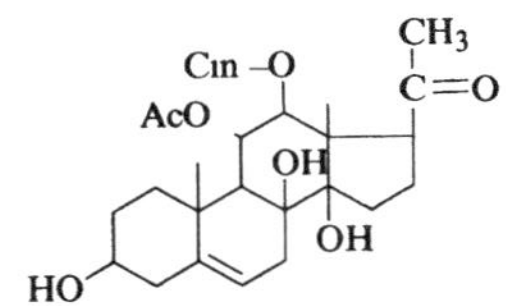

CONDURANGOGENIN E (*54*)

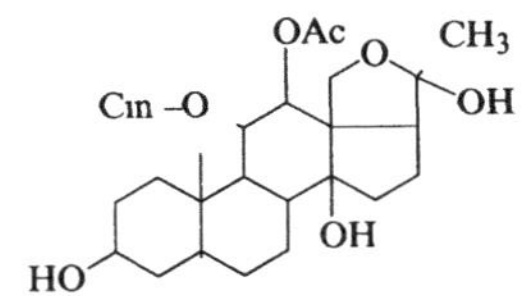

5α-H, 3β, 14β, 20-TRI-HYDROXY-11α-O-CINNAMOYL-12β-O-ACETYL (18, 20)-EPOXYPREGNANE (*130*)

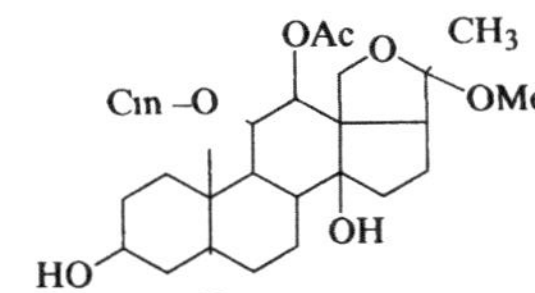

5α-H, 3β, 14β DIHYDROXY-11α-O-CINNAMOYL-12β-O-ACETYL (18, 20)-EPOXY-20-O-METHYLPREGNANE (*130*)

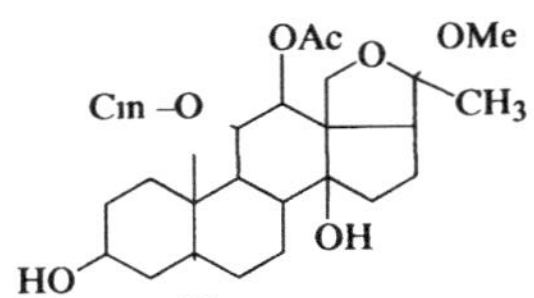

5α-H, 3β, 14β-DIHYDROXY 11α-O-CINNAMOYL-12β-O-ACETYL-(18,20)-EPOXY-20-ISO-O-METHYLPREGNANE (*130*)

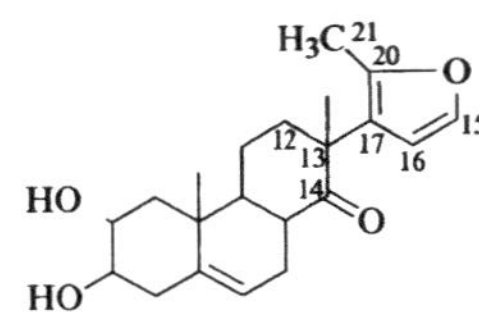

ATRATOGENIN-A (*58 290*)
69–73° [–88 2° MeOH]

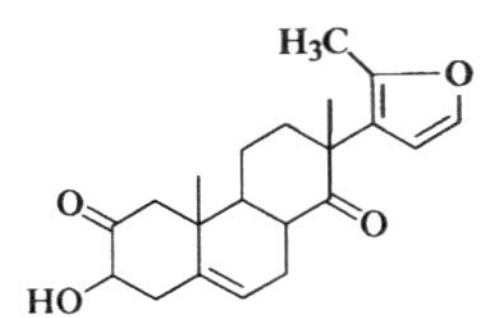

ATRATOGENIN-B (*58*)
Liq [– 87 4° $CHCl_3$]

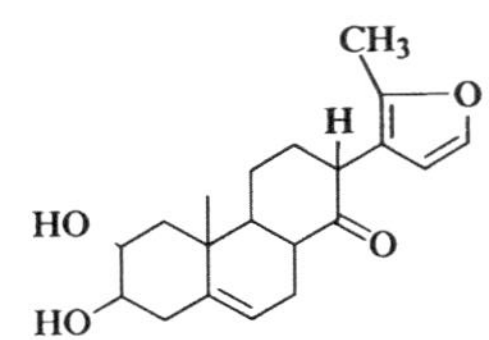

CYNAJAPOGENIN (*58 290*)
67–73° [– 42 5° $CHCl_3$]

Chart 2 (continued)

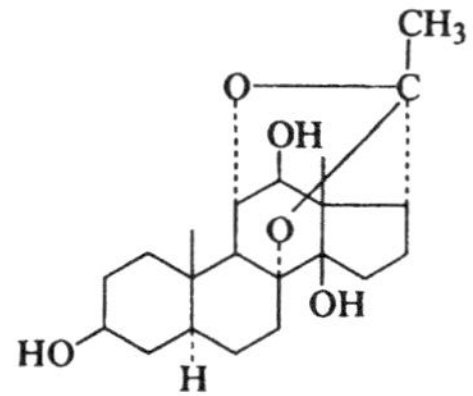

TENACIGENIN-A (*309, 311*)

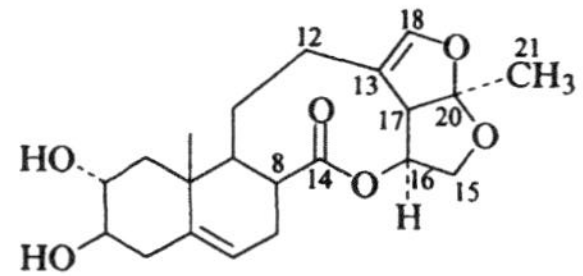

GLAUCOGENIN-A (*212, 253*)

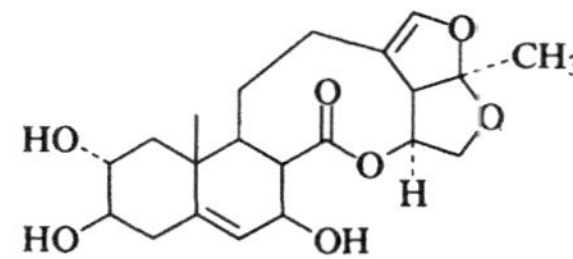

GLAUCOGENIN-B (79, *212, 253*)

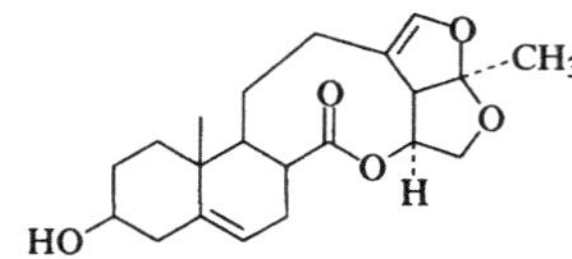

GLAUCOGENIN-C (*212, 253*)
205–206° [+ 83.2° MeOH]

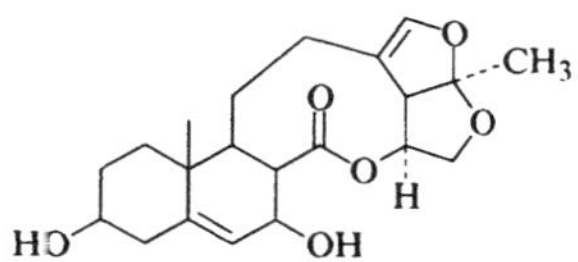

GLAUCOGENIN-D (*79*)

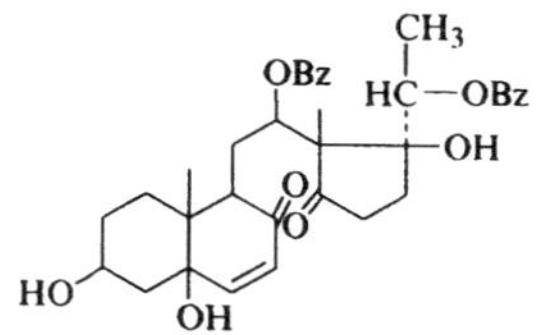

(20 S)-12β, 20 DIBENZOYL-OXY-3β,5β, 17-TRIHYDROXY-8.14-SECOPREGN-6-ENE 8,14-DIONE (*137*)

Ac = —C(=O)—CH_3

Bz = —C(=O)—Ph

Cin. = —C(=O)—CH=CHPh

Ikem. = —C(=O)—CH=C(Me)$CHMe_2$

tig. = —C(=O)—C(Me)=CHMe

isoval. = —C(=O)—CH_2CHMe_2

nic. = 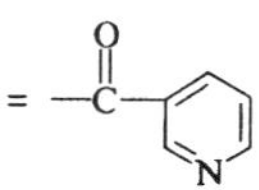

Chart 2 (continued)

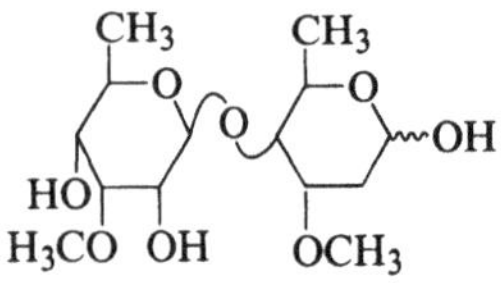
Asclepobiose

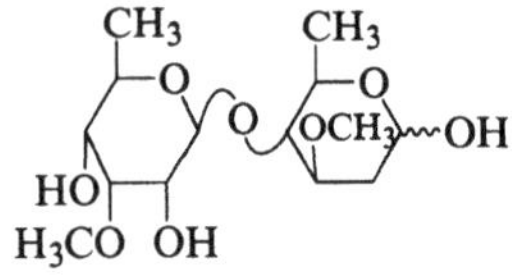
Pachybiose

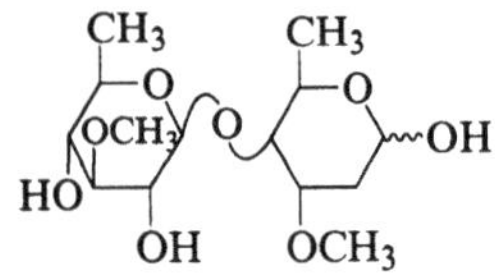
Lilacinabiose

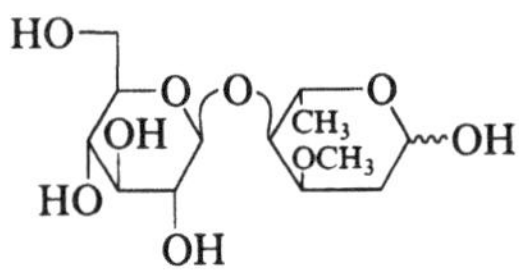
Glaucobiose

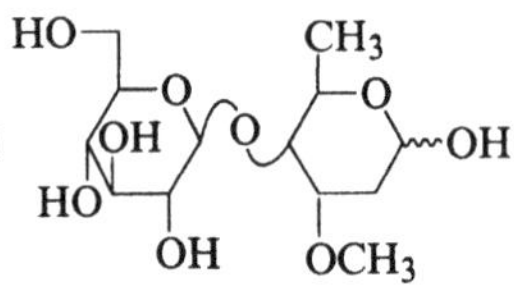
Strophanthobiose

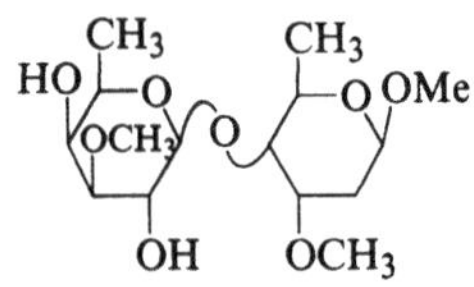
Methyl-β-D-Digitalo-pyranosyl-(1 → 4) β-D-Cymaropyranoside

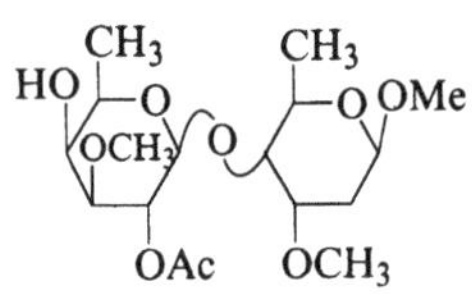
Methyl-4-O-(2-O-acetyl-β-D-Digitalopyranosyl)-β-D-Cymaropyranoside

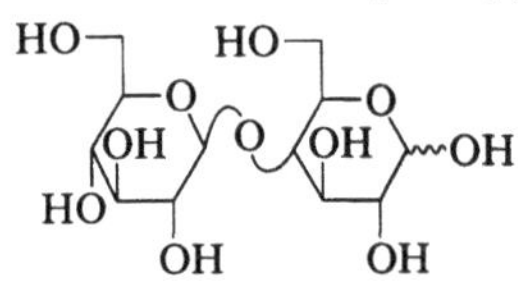
Cellobiose

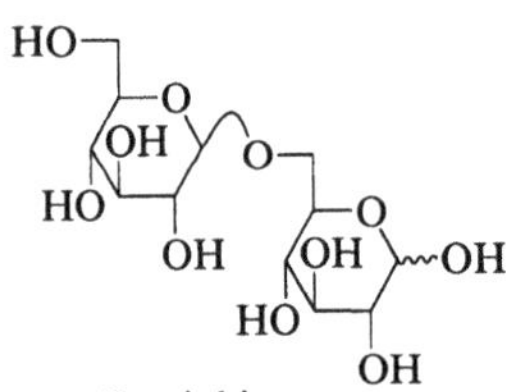
Gentiobiose

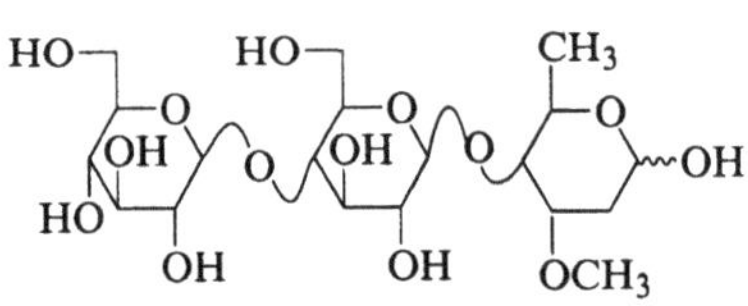
Leptatriose

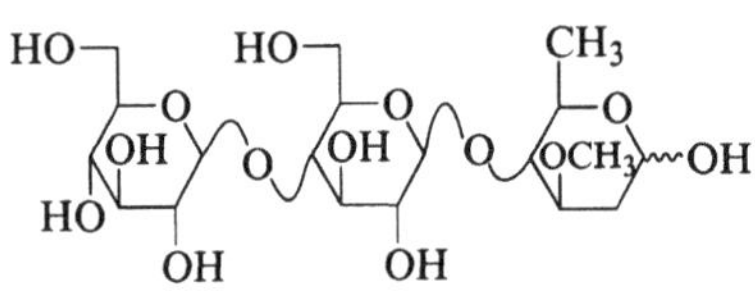
Cynanchotriose

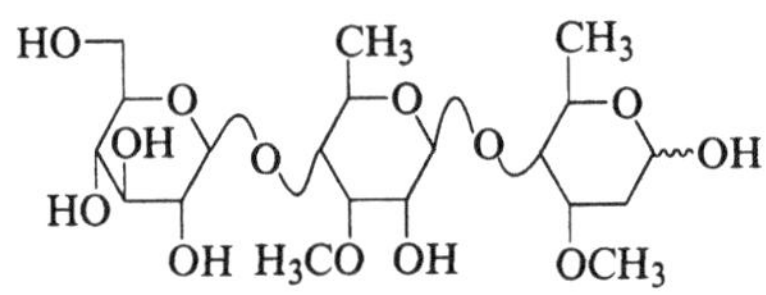
Dregeatriose

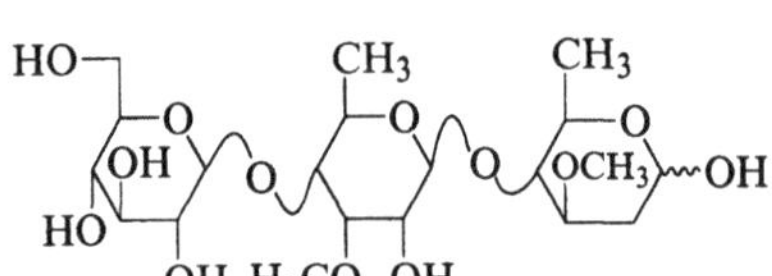
Neocondurangotriose

Chart 3

hydrolyse faster. Disaccharides from pregnane glycosides are listed in Chart 3.

The disaccharides pachybiose (*56, 195, 219–221*) and asclepobiose (*19, 56, 222–223*) are most frequently encountered in pregnane glycosides. Lilacinabiose (*62, 224*) is 3-O-methyl-6-deoxy-β-D-glucopyranosyl-(1→4)-D-cymaropyranose. Glaucobiose (*70, 225*) and strophanthobiose (*58, 76, 225*) differ from each other in that in the former the β-D-glucopyranosyl moiety is linked to L-cymarose by a (1→4) linkage whereas in the latter β-D-glucopyranosyl half is linked to D-cymarose by a (1→4) linkage. Two disaccharides, methyl β-D-digitalopyranosyl-(1→4) β-D-cymaropyranoside (*31*) and methyl-4-O-(2-O-acetyl-β-D-digitalopyranosyl)-β-D-cymaropyranoside (*162, 226*), have been obtained from the methanolic H_2SO_4 hydrolysate of pregnane glycosides of *Periploca sepium*. The disaccharides gentiobiose (*8, 34*) and cellobiose (*50, 53*) have also been found present in pregnane glycosides.

5.3. Trisaccharides from Pregnane Glycosides

Hydrolysates of pregnane glycosides have yielded four trisaccharides which are also listed in Chart 3. These contain a 2-deoxysugar at the reducing end which is linearly linked to two normal hexoses. Leptatriose obtained from *Leptadenia reticulata* by SRIVASTAV *et al.* (*50, 53*) has a cellobiose moiety linked to D-cymarose by a 1→4 β-glycosidic linkage whereas in cynanchotriose (*76, 227*) from *Cynanchum wallichi* the cellobiose moiety is linked to D-oleandrose by a 1→4-β-glycosidic linkage. In dregeatriose (*76, 215*) the terminal D-glucose is linked to 3-O-methyl-6-deoxy-D-allose which is in turn linked to D-cymarose. In the case of neocondurangotriose (*76, 129–130, 215*) the reducing end is made up of D-oleandrose while 3-O-methyl-6-deoxy-D-allose and D-glucose form the intermediate and terminal end respectively. In both these trisaccharides, the two normal hexoses are linked by (1→4) β-glycosidic linkages.

Interestingly, in pregnane glycosides **211** and **213** isolated from *Periploca sepium* (*228*) the sugar component contains an ortho-ester function which is rather uncommon in natural products. In glycosides **217**–**221** and **224**–**226** from the same source, the glycosidic linkage between O-4 of the first sugar, 2,6-dideoxyarabinohexopyranose and C-1 of the second O-cymarosyl is peroxide.

6. Biosynthesis of Pregnane Glycosides

Biosynthesis of pregnanes and their glycosides has been covered in depth by REICHSTEIN (*1*). In this context, it is of interest that a pregnane glycoside isolated from *Mandevilla pentlandiana* (*25*) has a 21-O-methoxy-20-one C-17 side chain and is biogenetically related to 3β,14β,21-trihydroxy-5β-pregnane-20-one, a precursor of a cardenolide (*25*). The isolation of this glycoside suggests a pregnane route for the biosynthesis of cardenolides (*229–230*). The 21-O-methylated compound possibly is a storage form of a 21-hydroxy-20-keto pregnane derivative (*25*). Another pregnane, *i.e.* pregnenolone (Δ^5-pregnen-3β-ol-20-one), which is a known biosynthetic precursor of cardenolides has also been isolated as a constituent of the glucosides (*34*) from the root and trunk bark of *Nerium odorum.*

7. Biological Activity

Pregnane ester glycosides* closely resemble cardiac glycosides (*193*) which are important in medicinal chemistry due to their digitalis-like effect on cardiac muscles and their application in the therapy of auricular fibrillation and in many types of congestive heart failure (*229–230*). Biogenetic studies have revealed that pregnane derivatives are biological precursors of cardiac glycosides (*1, 25*) and therefore these substances can be isolated from plants only in very small quantities. Using modern pharmacological methods some of these compounds have shown specific biological activity.

The crude drug condurango cortex, the bark of *Marsdenia condurango*, has been used as an avomatic bitter stomachic in popular medicine and also against cancer or syphilis in folk remedies (*129*). In anti-tumor screening by CCNSC the extract of this plant was not effective against sarcoma-180, adenocarcinoma 755, human sarcoma HS-1 and KB system (*231*). However, condurango glycosides (CG) A_O (**164**), CGB_O (**166**), CGC_O (**165**), CGD_O (**167**), 20-O-methyl CGD_O (**168**) and 20-iso-O-methyl-CGD_O (**169**) from *Marsdenia condurango* were found active against Ehrlich ascites carcinoma (*129–130*). Two other pregnane glycosides, viz. condurangoglycoside E_{O1} (**170**) and E_{O2} (**171**) obtained from *Marsdenia condurango*, have also shown anticarcinogenic activity (*232*). AHSAN reported that the polyoxypregnane glycoside amplexoside A (**36**) from *Asclepias amplexicaulis* showed cancer inhibitory activity in the KB assay (*233*). Generally members of Asclepiadaceae produce an abundance of esterified polyoxypregnane glycosides (*1, 2*) and can therefore, be a promising source of antitumor agents. Thirteen pregnane glycosides (*215, 234–*235) were isolated from *Dregea volubilis*; among them, dregeosides A_{P1} (**131**) and A_{O1} (**132**) showed antitumor activity against Ehrlich carcinoma (solid type), with dregeoside A_{o1} also being active against melanoma B-16

* Table 1 which follows lists pregnane glycosides isolated since 1967 and their sources, arranged by plant family. Structures are listed in Table 2.

Table 1. *Pregnane Glycosides and their Sources*

Plant	Glycoside (Glycoside no.) Molecular Formula mp°C $[\alpha]_D$	Genin	Sugar	References
		Family Apocynaceae		
Apocynum venetum var. basikurumon	Basikoside A (**1**) $C_{27}H_{44}O_7$ 260–265°C −101.8°	Teikagenin	-3-O-β-D-Fucp.	(*30*)
”	Basikoside B (**2**) $C_{29}H_{46}O_8$ 240–246°C −47.7°	Teikagenin	-3-O-3Ac-β-D-Fucp.	
”	Basikoside C (**3**) $C_{33}H_{54}O_{10}$ 215–220°C −92.2°	Teikagenin	-3-O-β-D-Fucp-20-O-β-D-Canp.	
”	Basikoside D (**4**) $C_{40}H_{66}O_{13}$ – −97.0°	Teikagenin	-3-O-β-D-Fucp-20-O-β-D-Digp-(1→3)-β-D-Canp.	
Holarrhena antidysenterica	Holantosine A (**5**) $C_{28}H_{47}O_6N$ – –	Holantogenin	3-O-β-D-Holp.	(*218*)
”	Holantosine B (**6**) $C_{28}H_{45}O_5N$ – –	14,20 Anhydro holantogenin	-3-O-β-D-Holp.	

Table 1 (*continued*)

Plant	Glycoside (Glycoside no.) Molecular Formula mp°C $[\alpha]_D$	Genin	Sugar	References
Holarrhena curtissi	N-Demethyl holacurtin (**7**) $C_{28}H_{47}O_5N$ – –	5α-Pregnan-3β, 14β-diol-20-one	-3-O-β-D-Holp.	(*241*)
Korolkowia sewertzovii	Sevkorine (**8**) $C_{34}H_{57}O_7N$ 236–238°C −41.1°	Sevkoridinine	-3-O-β-D-Glup.	(*242*)
Malouetia glandulifera	Conopharyngine (**9**) $C_{27}H_{45}O_6N$ – –	Pregn-5-ene-20-amino-3β-ol	-3-O-β-D-Glup.	(*243*)
Mandevilla pentlandiana	–(**10**) $C_{43}H_{72}O_{13}$ Amorph. 17.1°	3β, 14β-Dihydroxy-21-methoxy-5β-pregnan 20-one	-3-O-β-D-Digp-(1→4)-β-D-Cymp-(1→4)-β-D-Cymp.	(*25*)
Nerium odorum	Pregnenolone glucoside I (**11**) $C_{39}H_{62}O_{17}\cdot H_2O$ 231–235°C −20.5°	Δ^5-Pregnen-3β-ol-20-one	-bis-3-O-β-D-Glup-(1→2;1→6)-β-D-Glup.	(*34*)

”	Pregnenolone glucoside II (**12**) $C_{33}H_{52}O_{12}$ 255–259°C −11.9°	Δ^5-Pregnen-3β-ol-20-one	-3-O-β-D-Glup-(1→6)-β-D-Glup.	
”	Pregnenolone glucoside III (**13**) $C_{33}H_{52}O_{12}$ 252–256°C −7.2°	Δ^5-Pregnen-3β-ol-20-one	-3-O-β-D-Glup-(1→2)-β-D-Glup.	
”	Pregnenologlucoside IV (**14**) $C_{27}H_{42}O_7$ 269–271°C –	Δ^5-Pregnen-3β-ol-20-one	-3-O-β-D-Glup.	
Nerium indicum ”	–(**15**) $C_{33}H_{54}O_{13}$ Amorphous −15.8°	Δ^5-Pregnen-3β,14β-dihydroxy-20-one	-3-O-β-D-Glup-(1→6)-β-D-Glup.	(*8*)
”	–(**16**) $C_{27}H_{40}O_9$ Amorphous +22.6°	Δ^4-Pregn-14β,21-dihydroxy-3,20-dione	-21-O-β-D-Glup.	
Trachelospermum asiaticum	Teikaside A (**17**) $C_{48}H_{80}O_{18}$ – –	Pregn-6-ene-3β, 17α, 20α-triol	-3-O-β-D-Dgtp-20-O-β-D-Glup-(1→4)-β-D-Sarp-(1→4)-β-D-Sarp.	(*244*)
”	Teikaside A-Ia (**18**) $C_{34}H_{56}O_{10}$ 205–213°C −122.3°	Teikagenin	-3-O-β-D-Dgtp-20-O-β-D-Canp.	(*21*)

Table 1 (*continued*)

Plant	Glycoside (Glycoside no.) Molecular Formula mp°C $[\alpha]_D$	Genin	Sugar	References
Trachelospermum asiaticum	Teikaside A-Ib (**19**) $C_{35}H_{58}O_{10}$ 207–211°C −117.2°	Teikagenin	-3-O-β-D-Dgtp-20-O-β-D-Olep.	
”	Teikaside A-IIa (**20**) $C_{41}H_{68}O_{15} \cdot 1/2\ H_2O$ 265–280°C −113.1°	Teikagenin	-3-O-β-D-Dgtp-20-O-β-D-Glup-(1→4)-β-D-Digp.	
”	Teikaside A-IIb (**21**) $C_{41}H_{68}O_{15} \cdot 4.5\ H_2O$ 250–260°C −44.8°	Teikagenin	-3-O-β-D-Dgtp-20-O-β-D-Glup-(1→4)-β-D-Olep.	
”	Teikaside A-IIc (**22**) $C_{40}H_{66}O_{15} \cdot 3H_2O$ 285–295°C −95.9°	Teikagenin	-3-O-β-D-Dgtp-20-O-β-D-Glup-(1→4)-β-D-Canp.	
”	Teikaside A-IIIb (**23**) $C_{48}H_{80}O_{18}$ – −35°	Teikagenin	-3-O-β-D-Dgtp-20-O-β-D-Glup-(1→4)-β-D-Olep-(1→4)-β-D-Sarp.	
”	Teikaside A-IIIc (**24**) $C_{48}H_{80}O_{18}$ – −64.2°	Teikagenin	-3-O-β-D-Dgtp-20-O-β-D-Glup-(1→4)-β-D-Olep-(1→4)-β-D-Olep.	

References pp. 309–325

”	Teikaside A-IIId (**25**) $C_{48}H_{80}O_{18}$ – $-102.2°$	Teikagenin	-3-O-β-D-Dgtp-20-O-β-D-Glup-(1→4)-β-D-Digp-(1→4)-β-D-Sarp.	
”	Teikaside C-O (**26**) $C_{37}H_{60}O_{11}$ – $-105.8°$	Teikagenin	-3-O-4Ac-α-L-Sarp-(1→4)-β-D-Dgtp.	(55)
”	Teikaside C-IIa (**27**) $C_{50}H_{82}O_{19}$ – $-93.5°$	Teikagenin	-3-O-4Ac-α-L-Sarp-(1→4)-β-D-Dgtp-20-O-β-D-Glup-(1→4)-β-D-Digp.	
”	Teikaside C-IIb (**28**) $C_{50}H_{82}O_{19}$ – $-65.8°$	Teikagenin	-3-O-4Ac-α-L-Sarp-(1→4)-β-D-Dgtp-20-O-β-D-Glup-(1→4)-β-D-Olep.	
”	Teikaside C-IIc (**29**) $C_{49}H_{80}O_{19}$ – $-112.2°$	Teikagenin	-3-O-4Ac-α-L-Sarp-(1→4)-β-D-Dgtp-20-O-β-D-Glup-(1→4)-β-D-Canp.	
”	Teikaside C-IIIa (**30**) $C_{57}H_{94}O_{22}$ – $-85.9°$	Teikagenin	-3-O-4Ac-α-L-Sarp-(1→4)-β-D-Dgtp-20-O-β-D-Glup-(1→4)-β-D-Olep-(1→4)-β-D-Olep.	
”	Teikaside C-IVa (**31**) $C_{64}H_{106}O_{25}$ – $-83.7°$	Teikagenin	-3-O-4Ac-α-L-Sarp-(1→4)-β-D-Dgtp-20-O-β-D-Glup-(1→4)-β-D-Cymp-(1→4)-β-D-Olep-(1→4)-β-D-Olep.	
”	Teikaside B-IVa (**32**) $C_{62}H_{104}O_{24}$ – $-78.4°$	Teikagenin	-3-O-α-L-Sarp-(1→4)-β-D-Dgtp-20-O-β-D-Glup-(1→4)-β-D-Cymp-(1→4)-β-D-Olep-(1→4)-β-D-Olep.	

Table 1 (*continued*)

Plant	Glycoside (Glycoside no.) Molecular Formula mp°C $[\alpha]_D$	Genin	Sugar	References
Trachelospermum liukiuense	Teikaside AL-Ic (**33**) $C_{35}H_{58}O_{11}$ – $-86.5°$	Teikagenin	-3, 20-bis-O-β-D-Dgtp.	(*5*)
"	Teikaside AL-IId (**34**) $C_{41}H_{68}O_{16}$ – $-74.1°$	Teikagenin	-3-O-β-D-Dgtp-20-O-β-D-Glup-(1→4)-β-D-Dgtp.	
"	Teikaside BL-Ic (**35**) $C_{42}H_{70}O_{15}$ – $-79.1°$	Teikagenin	-3-O-β-D-Dgtp-(1→4)-β-D-Dgtp-20-O-β-D-Dgtp.	
		Family Asclepiadaceae		
Asclepias amplexicaulis	Amplexoside A (**36**) $C_{52}H_{76}O_{18}$ 258–260°C 183°	20-O-Acetyl-12β-O-cinnamoyl 5α-dihydro sarcostin	-3-O-α/β-[Dgxp and 3Me-6d-β-D-Allop(1→4)-D-Cymp].	(*233*)
Asclepias fruticosa	–(**242**) $C_{46}H_{74}O_{17} \cdot 5/2H_2O$ Amorphous $+13.4°$	Lineolon	-3-O-β-D-Cymp-(1→4)-β-D-Dgxp-(1→4)-β-D-Olip-(1→4)–β-D-Dgxp.	(*245*)

”	–(**243**) $C_{47}H_{76}O_{17} \cdot 2H_2O$ Amorphous +8.9°	Lineolon	-3-O-β-D-Cymp-(1→4)-β-D-Dgxp-(1→4)-β-D-Olep-(1→4)–β-D-Dgxp.	
”	–(**244**) $C_{48}H_{78}O_{17} \cdot H_2O$ Amorphous +16.0°	Lineolon	-3-O-β-D-Cymp-(1→4)-β-D-Cymp-(1→4)-β-D-Olep-(1→4)–β-D-Dgxp.	
”	–(**245**) $C_{48}H_{78}O_{17} \cdot H_2O$ Amorphous +16.0°	Lineolon	-3-O-β-D-Olep-(1→4)-β-D-Cymp-(1→4)-β-D-Olep-(1→4)–β-D-Dgxp.	
”	–(**246**) $C_{48}H_{78}O_{17} \cdot 5/2H_2O$ Amorphous 0°	Lineolon	-3-O-β-D-Cymp-(1→4)-β-D-Olep-(1→4)-β-D-Olep-(1→4)–β-D-Dgxp.	
”	–(**247**) $C_{46}H_{74}O_{17} \cdot 3/2H_2O$ Amorphous −8.8°	Lineolon	-3-O-β-D-Olep-(1→4)-β-D-Dgxp-(1→4)-β-D-Olip-(1→4)–β-D-Dgxp.	(*246*)
”	–(**248**) $C_{46}H_{76}O_{17} \cdot H_2O$ Amorphous −9.8°	Lineolon	-3-O-β-D-Olep-(1→4)-β-D-Dgxp-(1→4)-β-D-Olep-(1→4)–β-D-Dgxp.	
”	–(**249**) $C_{46}H_{76}O_{17}$ Amorphous +29.3°	Isolineolon	-3-O-β-D-Olep-(1→4)-β-D-Dgxp-(1→4)-β-D-Olep-(1→4)–β-D-Dgxp.	

Table 1 (*continued*)

Plant	Glycoside (Glycoside no.) Molecular Formula mp°C $[\alpha]_D$	Genin	Sugar	References
Asclepias fruticosa	–(**250**) $C_{47}H_{78}O_{17}$ Amorphous +36.8°	Isolineolon	-3-O-β-D-Olep-(1→4)-β-D-Cymp-(1→4)-β-D-Olep-(1→4)-β-D-Dgxp.	
"	–(**251**) $C_{62}H_{92}O_{23}$ – +16.1°	Ikemagenin	-3-O-β-D-Glup-(1→4)-β-D-Cymp-(1→4)-β-D-Dgxp-(1→4)-β-D-Olep-(1→4)-β-D-Dgxp.	(*247*)
"	–(**252**) $C_{62}H_{92}O_{23}$ – +9.6°	Ikemagenin	-3-O-β-D-Glup-(1→4)-β-D-Olep-(1→4)-β-D-Dgxp-(1→4)-β-D-Olep-(1→4)-β-D-Dgxp.	
"	–(**253**) $C_{63}H_{94}O_{23}$ – +20.8°	Ikemagenin	-3-O-β-D-Glup-(1→4)-β-D-Cymp-(1→4)-β-D-Cymp-(1→4)-β-D-Olep-(1→4)-β-D-Dgxp.	
"	–(**254**) $C_{63}H_{94}O_{23}$ – +24.08°	Ikemagenin	-3-O-β-D-Glup-(1→4)-β-D-Cymp-(1→4)-β-D-Dgxp-(1→4)-β-D-Olep-(1→4)-β-D-Cymp.	
"	–(**255**) $C_{64}H_{96}O_{23}$ – +20.0°	Ikemagenin	-3-O-β-D-Glup-(1→4)-β-D-Cymp-(1→4)-β-D-Olep-(1→4)-β-D-Olep-(1→4)-β-D-Cymp.	

”	–(**256**) $C_{64}H_{96}O_{23}$ – + 9.0°	Ikemagenin	-3-O-β-D-Glup-(1→4)-β-D-Olep-(1→4)-β-D-Olep-(1→4)-β-D-Olep-(1→4)-β-D-Cymp.
”	–(**257**) $C_{62}H_{92}O_{24}$ – + 24.3°	Kidjolanin	-3-O-β-D-Glup-(1→4)-β-D-Cymp-(1→4)-β-D-Dgxp-(1→4)-β-D-Olep-(1→4)-β-D-Dgxp.
”	–(**258**) $C_{63}H_{94}O_{24}$ – + 26.0°	Kidjolanin	-3-O-β-D-Glup-(1→4)-β-D-Cymp-(1→4)-β-D-Cymp-(1→4)-β-D-Olep-(1→4)-β-D-Dgxp.
”	–(**259**) $C_{63}H_{94}O_{24}$ – + 33.5°	Kidjolanin	-3-O-β-D-Glup-(1→4)-β-D-Cymp-(1→4)-β-D-Dgxp-(1→4)-β-D-Olep-(1→4)-β-D-Cymp.
”	–(**260**) $C_{64}H_{96}O_{24}$ – + 25.2°	Kidjolanin	-3-O-β-D-Glup-(1→4)-β-D-Cymp-(1→4)-β-D-Olep-(1→4)-β-D-Olep-(1→4)-β-D-Cymp.
”	–(**261**) $C_{64}H_{96}O_{24}$ – + 12.7°	Kidjolanin	-3-O-β-D-Glup-(1→4)-β-D-Olep-(1→4)-β-D-Olep-(1→4)-β-D-Olep-(1→4)-β-D-Cymp.
”	–(**262**) $C_{64}H_{96}O_{24}$ – + 32.0°	Kidjolanin	-3-O-β-D-Glup-(1→4)-β-D-Cymp-(1→4)-β-D-Olep-(1→4)-β-D-Cymp-(1→4)-β-D-Cymp.

Table 1 (*continued*)

Plant	Glycoside (Glycoside no.) Molecular Formula mp°C $[\alpha]_D$	Genin	Sugar	References
Boucerosia aucheriana	Bouceroside AI (**37**) $C_{62}H_{88}O_{21} \cdot 3H_2O$ 152–157.5°C +9°	Boucerogenin I	-3-O-β-D-Glup-(1→4)-3Me-6d-β-D-Allop-(1→4)-β-D-Olep-(1→4)-β-D-Cymp.	(*18*)
"	Bouceroside AII (**38**) $C_{62}H_{90}O_{21} \cdot H_2O$ 153–158.5°C +8.6°	Boucerogenin II	-3-O-β-D-Glup-(1→4)-3Me-6d-β-D-Allop-(1→4)-β-D-Olep-(1→4)-β-D-Cymp.	
"	Bouceroside BI (**39**) $C_{62}H_{88}O_{21} \cdot 4H_2O$ 161.5–168°C +22.4°	Boucerogenin I	-3-O-β-D-Glup-(1→4)-3Me-6d-β-D-Allop-(1→4)-β-D-Cymp-(1→4)-β-D-Cymp.	
"	Bouceroside BII (**40**) $C_{62}H_{90}O_{21} \cdot 9/2H_2O$ 157.5–165°C +21°	Boucerogenin II	-3-O-β-D-Glup-(1→4)-3Me-6d-β-D-Allop-(1→4)–β-D-Cymp-(1→4)-β-D-Cymp.	
"	Bouceroside ANC (**41**) $C_{49}H_{76}O_{15}$ 138.5–142.5°C −3.2°	12-O-Benzoyldihydro-boucerin	-3-O-3Me-6d–β-D-Allop-(1→4)-β-D-Cymp-(1→4)-β-D-Cymp.	(*19*)
"	Bouceroside ADC (**42**) $C_{49}H_{74}O_{15}$ 132–135.5°C −12.5°	12-O-Benzoylboucerin	-3-O-3Me-6d-β-D-Allop-(1→4)-β-D-Cymp-(1→4)-β-D-Cymp.	

”	Bouceroside ANO (**43**) $C_{49}H_{76}O_{15}$ 113.5–116°C −12.4°	12-O-Benzoyldihydro- boucerin	-3-O-3Me-6d-β-D-Allop- (1→4)-β-D-Olep-(1→4)-β-D-Cymp.
”	Bouceroside ADO (**44**) $C_{49}H_{74}O_{15}$ 107.5–111°C −11.8°	12-O-Benzoylboucerin	-3-O-3Me-6d-β-D-Allop- (1→4)-β-D-Olep-(1→4)-β-D-Cymp.
”	Bouceroside BNO (**45**) $C_{51}H_{78}O_{16}$ 133.5–137°C +2.4°	12-O-Benzoyl-20-O- acetyldihydroboucerin	-3-O-3Me-6d-β-D-Allop- (1→4)-β-D-Olep-(1→4)-β-D-Cymp.
”	Bouceroside BDO (**46**) $C_{51}H_{76}O_{16}$ 135.5–139°C −21.0°	12-O-Benzoyl-20-O- acetylboucerin	-3-O-3Me-6d-β-D-Allop- (1→4)-β-D-Olep-(1→4)-β-D-Cymp.
”	Bouceroside BNC (**47**) $C_{51}H_{78}O_{16}$ 138.5–141°C +18.5°	12-O-Benzoyl-20-O- acetyldihydroboucerin	-3-O-3Me-6d-β-D-Allop- (1→4)-β-D-Cymp-(1→4)-β-D-Cymp.
”	Bouceroside BDC (**48**) $C_{51}H_{76}O_{16}$ 103.5–106°C +2.1°	12-O-Benzoyl-20-O- acetylboucerin	-3-O-3Me-6d-β-D-Allop- (1→4)-β-D-Cymp-(1→4)-β-D-Cymp.
”	Bouceroside CNO (**49**) $C_{56}H_{80}O_{16}$ 143.5–147°C −7.5°	Boucerogenin II	-3-O-3Me-6d-β-D-Allop- (1→4)-β-D-Olep-(1→4)-β-D-Cymp.
”	Bouceroside CNC (**50**) $C_{56}H_{80}O_{16}$ 114–117.5°C +8°	Boucerogenin II	-3-O-3Me-6d-β-D-Allop- (1→4)-β-D-Cymp-(1→4)-β-D-Cymp.

Table 1 (*continued*)

Plant	Glycoside (Glycoside no.) Molecular Formula mp°C $[\alpha]_D$	Genin	Sugar	References
Calotropis gigantea	Calotroposide A (**51**) $C_{63}H_{96}O_{21}$ Amorphous +2.3°	12β-O-Benzoyllineolon	-3-O-β-D-Cymp-(1→4)-β-D-Olep-(1→4)-β-D-Olep-(1→4)-β-D-Cymp-(1→4)-β-D-Cymp.	(*77*)
"	Calotroposide B (**52**) $C_{63}H_{96}O_{22}$ Amorphous +12.2°	12β-O-Benzoyldeacetyl metaplexigenin	-3-O-β-D-Cymp-(1→4)-β-D-Olep-(1→4)-β-D-Olep-(1→4)-β-D-Cymp-(1→4)-β-D-Cymp.	
"	Calotroposide C (**53**) $C_{63}H_{96}O_{22}$ Amorphous −1.9°	12β-O-Benzoyldeacetyl metaplexigenin	-3-O-β-D-Olep-(1→4)-β-D-Olep-(1→4)-β-D-Olep-(1→4)-β-D-Cymp-(1→4)-β-D-Cymp.	(*51*)
"	Calotroposide D (**54**) $C_{63}H_{96}O_{21}$ Amorphous −17.6°	12β-O-Benzoyllineolon	-3-O-β-D-Olep-(1→4)-β-D-Olep-(1→4)-β-D-Olep-(1→4)-β-D-Cymp-(1→4)-β-D-Cymp.	
"	Calotroposide E (**55**) $C_{56}H_{84}O_{19}$ Amorphous −1.6°	12β-O-Benzoyldeacetyl metaplexigenin	-3-O-β-D-Olep-(1→4)-β-D-Olep-(1→4)-β-D-Cymp-(1→4)-β-D-Cymp.	
"	Calotroposide F (**56**) $C_{56}H_{84}O_{18}$ Amorphous −15.6°	12β-O-Benzoyllineolon	-3-O-β-D-Olep-(1→4)-β-D-Olep-(1→4)-β-D-Cymp-(1→4)-β-D-Cymp.	

”	Calotroposide G (**57**) $C_{49}H_{72}O_{15}$ Amorphous −17.4°	12β-O-Benzoyllineolon	-3-O-β-D-Olep-(1→4)-β-D-Cymp-(1→4)-β-D-Cymp.	
Caralluma tuberculata	Caratuberside A (**58**) $C_{34}H_{56}O_{12}$ 170–171°C +60°	3β, 14β-Dihydroxy-pregnan-20-one	-3-O-β-D-Glup-(1→4)-3Me-6d-β-D-Galp.	(*32*)
”	Caratuberside B (**59**) $C_{34}H_{58}O_{12}$ 182–185°C –	3β, 14β, 20-Trihydroxy-pregnane	-3-O-β-D-Glup-(1→4)-3Me-6d-β-D-Galp.	
Caralluma umbellata	Carumbelloside I (**60**) $C_{33}H_{52}O_{13}$ 254–256°C −46°	3β, 14β-Dihydroxy-pregn-5-en-20-one	-3-O-β-D-Glup-(1→6)-β-D-Glup.	(*86*)
”	Carumbelloside II (**61**) $C_{27}H_{42}O_8$ 274–276°C −17.3°	3β, 14β-Dihydroxy-pregn-5-en-20-one	-3-O-β-D-Glup.	
Cynanchum africanum	Cynafoside A (**62**) $C_{57}H_{86}O_{21} \cdot 4/3H_2O$ 142–144°C +14.8°	Cynafogenin	-3-O-β-D-Glup-(1→4)-α-L-Cymp-(1→4)-β-D-Cymp-(1→4)-β-D-Cymp.	(*248*)
”	Cynafoside B (**63**) $C_{56}H_{84}O_{21} \cdot H_2O$ 152–154°C +8.8°	Cynafogenin	-3-O-β-D-Glup-(1→4)-α-L-Cymp-(1→4)-β-D-Cymp-(1→4)-β-D-Dgxp.	
”	Cynafoside C (**64**) $C_{63}H_{96}O_{24} \cdot H_2O$ 88–90°C +7.6°	Cynafogenin	-3-O-β-D-Glup-(1→4)-α-L-Cymp-(1→4)-β-D-Olep-(1→4)-β-D-Dgxp-(1→4)-β-D-Cymp.	(*70*)

Table 1 (*continued*)

Plant	Glycoside (Glycoside no.) Molecular Formula mp°C $[\alpha]_D$	Genin	Sugar	References
Cynanchum africanum	Cynafoside D (**65**) $C_{62}H_{94}O_{24}\cdot 5/2H_2O$ 153–155°C +12.8°	Cynafogenin	-3-O-β-D-Glup-(1→4)-α-L-Cymp-(1→4)-β-D-Olep-(1→4)-β-D-Dgxp-(1→4)-β-D-Dgxp.	
Cynanchum atratum	Cynatratoside A (**66**) $C_{28}H_{40}O_8\cdot 1/2H_2O$ 209–210°C +15.5°	Glaucogenin-C	-3-O-β-D-Olep.	(*207*)
”	Cynatratoside B (**67**) $C_{41}H_{62}O_{14}\cdot 1/2H_2O$ 100–103°C −21.5°	Glaucogenin-C	-3-O-α-L-Cymp-(1→4)-β-D-Dgxp-(1→4)-β-D-Olep.	
”	Cynatratoside C (**68**) $C_{41}H_{62}O_{14}\cdot 1/2H_2O$ 104–108°C −7.2°	Glaucogenin-C	-3-O-α-D-Olep-(1→4)-β-D-Dgxp-(1→4)-α-D-Olep.	
”	Cynatratoside D (**69**) $C_{47}H_{72}O_{19}\cdot 1/2H_2O$ 140–145°C −25.8°	Glaucogenin-C	-3-O-β-D-Glup-(1→4)-α-L-Cymp-(1→4)-α-D-Dgxp-(1→4)-β-D-Olep.	
”	Cynatratoside E (**70**) $C_{47}H_{72}O_{19}\cdot 3/2H_2O$ 150–155°C −19.9°	Glaucogenin-C	-3-O-α-D-Glup-(1→4)-β-D-Olep-(1→4)-β-D-Dgxp-(1→4)-α-D-Olep.	

"	Cynatratoside F (**71**) $C_{42}H_{64}O_{15}$ 131–135°C −15.3°	2α-Hydroxy glaucogenin-C	-3-O-β-D-Cymp-(1→4)-α-L-Digp-(1→4)-β-D-Cymp.	*(208)*
"	Atratoside A (**72**) $C_{42}H_{64}O_{13} \cdot 3/2H_2O$ 105–110°C −65.9°	Atratogenin A	-3-O-β-D-Cymp-(1→4)-α-L-Digp-(1→4)-β-D-Cymp.	*(58)*
"	Atratoside B (**73**) $C_{48}H_{74}O_{18} \cdot 5/2H_2O$ 153–158°C −48.3°	Atratogenin A	-3-O-β-D-Glup-(1→4)-β-D-Cymp-(1→4)-α-L-Digp-(1→4)-β-D-Cymp.	
"	Atratoside C (**74**) $C_{48}H_{72}O_{18} \cdot 3H_2O$ 148–153°C −58.8°	Atratogenin B	-3-O-β-D-Glup-(1→4)-β-D-Cymp-(1→4)-α-L-Digp-(1→4)-β-D-Cymp.	
"	Atratoside D (**75**) $C_{40}H_{60}O_{13} \cdot H_2O$ 92–94°C −52.3°	Cynajapogenin A	-3-O-α-D-Olep-(1→4)-β-D-Dgxp-(1→4)-β-D-Cymp.	
Cynanchum auriculatum	Cynauricuoside A (**76**) $C_{64}H_{96}O_{24}$ 167–173°C −28.6°	Kidjoranin	-3-O-β-D-Glup-(1→4)-α-L-Cymp-(1→4)-β-D-Cymp-(1→4)-α-L-Digp-(1→4)-β-D-Cymp.	*(10)*
"	Cynauricuoside B (**77**) $C_{52}H_{82}O_{19}$ 137–142°C −64.96°	Metaplexigenin	-3-O-α-L-Cymp-(1→4)-β-D-Cymp-(1→4)-α-L-Digp-(1→4)-β-D-Cymp.	
"	Cynauricuoside C (**78**) $C_{68}H_{110}O_{29}$ 174–181°C −25.15°	Caudatin	-3-O-β-D-Glup-(1→4)-β-D-Glup-(1→4)-α-L-Cymp-(1→4)-β-D-Cymp-(1→4)-α-L-Digp-(1→4)-β-D-Cymp.	

Table 1 (*continued*)

Plant	Glycoside (Glycoside no.) Molecular Formula mp°C $[\alpha]_D$	Genin	Sugar	References
Cynanchum caudatum	Cynanchoside C_1 (**79**) $C_{49}H_{78}O_{15} \cdot 1/2H_2O$ 123.5–129°C +30.4°	Cynanchogenin	-3-O-β-D-Cymp-(1→4)-β-D-Cymp-(1→4)-β-D-Cymp.	(*249*)
”	Cynanchoside C_2 (**80**) $C_{49}H_{78}O_{15}$ 132.5–135.5°C −14.6°	Cynanchogenin	-3-O-β-D-Olep-(1→4)-β-D-Cymp-(1→4)-β-D-Cymp.	(*249, 250*)
”	–(**263**) $C_{28}H_{46}O_9 \cdot H_2O$ Amorphous +64.4°	Sarcostin	-3-O-β-D-Cymp.	(*251*)
”	–(**264**) $C_{49}H_{82}O_{18} \cdot 2H_2O$ – −2.9°	Sarcostin	-3-O-α-D-Cymp-(1→4)-β-D-Olep-(1→4)-β-D-Cymp-(1→4)-β-D-Cymp.	
”	–(**265**) $C_{49}H_{82}O_{18} \cdot 3H_2O$ Amorphous +40.1°	Sarcostin	-3-O-β-D-Olep-(1→4)-β-D-Olep-(1→4)-β-D-Cymp-(1→4)-β-D-Cymp.	
”	–(**266**) $C_{42}H_{70}O_{15} \cdot 2H_2O$ Amorphous +4.4°	Sarcostin	-3-O-α-D-Cymp-(1→4)-β-D-Cymp-(1→4)-β-D-Cymp.	

"	-(**267**) $C_{42}H_{70}O_{15}$ Amorphous +4.5°	Sarcostin	-3-O-α-D-Cymp-(1→4)-β-D-Olep-(1→4)-β-D-Cymp.	
"	-(**268**) $C_{42}H_{70}O_{15}$ Amorphous +32.0°	Sarcostin	-3-O-β-D-Olep-(1→4)-β-D-Olep-(1→4)-β-D-Cymp.	
"	-(**269**) $C_{42}H_{70}O_{15} \cdot 2H_2O$ Amorphous +51.8°	Sarcostin	-3-O-β-D-Olep-(1→4)-β-D-Cymp-(1→4)-β-D-Cymp.	
"	-(**270**) $C_{35}H_{58}O_{12} \cdot 2H_2O$ Amorphous +40.5°	Sarcostin	-3-O-β-D-Cymp-(1→4)-β-D-Cymp.	
"	-(**271**) $C_{35}H_{58}O_{12} \cdot 2H_2O$ Amorphous +41.4°	Sarcostin	-3-O-β-D-Olep-(1→4)-β-D-Cymp.	
"	-(**272**) $C_{49}H_{80}O_{18} \cdot 2H_2O$ Amorphous +8.2°	Deacylmetaplexigenin	-3-O-β-D-Olep-(1→4)-β-D-Olep-(1→4)-β-D-Cymp-(1→4)-β-D-Cymp.	
"	-(**273**) $C_{49}H_{78}O_{16} \cdot H_2O$ Amorphous +5.1°	Caudatin	-3-O-β-D-Olep-(1→4)-β-D-Cymp-(1→4)-β-D-Cymp.	(*252*)
"	-(**274**) $C_{51}H_{74}O_{15} \cdot 5/2H_2O$ Amorphous +9.3°	Ikemagenin	-3-O-β-D-Olep-(1→4)-β-D-Cymp-(1→4)-β-D-Cymp.	

Table 1 (*continued*)

Plant	Glycoside (Glycoside no.) Molecular Formula mp°C $[\alpha]_D$	Genin	Sugar	References
Cynanchum caudatum	–(**275**) $C_{51}H_{76}O_{16} \cdot 2H_2O$ Amorphous + 27.9°	Penupogenin	-3-O-β-D-Olep-(1→4)-β-D-Cymp-(1→4)-β-D-Cymp.	
”	–(**276**) $C_{49}H_{72}O_{16} \cdot 3/2H_2O$ Amorphous + 0°	12-O-Benzoyl-deacylmetaplexigenin	-3-O-β-D-Olep-(1→4)-β-D-Cymp-(1→4)-β-D-Cymp.	
”	–(**277**) $C_{49}H_{74}O_{16} \cdot 2H_2O$ Amorphous + 16.9°	12-O-Benzoyl-sarcostin	-3-O-β-D-Olep-(1→4)-β-D-Cymp-(1→4)-β-D-Cymp.	
”	–(**278**) $C_{49}H_{78}O_{15} \cdot 2H_2O$ Amorphous – 29.1°	Cynanchogenin	-3-O-β-D-Olep-(1→4)-β-D-Olep-(1→4)-β-D-Cymp.	
”	–(**279**) $C_{49}H_{78}O_{16} \cdot 3/2H_2O$ Amorphous – 8.9°	Caudatin	-3-O-β-D-Olep-(1→4)-β-D-Olep-(1→4)-β-D-Cymp.	
”	–(**280**) $C_{56}H_{90}O_{18} \cdot 3/2H_2O$ Amorphous – 13.1°	Cynanchogenin	-3-O-β-D-Olep-(1→4)-β-D-Olep-(1→4)-β-D-Cymp-(1→4)-β-D-Cymp.	

”	-(**281**) $C_{56}H_{90}O_{19} \cdot 3/2H_2O$ Amorphous $-0.45°$	Caudatin	-3-O-β-D-Olep-(1→4)-β-D-Olep-(1→4)-β-D-Cymp-(1→4)-β-D-Cymp.
”	-(**282**) $C_{56}H_{90}O_{18} \cdot H_2O$ Amorphous $+7.0°$	Cynanchogenin	-3-O-β-D-Cymp-(1→4)-β-D-Olep-(1→4)-β-D-Cymp-(1→4)-β-D-Cymp.
”	-(**283**) $C_{56}H_{90}O_{19} \cdot 3/2H_2O$ Amorphous $-0.74°$	Caudatin	-3-O-β-D-Olep-(1→4)-β-D-Olep-(1→4)-β-D-Olep-(1→4)-β-D-Cymp.
”	-(**284**) $C_{63}H_{102}O_{21} \cdot 3/2H_2O$ Amorphous $-18.9°$	Cynanchogenin	-3-O-β-D-Olep-(1→4)-β-D-Olep-(1→4)-β-D-Olep-(1→4)-β-D-Cymp-(1→4)-β-D-Cymp.
”	-(**285**) $C_{63}H_{102}O_{22} \cdot H_2O$ Amorphous $-5.9°$	Caudatin	-3-O-β-D-Olep-(1→4)-β-D-Olep-(1→4)-β-D-Olep-(1→4)-β-D-Cymp-(1→4)-β-D-Cymp.
”	-(**286**) $C_{63}H_{102}O_{21} \cdot H_2O$ Amorphous $-4.5°$	Cynanchogenin	-3-O-β-D-Cymp-(1→4)-β-D-Olep-(1→4)-β-D-Olep-(1→4)-β-D-Cymp-(1→4)-β-D-Cymp.
”	-(**287**) $C_{63}H_{102}O_{22} \cdot H_2O$ Amorphous $+15.8°$	Caudatin	-3-O-β-D-Cymp-(1→4)-β-D-Olep-(1→4)-β-D-Olep-(1→4)-β-D-Cymp-(1→4)-β-D-Cymp.

Table 1 (*continued*)

Plant	Glycoside (Glycoside no.) Molecular Formula mp°C $[\alpha]_D$	Genin	Sugar	References
Cynanchum caudatum	–(**288**) $C_{63}H_{102}O_{21} \cdot 1/2H_2O$ Amorphous $-8.1°$	Cynanchogenin	-3-O-β-D-Olep-(1→4)-β-D-Cymp-(1→4)-β-D-Olep-(1→4)-β-D-Cymp-(1→4)-β-D-Cymp.	
”	–(**289**) $C_{63}H_{102}O_{22} \cdot 2H_2O$ Amorphous $+5.9°$	Caudatin	-3-O-β-D-Olep-(1→4)-β-D-Cymp-(1→4)-β-D-Olep-(1→4)-β-D-Cymp-(1→4)-β-D-Cymp.	
Cynanchum formosanum	Cynanformoside A (**81**) $C_{28}H_{46}O_8$ 183°C $-15.5°$	Utendin	-3-O-β-D-Olep.	(*22*)
”	Cynanformoside B (**82**) $C_{30}H_{48}O_9$ 212–214°C $-17.5°$	20-O-Acetylutendin	-3-O-β-D-Olep.	
Cynanchum forrestii	Cynaforroside A (**83**) $C_{42}H_{64}O_{14}$ 122–126°C $-25.83°$	Glaucogenin C	-3-O-α-L-Cymp-(1→4)-β-L-Cymp-(1→4)-β-D-Olep.	(*84*)
Cynanchum glaucescens	Glaucoside A (**84**) $C_{28}H_{40}O_9$ 112–117°C $+7.17°$	Glaucogenin A	-3-O-β-D-Olep.	(*209*)

References pp. 309–325

”	Glaucoside B (**85**) $C_{42}H_{64}O_{15}$ 115–120°C $-1.83°$	Glaucogenin A	-3-O-α-L-Cymp-(1→4)-β-L-Cymp-(1→4)-β-L-Cymp.	
”	Glaucoside C (**86**) $C_{41}H_{62}O_{15} \cdot 1/2H_2O$ 127–133°C $-14.6°$	Glaucogenin A	-3-O-α-L-Cymp-(1→4)-β-D-Dgxp-(1→4)-β-L-Cymp.	
”	Glaucoside D (**87**) $C_{41}H_{62}O_{15} \cdot 1/2H_2O$ 118–124°C $-28.3°$	Glaucogenin A	-3-O-α-L-Cymp-(1→4)-β-D-Dgxp-(1→4)-β-D-Olep.	
”	Glaucoside E (**88**) $C_{42}H_{64}O_{15} \cdot 1/2H_2O$ 100–106°C $-21.4°$	Glaucogenin C	-3-O-α-L-Cymp-(1→4)-β-L-Cymp-(1→4)-β-D-Thevp.	
”	Glaucoside F(**89**) $C_{42}H_{64}O_{15} \cdot H_2O$ 110–113°C $-17.4°$	Glaucogenin A	-3-O-α-L-Cymp-(1→4)-β-L-Cymp-(1→4)-β-D-Olep.	(*210*)
”	Glaucoside G(**90**) $C_{41}H_{62}O_{15}$ 117–123°C $-29.6°$	Glaucogenin C	-3-O-α-L-Cymp-(1→4)-β-D-Dgxp-(1→4)-β-D-Thevp.	
”	Glaucoside H (**91**) $C_{47}H_{72}O_{20} \cdot 2H_2O$ 156–159°C $-26.8°$	Glaucogenin A	-3-O-β-D-Glup-(1→4)-α-L-Cymp-(1→4)-β-D-Dgxp-(1→4)-β-L-Cymp.	(*211*)
”	Glaucoside I (**92**) $C_{48}H_{74}O_{20} \cdot 2H_2O$ 150–152°C $-19.6°$	Glaucogenin A	-3-O-β-D-Glup-(1→4)-α-L-Cymp-(1→4)-β-L-Cymp-(1→4)-β-L-Cymp.	

Table 1 (*continued*)

Plant	Glycoside (Glycoside no.) Molecular Formula mp°C $[\alpha]_D$	Genin	Sugar	References
Cynanchum glaucescens	Glaucoside J (**93**) $C_{47}H_{72}O_{21} \cdot H_2O$ 134–139°C −25.3°	Glaucogenin B	-3-O-β-D-Glup-(1→4)-α-L-Cymp-(1→4)-β-D-Dgxp-(1→4)-β-D-Olep.	
”	Glaucogenin C mono-D-thevetoside (**94**) $C_{28}H_{40}O_9$ 187–190.5°C +27.4°	Glaucogenin C	-3-O-β-D-Thevp.	(*212, 253*)
Cynanchum hancockianum	Hancoside (**95**) $C_{44}H_{62}O_8$ 185–187°C −12.31°	3β, 14β, 15β-Trihydroxy-pregn-5-en-20-one	-3-O-6Sin-β-D-Glup-(1→2)-β-D-Glup.	(*39*)
Cynanchum otophyllum	Otophylloside A (**96**) $C_{49}H_{72}O_{17}$ – –	12β-O-p-Hydroxybenzoyldeacetylmetaplexigenin	-3-O-β-D-Olep-(1→4)-β-D-Cymp-(1→4)-β-D-Cymp.	(*238*)
”	Otophylloside B (**97**) $C_{49}H_{78}O_{16}$ – –	12β-O-Ikemoyldeacetylmetaplexigenin	-3-O-β-D-Olep-(1→4)-β-D-Cymp-(1→4)-β-D-Cymp.	
Cynanchum paniculatum	Cynapanoside A (**98**) $C_{28}H_{40}O_9$ 114–117°C +21.3°	Glaucogenin D	-3-O-β-D-Olep.	(*79*)

”	Cynapanoside B (**99**) $C_{41}H_{62}O_{15}$ 125–126.5°C +39.4°	Glaucogenin D	-3-O-α-L-Cymp-(1→4)-β-D-Dgxp-(1→4)-β-D-Olep	
”	Cynapanoside C (**100**) $C_{41}H_{62}O_{15}$ 136–138°C −11.2°	Glaucogenin D	-3-O-α-D-Olep-(1→4)-β-D-Dgxp-(1→4)-β-D-Olep	
Cynanchum sibiricum and *C. maximowiczii*	Sibiricoside D (**101**) $C_{62}H_{100}O_{23}$ – −26.6°	Sibirigenin/Cynanchogenin	-3-O-β-D-Glup-(1→4)-β-D-Cymp-(1→4)-β-D-Cymp-(1→4)-β-D-Cymp-(1→4)-β-D-Olep	(*202, 254*)
Cynanchum sibiricum	Sibiricoside E (**102**) $C_{68}H_{110}O_{28}$ – −18.5°	Sibirigenin/Cynanchogenin	-3-O-β-D-Glup-(1→4)-β-D-Glup-(1→4)-β-D-Cymp-(1→4)-β-D-Cymp-(1→4)-β-D-Cymp-(1→4)-β-D-Olep	(*202*)
Cynanchum wilfordi	Wilfoside C1N (**103**) $C_{56}H_{90}O_{19} \cdot 2/3H_2O$ 140–142.5°C −44.7°	Caudatin	-3-O-α-L-Cymp-(1→4)-β-D-Cymp-(1→4)-α-L-Digp-(1→4)-β-D-Cymp	(*255*)
”	Wilfoside C2N (**104**) $C_{55}H_{88}O_{19} \cdot 3/2H_2O$ 142–143°C −50.3°	Caudatin	-3-O-α-L-Cymp-(1→4)-β-D-Cymp-(1→4)-α-L-Digp-(1→4)-β-D-Dgxp	
”	Wilfoside C3N (**105**) $C_{49}H_{78}O_{16}$ 124–126.5°C +14.8°	Caudatin	-3-O-β-D-Cymp-(1→4)-α-L-Digp-(1→4)-β-D-Cymp	
”	Wilfoside C1G (**106**) $C_{62}H_{100}O_{24}$ 143–147°C −31.8°	Caudatin	-3-O-β-D-Glup-(1→4)-α-L-Cymp-(1→4)-β-D-Cymp-(1→4)-α-L-Digp-(1→4)-β-D-Cymp	

Table 1 (*continued*)

Plant	Glycoside (Glycoside no.) Molecular Formula mp°C $[\alpha]_D$	Genin	Sugar	References
Cynanchum wilfordi	Wilfoside C2G (**107**) $C_{61}H_{98}O_{24} \cdot 3/2H_2O$ 135–138°C −37.8°	Caudatin	-3-O-β-D-Glup-(1→4)-α-L-Cymp-(1→4)-β-D-Cymp-(1→4)-α-L-Digp-(1→4)-β-D-Dgxp.	
"	Wilfoside C3G (**108**) $C_{55}H_{88}O_{21}$ 163–167°C +5.9°	Caudatin	-3-O-β-D-Glup-(1→4)-β-D-Cymp-(1→4)-α-L-Digp-(1→4)-β-D-Cymp.	
"	Wilfoside D1N (**109**) $C_{56}H_{84}O_{19} \cdot H_2O$ 143–145°C −46.9°	Cynanforidine	-3-O-α-L-Cymp-(1→4)-β-D-Cymp-(1→4)-α-L-Digp-(1→4)-β-D-Cymp.	(*85*)
"	Wilfoside G1G (**110**) $C_{70}H_{101}NO_{25} \cdot 1/2H_2O$ 164–167°C +28.2°	Gagaminine	-3-O-β-D-Glup-(1→4)-α-L-Cymp-(1→4)-β-D-Cymp-(1→4)-α-L-Digp-(1→4)-β-D-Cymp.	
"	Wilfoside F1N (**111**) $C_{65}H_{98}O_{20} \cdot 2/3H_2O$ 140–144°C +32.6°	Cynanforine	-3-O-α-L-Cymp-(1→4)-β-D-Cymp-(1→4)-α-L-Digp-(1→4)-β-D-Cymp.	
"	Wilfoside K1N (**112**) $C_{58}H_{86}O_{19} \cdot 1/3H_2O$ 183–187°C −23.2°	Kidjoranine	-3-O-α-L-Cymp-(1→4)-β-D-Cymp-(1→4)-α-L-Digp-(1→4)-β-D-Cymp.	

”	Wilfoside M1N (**113**) $C_{49}H_{80}O_{18}$ 141–143°C −40.3°	Deacylmetaplexigenin	-3-O-α-L-Cymp-(1→4)-β-D-Cymp-(1→4)-α-L-Digp-(1→4)-β-D-Cymp.	
”	Wilfoside W1N (**114**) $C_{63}H_{94}O_{20}\cdot 1/2H_2O$ 143.5–146°C +33°	Wilforidine	-3-O-α-L-Cymp-(1→4)-β-D-Cymp-(1→4)-α-L-Digp-(1→4)-β-D-Cymp.	
”	Wilfoside W3N (**115**) $C_{56}H_{82}O_{17}$ 120–123°C −43.3°	Wilforidine	-3-O-β-D-Cymp-(1→4)-α-L-Digp-(1→4)-β-D-Cymp.	
Cynanchum wallichii	Wallicoside (**116**) $C_{61}H_{98}O_{26}$ 194–196°C +22°	Caudatin	-3-O-β-D-Glup-(1→4)-β-D-Glup-(1→4)-β-D-Olep-(1→4)-β-D-Cymp-(1→4)-β-D-Cymp.	*(227)*
Dregea abyssinica	Drebyssoside 1 (**117**) $C_{49}H_{78}O_{17}$ 141–143°C +24.9°	Drevogenin A	-3-O-3Me-6d-β-D-Allop-(1→4)-β-D-Olep-(1→4)-β-D-Cymp.	*(256)*
”	Drebyssoside 2 (**118**) $C_{49}H_{78}O_{17}$ Amorphous +38.0°	Drevogenin A	-3-O-3Me-6d-β-D-Allop-(1→4)-β-D-Cymp-(1→4)-α/β-D-Cymp.	
”	Drebyssoside 3 (**119**) $C_{49}H_{78}O_{18}$ 176–178°C +42.3°	Drebyssogenin F	-3-O-3Me-6d-β-D-Allop-(1→4)-β-D-Olep-(1→4)-β-D-Cymp.	

Table 1 (*continued*)

Plant	Glycoside (Glycoside no.) Molecular Formula mp°C $[\alpha]_D$	Genin	Sugar	References
Dregea lanceolata	Drelin (**120**) $C_{43}H_{68}O_{16}$ 151°C +16.27°	11α-O-Acetylmarsdenin	-3-O-β-D-Bovp-(1→4)-β-D-Cymp-(1→4)-β-D-Cymp.	(*195*)
"	Ceolin (**121**) $C_{44}H_{70}O_{16}$ 146–148°C +4.38°	11α-O-Acetylmarsdenin	-3-O-β-D-Cymp-(1→4)-3Me-6d-β-D-Allop-(1→4)-β-D-Olep.	
"	Lanceolin (**122**) $C_{51}H_{84}O_{19}$ 108–110°C +24°	11α-O-Acetylmarsectohexol	-3-O-α-L-Digp-(1→4)-α-L-Digp-(1→4)-β-D-Cymp-(1→4)-β-D-Olep.	(*15*)
"	Lancin (**123**) $C_{35}H_{58}O_{12}$ 118–120°C −12.3°	Marsectohexol	-3-O-β-D-Cymp-(1→4)-β-D-Cymp.	
"	Lancinin (**124**) $C_{35}H_{56}O_{12}$ 95–97°C +16.04°	Marsdenin	-3-O-β-D-Cymp-(1→4)-β-D-Olep.	
"	Dregealin (**125**) $C_{51}H_{82}O_{19}$ 115°C +28.57°	11α-O-Acetylmarsdenin	-3-O-α-L-Digp-(1→4)-α-L-Digp-(1→4)-β-D-Cymp-(1→4)-β-D-Olep.	(*43*)

Dregea sinensis var. corrugata	Dregeoside (**126**) $C_{49}H_{76}O_{16}$ 125–128°C +43.2°	12β-O-Benzoyl-drevogenin	-3-O-β-D-Olep-(1→4)-β-D-Cymp-(1→4)-β-D-Cymp.	(*80*)
”	Dregeoside A (**127**) $C_{56}H_{94}O_{21}$ 145–148°C +28.5°	Drevogenin A	-3-O-3Me-6d-β-D-Allop-(1→4)-β-D-Olep-(1→4)-β-D-Cymp-(1→4)-β-D-Cymp.	(64)
”	Dregeoside B (**128**) $C_{53}H_{90}O_{22}$ 135–138°C +12.5°	12β-O-Isovaleryldi-hydrosarcostin	-3-O-β-D-Glup-(1→4)-3Me-6d-β-D-Allop-(1→4)-β-D-Olep-(1→4)-β-D-Cymp.	(*16*)
”	Dregeoside C (**129**) $C_{51}H_{78}O_{16}$ 143–146°C +37.5°	12β-O-Acetyl-20-O-benzoyltomento-genin	-3-O-β-D-Olep-(1→4)-β-D-Cymp-(1→4)-β-D-Cymp.	
Dregea volubilis	Dregoside A (**130**) $C_{35}H_{54}O_{10}$ 149–151.5°C +43.9°	Drevogenin A	-3-O-α/β-D-Cymp.	(*234*)
”	Dregeoside A_{pl} (**131**) $C_{56}H_{90}O_{20} \cdot H_2O$ 118–120°C +25.3°	Drevogenin A	-3-O-3Me-6d-β-D-Allop-(1→4)-β-D-Olep-(1→4)-β-D-Cymp-(1→4)-β-D-Cymp.	(*215*)
”	Dregeoside A_{ol} (**132**) $C_{62}H_{100}O_{20} \cdot 5/2H_2O$ 149–151.5°C +24.8°	Drevogenin A	-3-O-β-D-Glup-(1→4)-3Me-6d-β-D-Allop-(1→4)-β-D-Olep-(1→4)-β-D-Cymp-(1→4)-β-D-Cymp.	

Table 1 (*continued*)

Plant	Glycoside (Glycoside no.) Molecular Formula mp°C $[\alpha]_D$	Genin	Sugar	References
Dregea volubilis	Dregeoside A_{a1} (**133**) $C_{49}H_{78}O_{17} \cdot 3/2H_2O$ 130.5–133°C +35.2°	Drevogenin A	-3-O-3Me-6d-β-D-Allop-(1→4)-β-D-Cymp-(1→4)-β-D-Cymp.	
"	Dregeoside A_{11} (**134**) $C_{55}H_{88}O_{22} \cdot 3/2H_2O$ 162.5–165°C +33°	Drevogenin A	-3-O-β-D-Glup-(1→4)-3Me-6d-β-D-Allop-(1→4)-β-D-Cymp-(1→4)-β-D-Cymp.	
"	Dregeoside C_{11} (**135**) $C_{59}H_{86}O_{22} \cdot 2H_2O$ 142–145°C +52.1°	Drevogenin C	-3-O-β-D-Glup-(1→4)-3Me-6d-β-D-Allop-(1→4)-β-D-Cymp-(1→4)-β-D-Cymp.	
"	Dregeoside K_{p1} (**136**) $C_{54}H_{90}O_{19} \cdot H_2O$ 125.5–128°C +13.2°	Drebyssogenin K_2	-3-O-3Me-6d-β-D-Allop-(1→4)-β-D-Olep-(1→4)-β-D-Cymp-(1→4)-β-D-Cymp.	
"	Dregeoside K_{a1} (**137**) $C_{47}H_{78}O_{16} \cdot 3/2H_2O$ 131.5–135°C +15.5°	Drebyssogenin K_2	-3-O-3Me-6d-β-D-Allop-(1→4)-β-D-Cymp-(1→4)-β-D-Cymp.	
"	Dregeoside D_{p1} (**138**) $C_{49}H_{82}O_{18} \cdot H_2O$ 136.5–139°C +0.78°	Drevogenin D	-3-O-3Me-6d-β-D-Allop-(1→4)-β-D-Olep-(1→4)-β-D-Cymp-(1→4)-β-D-Cymp.	(*235*)

"	Dregeoside D_{al} (**139**) $C_{42}H_{70}O_{15}\cdot 1/2H_2O$ 139.5–143°C +2.13°	Drevogenin D	-3-O-3Me-6d-β-D-Allop-(1→4)-β-D-Cymp-(1→4)-β-D-Cymp.	
"	Dregeoside G_{pl} (**140**) $C_{56}H_{92}O_{20}\cdot H_2O$ 105–108°C +23.3°	Drebyssogenin G	-3-O-3Me-6d-β-D-Allop-(1→4)-β-D-Olep-(1→4)-β-D-Cymp-(1→4)-β-D-Cymp.	
"	Dregeoside G_{al} (**141**) $C_{49}H_{80}O_{17}\cdot 5/4H_2O$ 126.5–129°C +25.3°	Drebyssogenin G	-3-O-3Me-6d-β-D-Allop-(1→4)-β-D-Cymp-(1→4)-β-D-Cymp.	
"	Dregeoside H (**142**) $C_{41}H_{68}O_{16}\cdot 1/4H_2O$ 147–150°C +34.2°	Marsectohexol	-3-O-3Me-6d-β-D-Allop-(1→4)-β-D-Cymp-(1→4)-β-D-Dgxp.	
Folotsia sarcostemmoides	Folotsoside A (**143**) $C_{49}H_{72}O_{16}$ 209°C +19.0°	12-O-Benzoyllineolon	-3-O-3Me-6d-β-D-Allop-(1→4)-β-D-Cymp-(1→4)-β-D-Cymp.	(*124*)
Gymnema yunnanense	Gymnemaroside A (**144**) $C_{57}H_{86}O_{22}$ – –	Penupogenin	-3-O-β-D-Glup-(1→4)-3Me-6d-β-D-Allop-(1→4)-β-D-Cymp-(1→4)-β-D-Cymp.	(*257*)
"	Gymnemaroside B (**145**) – – –	Gymnemarogenin	-3-O-β-D-Glup-(1→4)-3Me-6d-β-D-Allop-(1→4)-β-D-Cymp-(1→4)-β-D-Cymp.	

Table 1 (*continued*)

Plant	Glycoside (Glycoside no.) Molecular Formula mp°C $[\alpha]_D$	Genin	Sugar	References
Hemidesmus indicus	Desinine (**146**) $C_{37}H_{58}O_{12}$ 115–118°C 0°	Drevogenin B	-3-O-β-D-Olep-(1→4)-β-D-Olep.	(*174*)
"	Indicine (**147**) $C_{27}H_{44}O_6$ 230–233°C −37°	Calogenin	-3-O-β-D-Dgxp.	(*63*)
"	Hemidine (**148**) $C_{27}H_{44}O_6$ 134–140°C −24°	Calogenin	-3-O-β-D-Bovp.	
"	Indicusin (**149**) $C_{46}H_{72}O_{18}$ 127–130°C −10.67°	11α, 12β-Di-O-acetyl-orgogenin	-3-O-β-D-Cymp-(1→4)-β-D-Cymp-(1→4) β-D-Cymp.	(*61*)
"	Hemidescine (**150**) $C_{36}H_{58}O_{10}$ 158°C +13.33°	20-O-Acetylcalogenin	-3-O-β-D-Dgxp-(1→4)-β-D-Olep.	(*3*)
"	Emidine (**151**) $C_{39}H_{64}O_{12}$ 192–196°C +10.3°	Calogenin	-3-O-β-D-Dgxp-(1→4)-β-D-Dgxp-(1→4)-β-D-Dgxp.	

”	Medidesmine (**152**) $C_{40}H_{66}O_{17}$ 116–118°C −27.6°	Sarcostin	-3-O-α-D-Glup-(1→4)-β-D-Dgxp-(1→4)-β-D-Olep.	(*258*)
”	Hemisine (**153**) $C_{48}H_{80}O_{19}$ 128–130°C −52.5°	Calogenin	-3-O-β-D-Cymp-(1→4)-3Me-β-D-Glup-(1→4)-β-D-Glup-(1→4)-β-D-Cymp.	
”	Desmisine (**154**) $C_{43}H_{70}O_{17}$ 98–100°C +205.3°	Calogenin	-3-O-β-D-Xylp-(1→4)-β-D-Dgxp-(1→4)-β-D-Xylp-(1→4)-β-D-Dgxp.	
Kanalia laniflora	Kalanoside H (**155**) $C_{43}H_{68}O_{15}$ Amorphous +1.4°	12-O-Acetyl-17-isolineolon	-3-O-α/β-D-Olep-α/β-D-Olep-α/β-D-Dgxp/Canp.	(*259*)
”	Kalanoside K (**156**) $C_{42}H_{66}O_{15}$ 165–169°C −1.9°	12-O-Acetyl-17-isolineolon	-3-O-β-D-Olep-(1→3/4)-β-D-Dgxp-(1→3/4)-β-D-Canp.	
Leptadenia hastata	–(**290**) $C_{44}H_{56}O_{11}$ – +73°	12-O-Benzoyl-20-O-cinnamoylsarcostin	-3-O-β-D-Cymp.	(*114*)
”	–(**291**) $C_{44}H_{64}O_{13}$ – +80°	Penupogenin	-3-O-β-D-Olep-(1→4)-β-D-Cymp.	

Table 1 (*continued*)

Plant	Glycoside (Glycoside no.) Molecular Formula mp°C $[\alpha]_D$	Genin	Sugar	References
Leptadenia hastata	–(**292**) $C_{51}H_{68}O_{14}$ – +88°	12-O-Benzoyl-20-O–cinnamoylsarcostin	-3-O-β-D-Olep-(1→4)-β-D-Cymp.	
Leptadenia reticulata	Reticulin (**157**) $C_{48}H_{80}O_{17}$ 119–122°C −7.1°	Calogenin	-3-O-β-D-Cymp-(1→4)-3Me-α-D-Galp-(1→4)-β-D-Dgxp-(1→4)-β-D-Cymp.	(*53*)
"	Deniculatin (**158**) $C_{34}H_{56}O_{11}$ 124–127°C −19.4°	Calogenin	-3-O-3Me-α-D-Galp-(1→4)-β-D-Dgxp.	
"	Leptaculatin (**159**) $C_{40}H_{66}O_{16}$ 107–110°C −5.8°	Calogenin	-3-O-β-D-Glup-(1→4)-β-D-Glup-(1→4)-β-D-Cymp.	
Marsdenia condurango	Condurangoglycoside A (**160**) $C_{53}H_{78}O_{17}$ – +39.4°	Condurangogenin A	-3-O-3Me-6d-β-D-Allop-(1→4)-β-D-Olep-(1→4)-β-D-Cymp.	(*54, 128*)

”	Condurangoglycoside C (**161**) $C_{53}H_{80}O_{17}$ – + 12°	Condurangogenin C	-3-O-3Me-6d-β-D-Allop-(1→4)-β-D-Olep-(1→4)-β-D-Cymp.	
”	Condurangoglycoside A_1 (**162**) $C_{65}H_{98}O_{27}$ – + 38°	Condurangogenin A	-3-O-β-D-Glup-(1→4)-β-D-Glup-(1→2/4)-3Me-6d-β-D-Allop-(1→4)-α/β-D-Olep-(1→4)-β-D-Cymp.	
”	Condurangoglykoside C_1 (**163**) $C_{65}H_{100}O_{27}$ – + 23°	Condurangogenin C	-3-O-β-D-Glup-(1→4)-β-D-Glup-(1→2/4)-3Me-6d-β-D-Allop-(1→4)-α/β-D-Olep-(1→4)-β-D-Cymp.	
”	Condurangoglycoside A_o (**164**) $C_{59}H_{88}O_{22}$ 170–174°C + 43.9°	Condurangogenin A	-3-O-β-D-Glup-(1→4)-3Me-6d-β-D-Allop-(1→4)-β-D-Olep-(1→4)-β-D-Cymp.	(*54, 129*)
”	Condurangoglycoside C_o (**165**) $C_{59}H_{90}O_{22}$ 160–170°C + 25.9°	Condurangogenin C	-3-O-β-D-Glup-(1→4)-3Me-6d-β-D-Allop-(1→4)-β-D-Olep-(1→4)-β-D-Cymp.	
”	Condurangoglycoside B_o (**166**) $C_{59}H_{86}O_{22}\cdot 2H_2O$ 170–180°C + 11.5°	Condurangogenin B	-3-O-β-D-Glup-(1→4)-3Me-6d-β-D-Allop-(1→4)-β-D-Olep-(1→4)-β-D-Cymp.	(*54, 130*)

Table 1 (*continued*)

Plant	Glycoside (Glycoside no) Molecular Formula mp°C $[\alpha]_D$	Genin	Sugar	References
Marsdenia condurango	Condurangoglycoside D_o (**167**) $C_{59}H_{88}O_{23}$ $4H_2O$ 183–188°C +13 5°	14β,20-Dihydroxycondurangogenin B hemiketal	-3-O-β-D-Glup-(1→4)-3Me-6d-β-D-Allop-(1→4)-β-D-Olep-(1→4)-β-D-Cymp	
"	20-O-Methyl-condurangoglycoside D_o (**168**) $C_{60}H_{90}O_{23}$ $4H_2O$ 180–190°C −8 76°	14β Hydroxy-20-O-methylcondurangogenin B hemiketal	-3-O-β-D-Glup-(1→4)-3Me-6d-β-D-Allop-(1→4)-β-D-Olep-(1→4)-β-D-Cymp	
"	20-Iso-O-methylcondurangoglycoside D_o (**169**) $C_{60}H_{90}O_{23}$ $4H_2O$ 168–173°C −19°	14β Hydroxy-20-iso-O-methylcondurangogenin B hemiketal	-3-O-β-D-Glup-(1→4)-3Me-6d-β-D-Allop-(1→4)-β-D-Olep-(1→4)-β-D-Cymp	
"	Condurangoglycoside E_{01} (**170**) $C_{66}H_{98}O_{26}$ – –	11α-O-Cinnamoyl-12β-O-acetyl-3β,8β,14β-trihydroxypregn-5-ene-20-one	-3-O-β-D-Glup-(1→4)-3Me-6d-β-D-Allop-(1→4)-β-D-Olep-(1→4)-β-D-Cymp-(1→4)-β-D-Cymp	(*54,232*)
"	Condurangoglycoside E_{02} (**171**) $C_{59}H_{86}O_{23}$ – –	11α-O-Cinnamoyl-12β-O-acetyl-3β,8β,14β-trihydroxypregn-5-ene-20-one	-3-O-β-D-Glup-(1→4)-3Me-6d-β-D-Allop-(1→4)-β-D-Olep-(1→4)-β-D-Cymp	

References pp 309–325

”	Condurangoglycoside E (**172**) $C_{53}H_{76}O_{18}$ 129–133°C +68.5°	Condurangogenin E	3-O-3Me-6d-β-D-Allop-(1→4)-β-D-Olep-(1→4)-β-D-Cymp.	*(54)*
”	Condurangoglycoside E_0 (**173**) $C_{59}H_{86}O_{23}$ 165–169°C +69°	Condurangogenin E	-3-O-β-D-Glup-(1→4)-3Me-6d-β-D-Allop-(1→4)-β-D-Olep-(1→4)-β-D-Cymp.	
”	Condurangoglycoside E_2 (**174**) $C_{60}H_{88}O_{21}$ 139–142°C +81.5°	Condurangogenin E	-3-O-3Me-6d-β-D-Allop-(1→4)-β-D-Olep-(1→4)-β-D-Cymp-(1→4)-β-D-Cymp.	
”	Condurangoglycoside E_3 (**175**) $C_{66}H_{98}O_{26}$ 168–172°C +68°	Condurangogenin E	-3-O-β-D-Glup-(1→4)-3Me-6d-β-D-Allop-(1→4)-β-D-Olep-(1→4)-β-D-Cymp-(1→4)-β-D-Cymp.	
”	Condurangoside A (**293**) $C_{46}H_{74}O_{17}\cdot 3H_2O$ Amorphous 24.5°	Gagaimogenin A	-3-O-3Me-6d-β-D-Allop-(1→4)-β-D-Olep-(1→4)-β-D-Cymp.	*(240)*
”	Condurangoside A_o (**294**) $C_{52}H_{84}O_{22}\cdot 3H_2O$ Amorphous 11.4°	Gagaimogenin A	-3-O-β-D-Glup-(1→4)-3Me-6d-β-D-Allop-(1→4)-β-D-Olep-(1→4)-β-D-Cymp.	
”	Condurangoside B (**295**) $C_{51}H_{76}O_{17}\cdot H_2O$ Amorphous 44.8°	Gagaimogenin B	-3-O-3Me-6d-β-D-Allop-(1→4)-β-D-Olep-(1→4)-β-D-Cymp.	

Table 1 (*continued*)

Plant	Glycoside (Glycoside no.) Molecular Formula mp°C $[\alpha]_D$	Genin	Sugar	References
Marsdenia condurango	Condurangoside C (**296**) $C_{51}H_{78}O_{17} \cdot H_2O$ Amorphous 32.0°	Gagaimogenin C	-3-O-3Me-6d-β-D-Allop-(1→4)-β-D-Olep-(1→4)-β-D-Cymp.	
"	Condurangoside B_o (**297**) $C_{57}H_{86}O_{22} \cdot 3H_2O$ Amorphous –	Gagaimogenin B	-3-O-β-D-Glup-(1→4)-3Me-6d-β-D-Allop-(1→4)-β-D-Olep-(1→4)-β-D-Cymp.	
"	Condurangoside C_o (**298**) $C_{57}H_{88}O_{22} \cdot 9/2H_2O$ Amorphous –	Gagaimogenin C	-3-O-β-D-Glup-(1→4)-3Me-6d-β-D-Allop-(1→4)-β-D-Olep-(1→4)-β-D-Cymp.	
"	Condurangoside D_{ol} (**299**) $C_{55}H_{90}O_{24} \cdot 2H_2O$ Amorphous –	Marsdenin	-3-O-β-D-Glup-(1→4)-3Me-6d-β-D-Allop-(1→4)-β-D-Olep-(1→4)-β-D-Cymp-(1→4)-β-D-Cymp.	
Marsdenia formosana	**MF-A (176)** $C_{35}H_{54}O_{10}$ 196–198°C +43.5°	Dehydrotomentosin	-3-O-β-D-Cymp.	(*260*)
"	MF-C (**177**) $C_{28}H_{44}O_8$ 245-248°C –	Pergularin	-3-O-β-D-Cymp.	

”	MF-D (**178**) $C_{28}H_{46}O_8$ 249–252°C –	Utendin	-3-O-β-D-Cymp.	
”	Marsformosadin-3-O-β-D-cymaropyranoside (**179**) $C_{39}H_{48}O_{10}$ – –	Marsformosadin	-3-O-β-D-Cymp.	(*261*)
”	Marsformoside (**180**) $C_{41}H_{64}O_{14}$ – –	12β-O-Tigloyl-20-O-acetylpregn-5-ene-3β, 14β,17-triol	-3-O-β-D-Quip-(1→4)-β-D-Cymp.	
”	Deacetyl marsformoside (**181**) $C_{39}H_{62}O_{13}$ – –	12β-O-Tigloyl-pregn-5-ene-3β, 14β, 17, 20-tetrol	-3-O-β-D-Quip-(1→4)-β-D-Cymp.	
Marsdenia incisa	Neomarinogenin (**182**) $C_{43}H_{72}O_{16}$ – –	3β, 5β, 14β, 17β, 20-Pentahydroxypregn-7β-al	-3-O-3Me-6d-β-D-Glup-(1→4)-β-D-Cymp-(1→4)-β-D-Cymp.	(*262*)
Marsdenia koi	Marsdekoiside A (**183**) $C_{51}H_{78}O_{17}$ 166–168°C +22.7°	12β-O-Cinnamoyl-dihydrosarcostin	-3-O-3Me-6d-β-D-Allop-(1→4)-β-D-Olep-(1→4)-β-D-Cymp.	(*17, 239*)
”	Marsdekoiside B (**184**) $C_{49}H_{76}O_{17}$ – –	12β-O-Benzoyldihydrosarcostin	-3-O-3Me-6d-β-D-Allop-(1→4)-β-D-Olep-(1→4)-β-D-Cymp.	(*239*)

Table 1 (*continued*)

Plant	Glycoside (Glycoside no.) Molecular Formula mp°C $[\alpha]_D$	Genin	Sugar	References
Marsdenia koi	Marsdekoiside D(**185**) $C_{42}H_{72}O_{16}$ – –	Dihydrosarcostin	-3-O-3Me-6d-β-D-Allop-(1→4)-β-D-Olep-(1→4)-β-D-Cymp.	(*263*)
"	Marsdekoiside E (**186**) $C_{51}H_{78}O_{16}$ – –	20-O-Cinnamoyl-dihydrosarcostin	-3-O-3Me-6d-β-D-Allop-(1→4)-β-D-Olep-(1→4)-β-D-Cymp.	(*264*)
Marsdenia oreophila	Marsdeoreophiside A (**187**) $C_{48}H_{82}O_{21}$ – –	Dihydrosarcostin	-3-O-β-D-Glup-(1→4)-3Me-6d-β-D-Allop-(1→4)-β-D-Olep-(1→4)-β-D-Cymp.	(*265*)
Marsdenia tenacissima	Tenacissoside A (**188**) $C_{48}H_{74}O_{19}$ 139.5–140.5°C −16.3°	Tenacigenin B-I	-3-O-β-D-Glup-(1→4)-3Me-6d-β-D-Allop-(1→4)-β-D-Olep.	(*266*)
"	Tenacissoside B (**189**) $C_{51}H_{78}O_{19}$ 132.5–134.5°C +11°	Tenacigenin B-II	-3-O-β-D-Glup-(1→4)-3Me-6d-β-D-Allop-(1→4)-β-D-Olep.	
"	Tenacissoside C (**190**) $C_{53}H_{76}O_{19}$ 128–132.5°C +16.3°	Tenacigenin B-III	-3-O-β-D-Glup-(1→4)-3Me-6d-β-D-Allop-(1→4)-β-D-Olep.	

"	Tenacissoside D (**191**) $C_{51}H_{80}O_{19}$ 137–140 5°C +16 4°	Tenacigenin B-IV	-3-O-β-D-Glup-(1→4)-3Me-6d-β-D-Allop-(1→4)- β-D-Olep	
"	Tenacissoside E (**192**) $C_{53}H_{78}O_{19}$ 140 5–142 5°C +26 2°	Tenacigenin B-V	-3-O-β-D-Glup-(1→4)-3Me-6d-β-D-Allop-(1→4)- β-D-Olep	
Metaplexis hemsleyana	Hemoside (**193**) $C_{35}H_{42}O_{10}$ – –	12β-O-Benzoyldea- cetylmetaplexigenin	-3-O-β-D-Cymp	*(267)*
Orthenthera viminea	Orthenine (**194**) $C_{58}H_{88}O_{19}$ 120–124°C +115°	12β-O-Cinnamoylsar- costin	-3-O-α-L-Olep-(1→4)-α-L-Olep-(1→4)-α-L-Olep- (1→4)-β-D-Cymp	*(217)*
"	Orine (**195**) $C_{46}H_{58}O_{11}$ 58–62°C +85 14°	12, 20-Di-O-cinnamoyl- sarcostin	-3-O-β-D-Cymp	*(268)*
"	Ornine (**196**) $C_{53}H_{70}O_{14}$ 124°C +141 3°	12, 20-Di-O-cinnamoyl- sarcostin	-3-O-α-L-Olep-(1→4)-β-D-Cymp	*(196)*
Oxystelma esculentum	Oxystine (**197**) $C_{57}H_{84}O_{20}$ 145–150°C −10 9°	12β-O-Cinnamoyl- deacylmetaplexigenin	-3-O-β-D-Cymp-(1→4)-β-D-Thevp-(1→4)-β-D- Cymp-(1→4)-β-D-Dgxp	*(269)*

Table 1 (*continued*)

Plant	Glycoside (Glycoside no.) Molecular Formula mp°C $[\alpha]_D$	Genin	Sugar	References
Oxystelma esculentum	Oxysine (**198**) $C_{48}H_{80}O_{16}$ 120–122°C −17.5°	Calogenin	-3-O-β-D-Olep-(1→4)-β-D-Thevp-(1→4)-β-D-Cymp-(1→4)-β-D-Dgxp.	(*60*)
”	Esculentin (**199**) $C_{42}H_{68}O_{17}$ 118–120°C +5°	Sarcogenin	-3-O-β-D-Thevp-(1→4)-β-D-Cymp-(1→4)-β-D-Olep.	(*62*)
Pergularia pallida	Pallidine (**200**) $C_{42}H_{54}O_{11}$ 108–112°C +20°	12,20-Di-O-benzoylsar-costin	-3-O-β-D-Olep.	(*270*)
”	Pallidinine (**201**) $C_{49}H_{66}O_{14}$ 118–122°C +88°	12,20-Di-O-benzoylsar-costin	-3-O-β-D-Cymp-(1→4)-β-D-Olep.	
Periploca calophylla	Calocin (**202**) $C_{27}H_{44}O_6$ 243–247°C −60.7°	Δ^5-Pregnene-3β,14β,20-triol	-3/20-β-D-Canp.	(*271*)
”	Plocin (**203**) $C_{49}H_{66}O_{13}$ 148–150°C +40°	12,20-Di-O-benzoyldre-vogenin D	-3-O-β-D-Olep-(1→4)-β-D-Olep.	(*272*)

"	Plocinine (**204**) $C_{53}H_{70}O_{14}$ 144–148°C +37°	12,20-Di-O-cinnamoyl-sarcostin	-3-O-α-L-Olep-(1→4)-α-L-Olep.	(*57*)
"	Locin (**205**) $C_{27}H_{44}O_7$ 110–115°C +20°	Boucerin	-3-O-β-D-Dgxp.	(*192*)
"	Calocinin (**206**) $C_{27}H_{44}O_6$ 250–255°C +16°	Calogenin	-3-O-2d-β-L-Fucp.	(*59*)
Periploca sepium	Glycoside K (**207**) $C_{40}H_{66}O_{24}$ 240–241°C −27.58°	Δ^5-Pregnene-3β,20α-diol	-20-O-β-D-Glup-(1→6)-β-D-Glup-(1→2)-β-D-Dgtp.	(*273, 274*)
"	Glycoside H_1(**208**) $C_{56}H_{92}O_{24}$ 182°C −22.83°	Δ^5-Pregnene-3β,20α-diol	-3-O-2Ac-β-D-Dgtp-(1→4)-β-D-Cymp-20-O-β-D-Glup-(1→6)-β-D-Glup-(1→2)-β-D-Dgtp.	(*275*)
"	Glycoside E_1(**209**) $C_{27}H_{44}O_6$ 239–240°C −69.9°	Δ^5-Pregnene-3β,17α,20α-triol	-20-O–β-D-Canp.	(*276*)
"	Glycoside H_2 (**210**) $C_{56}H_{92}O_{25}$ 191–192°C −25.9°	Δ^5-Pregnene-3β,16α,20α-triol	-3-O-2Ac-β-D-Dgtp-(1→4)-β-D-Cymp-20-O-β-D-Glup-(1→6)-β-D-Glup-(1→2)-β-D-Dgtp.	(*277*)

Table 1 (*continued*)

Plant	Glycoside (Glycoside no.) Molecular Formula mp°C $[\alpha]_D$	Genin	Sugar	References
Periploca sepium	Periploside A (**211**) $C_{65}H_{106}O_{24}$ 225–226°C +15.7°	Δ^5-Pregnene-3β,17α, 20α-triol	-20-O-2Ac-β-D-Dgtp-(1→4)-β-D-Cymp-(1→4)-β-D-Cymp-(1→4)-β-D-Cymp-(1→4)-D-Olep-{(1→3)(-CH_2O-)}(1→4)-β-D-Canp.	(*228*)
”	Periploside B (**212**) $C_{34}H_{52}O_9$ 146–147°C −71.6°	Δ^5-Pregnene-3β,17α, 20α-triol	-3-O-4′,6′-did-3′Me-$\Delta^{3'}$-D-2′Hex-20-O-β-D-Canp.	
”	Periploside C (**213**) $C_{72}H_{114}O_{27}$ 194–195°C +1.3°	Δ^5-Pregnene-3β,17α, 20α-triol	-3-O-4′,6′-did-3′Me-$\Delta^{3'}$-D-2′Hex-20-O-2Ac-β-D-Dgtp-(1→4)-β-D-Cymp-(1→4)-β-D-Cymp-(1→4)-β-D-Cymp-(1→4)-D-Olep-{(1→3)(-CH_2O-)}(1→4)-β-D-Canp.	
”	S-4a (**214**) $C_{56}H_{92}O_{25}$ 182–184°C −16.24°	Δ^5-Pregnene-3β,16β, 20(R)-triol	-3-O-2Ac-β-D-Dgtp-(1→4)-β-D-Cymp-20-O-β-D-Glup-(1→6)-β-D-Glup-(1→2)-β-D-Dgtp.	(*187*)
”	S-5 (**215**) $C_{54}H_{90}O_{27}$ 175–177°C −25.22°	Δ^5-Pregnene-3β, 20(S)-diol	-3-O-β-D-Dgtp-(1→4)-β-D-Cymp-20-O-β-D-Glup-(1→6)-β-D-Glup-(1→2)-β-D-Dgtp.	
”	S-10 (**216**) $C_{40}H_{66}O_{17}$ 167–169°C −2.6°	Δ^5-Pregnene-3β,16β, 20(R)-triol	-20-O-β-D-Glup-(1→6)-β-D-Glup-(1→2)-β-D-Dgtp.	
”	Periplocoside A (**217**) $C_{72}H_{114}O_{27}\cdot 2H_2O$	Δ^5-Pregnene-3β,17α, 20(S)-triol	-3-O-4′,6′-did-3′Me-$\Delta^{3'}$-D-2′Hex-20-O-2Ac-β-D-Dgtp-(1→4)-β-D-Cymp-(1→4)-β-D-Cymp-(1→4)-	(*162*)

	174-176°C −1.2°		β-D-Cymp-(1→5)-3, 7-did-4Me-α-D-gluco-2-Hepp-(2→4)-dioxy-(1→3)-β-D-Canp.	
"	Periplocoside B (**218**) $C_{56}H_{88}O_{19}$ 136–138°C +1.9°	Δ^5-Pregnene-3β, 17α, 20(S)-triol	-3-O-4′, 6′-did-3′Me-Δ^3-D-2′ Hex-20-O–β-D-Cymp-(1→4)-β-D-Cymp-(1→5)-3, 7-did-4Me-α-D-gluco-2-Hepp-(2→4)-dioxy-(1→3)-β-D-Canp.	
"	Periplocoside C (**219**) $C_{49}H_{76}O_{16}$ 180–182°C −8.4°	Δ^5-Pregnene-3β, 17α, 20(S)-triol	-3-O-4′, 6′-did-3′Me-Δ^3-D-2′ Hex-20-O-β-D-Cymp-(1→5)-3,7-did-4Me-α-D-gluco-2-Hepp-(2→4)-dioxy-(1→3)-β-D-Canp.	
"	Periplocoside D (**220**) $C_{70}H_{112}O_{26}$ 191–193°C −3.08°	Δ^5-Pregnene-3β, 17α, 20(S)-triol	-3-O-4′, 6′-did-3′Me-$\Delta^{3'}$-D-2′ Hex-20-O-β-D-Dgtp-(1→4)-β-D-Cymp-(1→4)-β-D-Cymp-(1→4)-β-D-Cymp-(1→5)-3,7-did-4Me-α-D-gluco-2-Hepp-(2→4)-dioxy-(1→3)-β-D-Canp.	*(69)*
"	Periplocoside E (**221**) $C_{65}H_{106}O_{24}$ 183–185°C −7.5°	Δ^5-Pregnene-3β, 17α, 20(S)-triol	-20-O-2Ac-β-D-Dgtp-(1→4)-β-D-Cymp-(1→4)-β-D-Cymp-(1→4)-β-D-Cymp-(1→5)-3, 7-did-4Me-α-D-gluco-2-Hepp-(2→4)-dioxy-(1→3)-β-D-Canp.	
"	Periplocoside L (**222**) $C_{28}H_{48}O_7$. 238–240°C −53.3°	Δ^5-Pregnene-3β, 17α, 20(S)-triol	-3-O-β-D-Dgtp.	
"	Periplocoside M (**223**) $C_{34}H_{52}O_9$. 195–197°C −89.91°	Δ^5-Pregnene-3β, 17α, 20(S)-triol	-3-O-4′, 6′-did-3′Me-$\Delta^{3'}$-D-2′ Hex-20-O-β-D-Canp.	
"	Periplocoside J (**224**) $C_{61}H_{100}O_{23}$ 178–181°C +24.13°	Δ^5-Pregnene-3β, 17α, 20(S)-triol	-20-O-β-D-Dgtp-(1→4)-β-D-Cymp-(1→4)-β-D-Canp-(1→4)-β-D-Dgxp-(1→5)-3, 7-did-4Me-α-D-gluco-2-Hepp-(2→4)-dioxy-(1→3)-β-D-Canp.	*(31)*

Table 1 (*continued*)

Plant	Glycoside (Glycoside no.) Molecular Formula mp°C $[\alpha]_D$	Genin	Sugar	References
Periploca sepium	Periplocoside K (**225**) $C_{68}H_{108}O_{26}$ 208–212°C −4.76°	Δ^5-Pregnene-3β,17α, 20(S)-triol	-3-O-4′,6′-did-3′Me-Δ^3-D-2′ Hex-20-O-β-D-Dgtp-(1→4)-β-D-Cymp-(1→4)-β-D-Canp-(1→4)-β-D-Dgxp-(1→5)-3,7-did-4Me-α-D-gluco-2-Hepp-(2→4)-dioxy-(1→3)-β-D-Canp.	
"	Periplocoside F (**226**) $C_{63}H_{104}O_{23}$ 195–198°C +8.1°	Δ^5-Pregnene-3β,17α, 20(S)-triol	-20-O-β-D-Dgtp-(1→4)-β-D-Cymp-(1→4)-β-D-Cymp-(1→4)-β-D-Cymp-(1→5)-3,7-did-4Me-α-D-gluco-2-Hepp-(2→4)-dioxy-(1→3)-β-D-Canp.	
"	Periplocoside O (**227**) $C_{36}H_{56}O_{10}$ 103–106°C +84.0°	Δ^5-Pregnene-3β,17α, 20(S)-triol	-3-O-4′, 6′-did-3′Me-Δ^3-D-2′ Hex-20-O-3MeMe-β-D-Canp.	
Sarcostemma brevistigma	Brevinine (**228**) $C_{42}H_{60}O_{14}$ 260–262°C +27°	11-O-Benzoylsar-cogenin	-3-O-α-L-Digp-(1→4)-α-L-Digp.	(*278*)
"	**Brevine (229)** $C_{49}H_{72}O_{17}$ 100–105°C + 21.2°	11-O-Benzoylsar-cogenin	-3-O-α-L-Digp-(1→4)-α-L-Digp-(1→4)-α-L-Digp.	(*173*)
Sarcostemma viminale	Sarcovimiside A (**230**) $C_{49}H_{72}O_{16}\cdot 2.5H_2O$ 137–140°C –	Cynanforidine	-3-O-α-L-Cymp-(1→4)-β-D-Cymp-(1→4)-β-D-Cymp.	(*137*)

”	Sarcovimiside B (**231**) $C_{55}H_{74}O_{18} \cdot 2H_2O$ 129–132°C –	(20S)-12β, 20-Dibenzoyloxy-3β, 5, 17-trihydroxy-8, 14-seco-5β, 17α-pregn-6-ene-8, 14-dione	-3-O-α-L-Cymp-(1→4)-β-D-Cymp-(1→4)-β-D-Dgtp.	
”	Sarcovimiside C (**232**) $C_{61}H_{84}O_{23} \cdot 4H_2O$ 158–160°C –	(20S)-12β, 20-Dibenzoyloxy-3β, 5, 17-trihydroxy-8, 14-seco-5β, 17α-pregn-6-ene-8, 14-dione	-3-O-β-D-Glup-(1→4)-α-L-Cymp-(1→4)-β-D-Cymp-(1→4)-β-D-Dgtp.	
		Family Compositae		
Carthamus tinctorius	–(**233**) $C_{33}H_{52}O_{13}$ – –	15α, 20-Dihydroxy-Δ^4-pregnen-3-one	-20-O-β-D-Glup-(1→4)-β-D-Glup.	(*4*)
		Family Liliaceae		
Paris polyphylla	–(**234**) $C_{39}H_{60}O_{15}$ 260–262°C −72.2°	Pregn-5,16-diene-3β-hydroxy-20-one	-3-O-α-L-Rhap-(1→2)-[α-L-Rhap-(1→4)-β-D-Glup].	(*216*)
		Family Meliaceae		
Melia toosendan	Toosendanoside (**235**) $C_{27}H_{46}O_9$ 265.5–268.5°C −8.1°	5α-Pregnane-2α, 3α, 16β, 20(R)-tetrol	-2-O-β-D-Glup	(*6*)

Table 1 (*continued*)

Plant	Glycoside (Glycoside No.) Molecular Formula mp°C $[\alpha]_D$	Genin	Sugar	References
		Family Moraceae		
Streblus asper	Sioraside (**236**) $C_{28}H_{46}O_8$ 214–217°C −1.3°	3β, 14β-Dihydroxypregn-20-one	-3-O-3Me-β-D-Glup.	(*9*)
		Family Palmae		
Sabal causiarum	Causiaroside II (**237**) $C_{57}H_{92}O_{28}$ 172–175°C −47.2°	3β, 16β-Dihydroxypregn-5-en-20-one	-3-O-[α-L-Rhap-(1→4)-β-D-(1→4)] [α-L-Rhap-(1→2)]-β-D-Glup-16-O-[δ-(β-D-Glup-oxy)-γ-Me] valerate.	(*23*)
		Animal Sources		
Eunicella verrucosa	Verrucoside (**238**) $C_{30}H_{48}O_7$ – −30°	5β-H-Pregn-20-ene-3β, 4β-diol	-4-O-2Ac-α-L-Dgtp.	(*7*)
Menduca sexta	–(**239**) $C_{39}H_{64}O_{17}$ (205–209°C) 265–279°C –	3β, 20(R)-Dihydroxy-pregn-5-ene	-3-O-β-D-Glup-(1→2)-β-D-Glup-20-O-β-D-Glup.	(*198*)

”	–(**240**) $C_{33}H_{54}O_{12}$ – –	3β, 20(R)-Dihydroxy, pregn-5-ene	-3-O-β-D-Glup-20-O-β-D-Glup.	
Pseudoplexaura wagenaari	–(**241**) $C_{27}H_{42}O_6$ 268–270°C –	3β-Hydroxypregna-5, 20-diene	-3-O-β-D-Galp.	(*113*)

Allo = Allose; Bov = Boivinose; Can = Canarose; Cym = Cymarose; Dgt = Digitalose; Dgx = Digitoxose; Dig = Diginose; Fuc = Fucose; Gal = Galactose; Glu = Glucose; Hepp = Heptulose; Hex = Hexosulose; Hol = Holosamine; Ole = Oleandrose; Oli = Olivose; Qui = Quinovose; Rha = Rhamnose; Sar = Sarmentose; Thev = Thevetose; 2d = 2deoxy; 6d = 6deoxy; did = dideoxy; 2Ac = 2-O-Acetyl; 4Ac = 4-O-Acetyl; 3Me = 3-O-Methyl; 4Me = 4-O-Methyl; 6Sin = 6-O-Sinapoyl

Table 2. *Structures of Pregnane Glycosides*

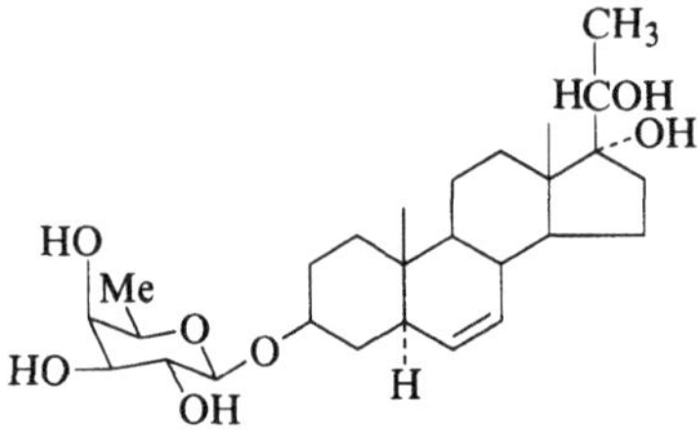

(1)

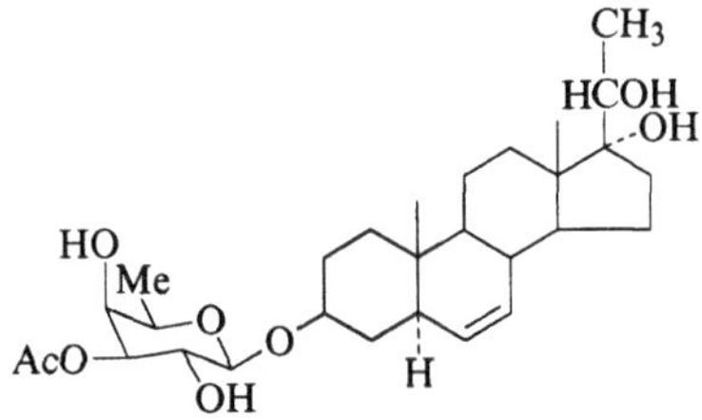

(2)

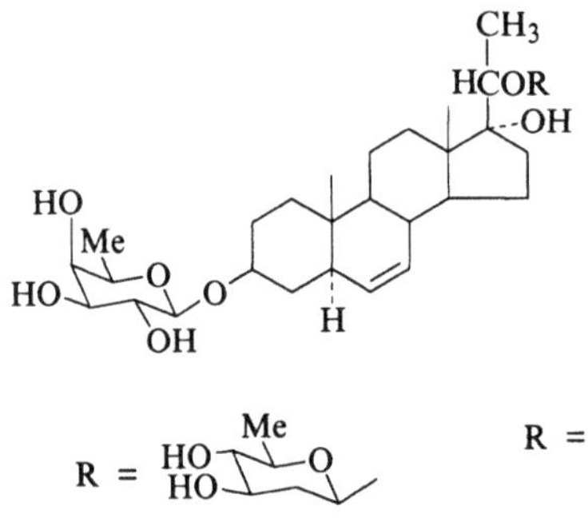

(3)

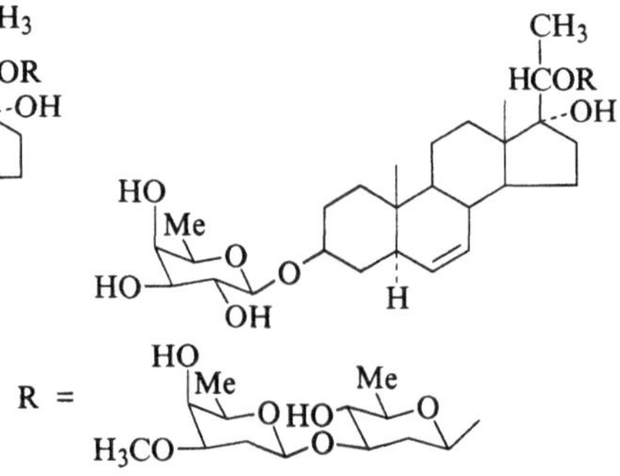

(4)

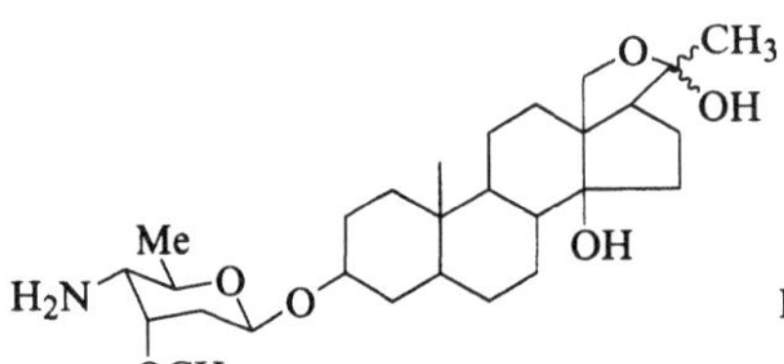

(5)

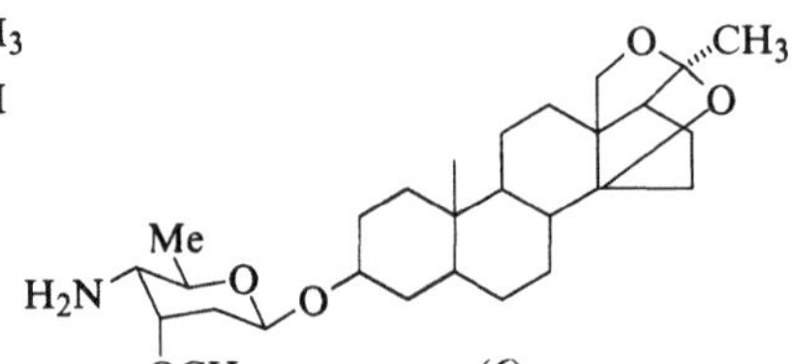

(6)

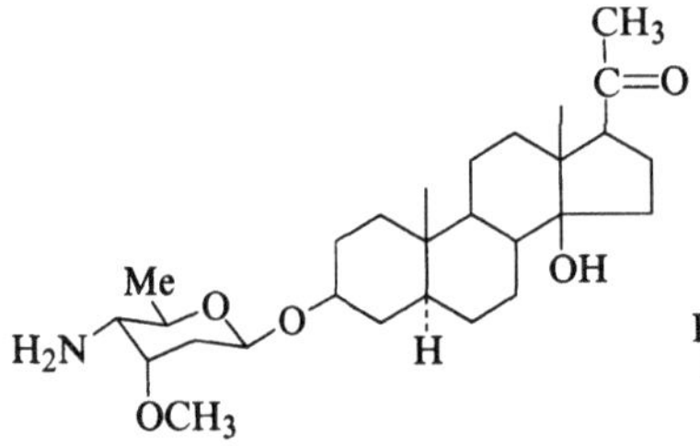

(7)

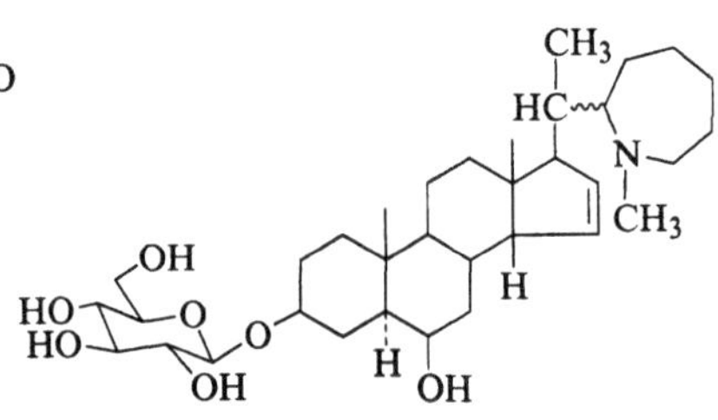

(8)

Table 2 (*continued*)

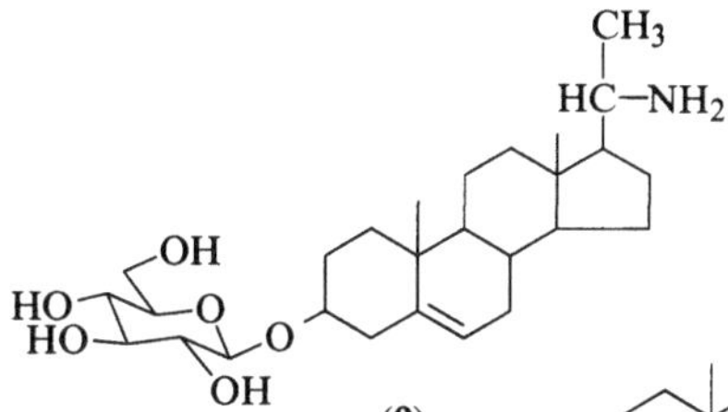

(**9**)

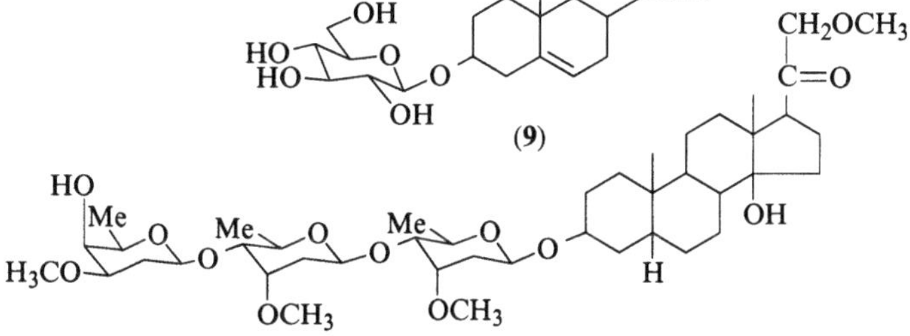

(**10**)

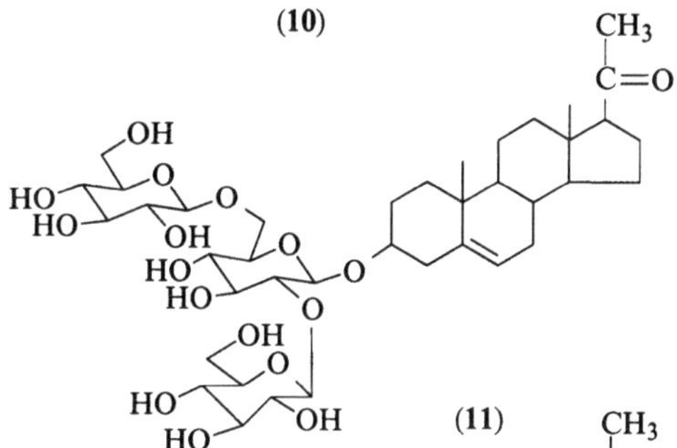

(**11**)

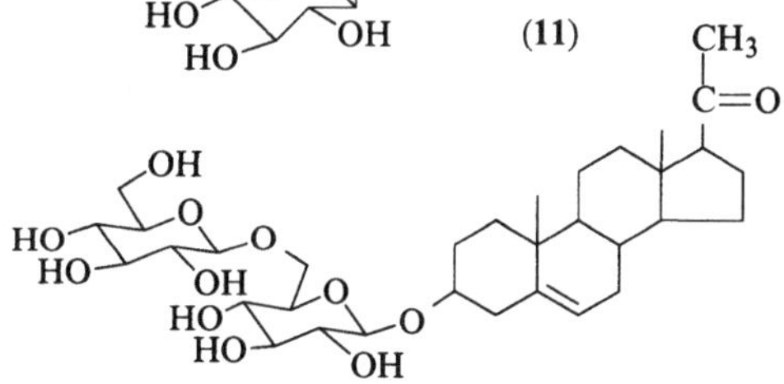

(**12**)

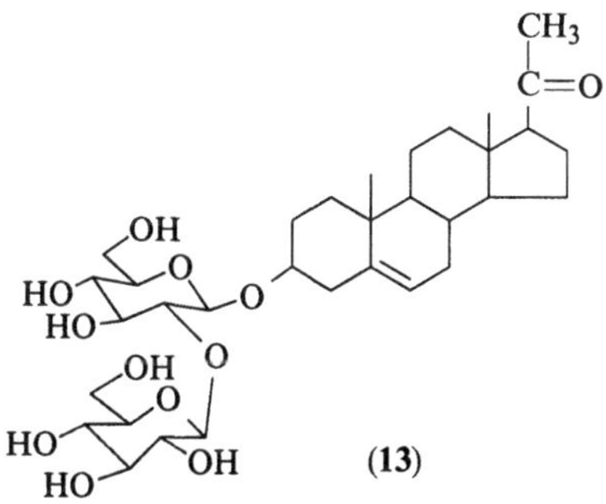

(**13**)

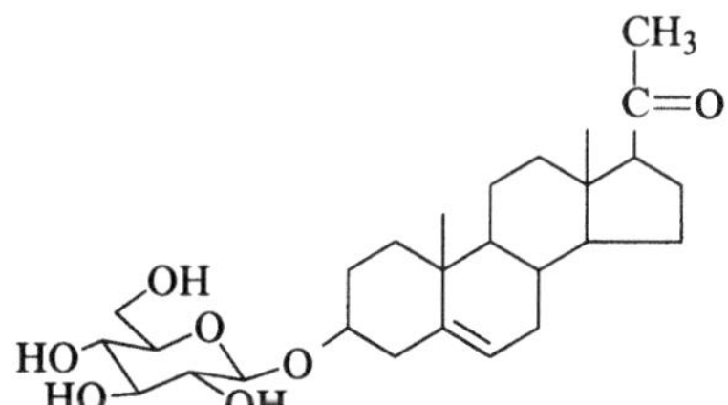

(**14**)

Table 2 (*continued*)

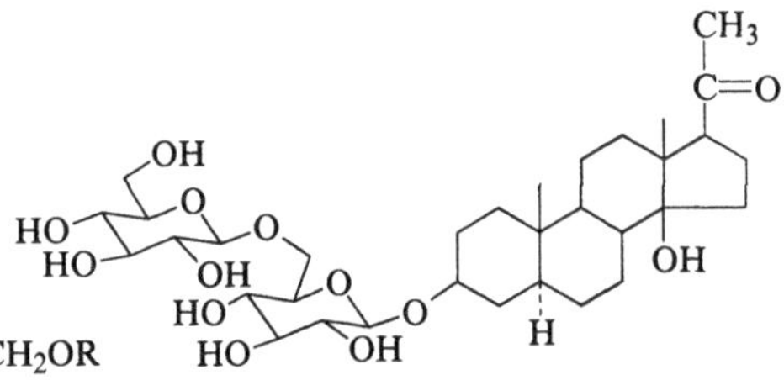

(**15**)

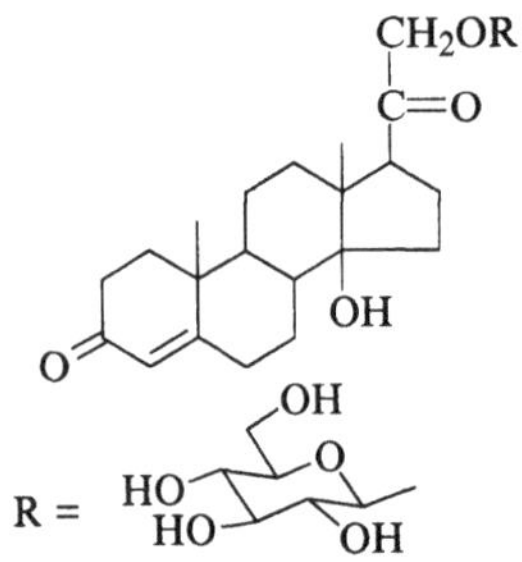

(**16**)

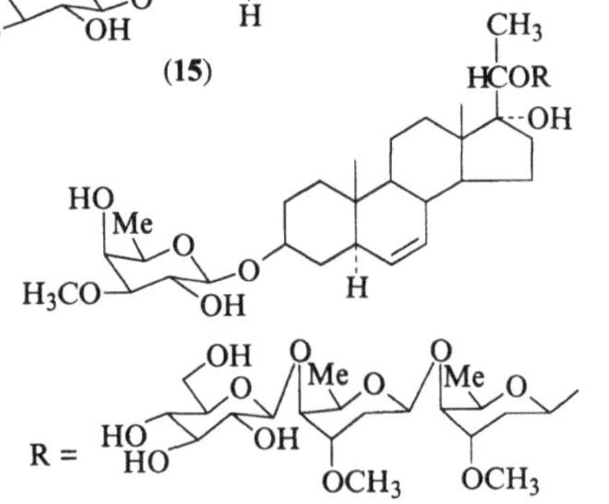

(**17**)

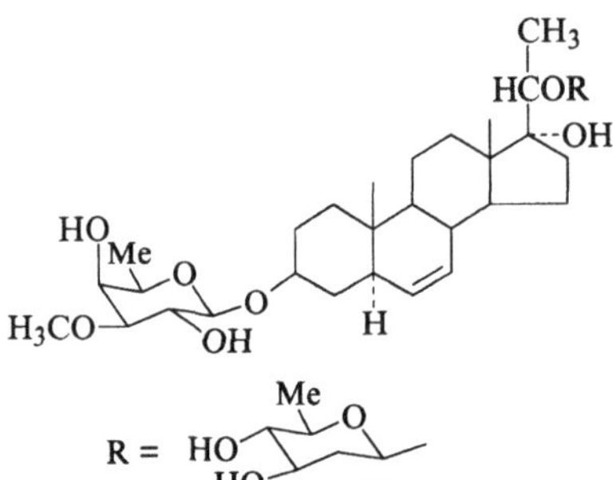

(**18**)

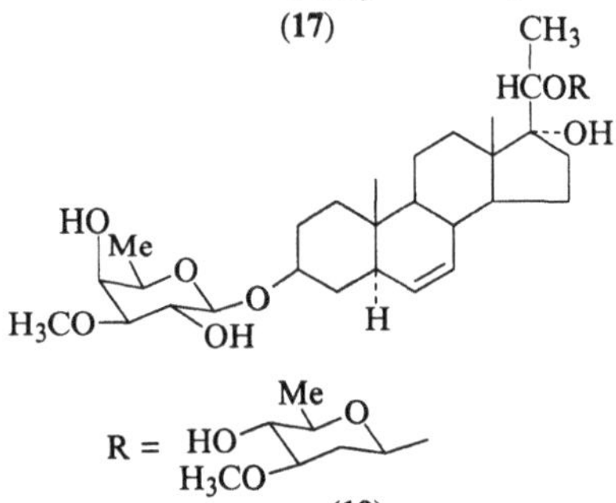

(**19**)

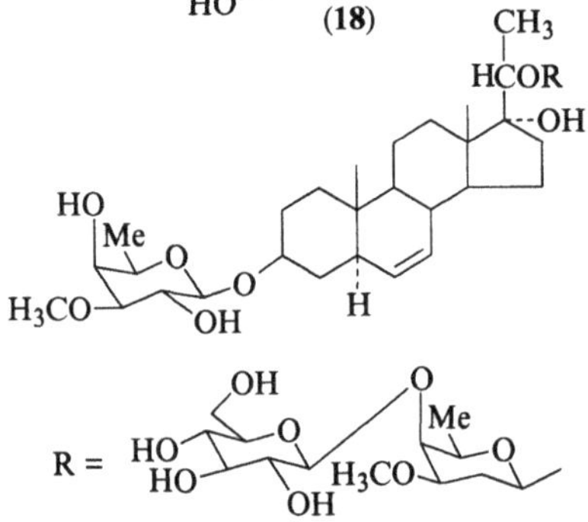

(**20**)

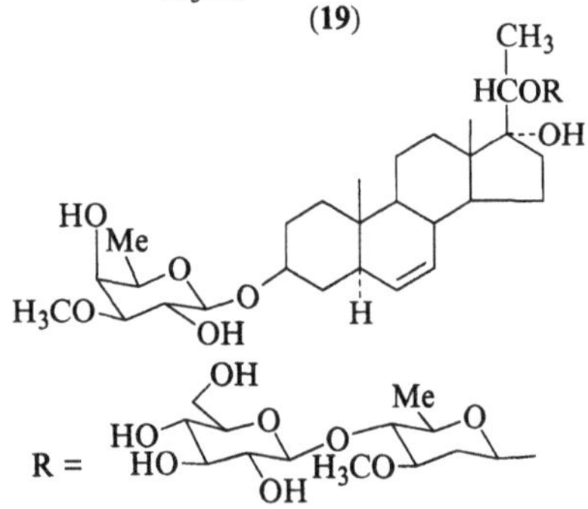

(**21**)

References, pp. 309–325

Table 2 (*continued*)

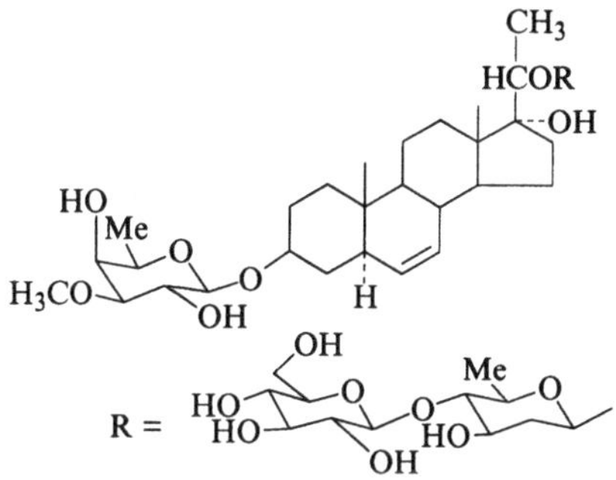

(**22**)

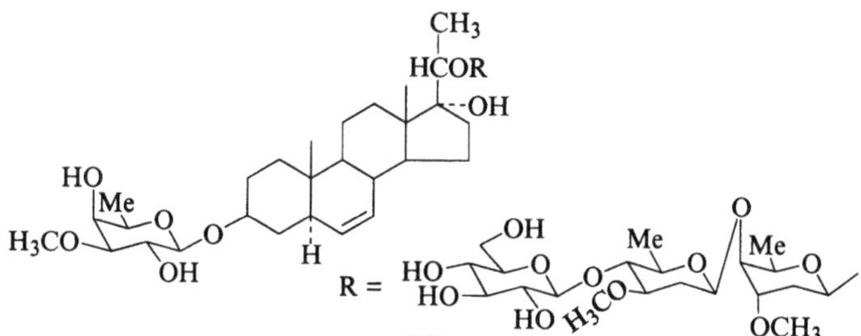

(**23**)

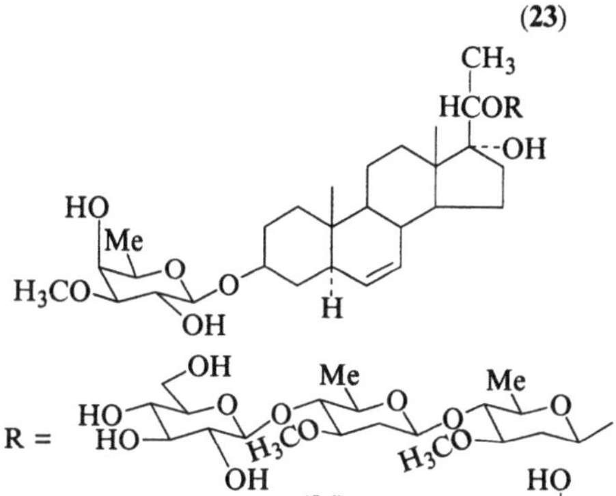

(**24**)

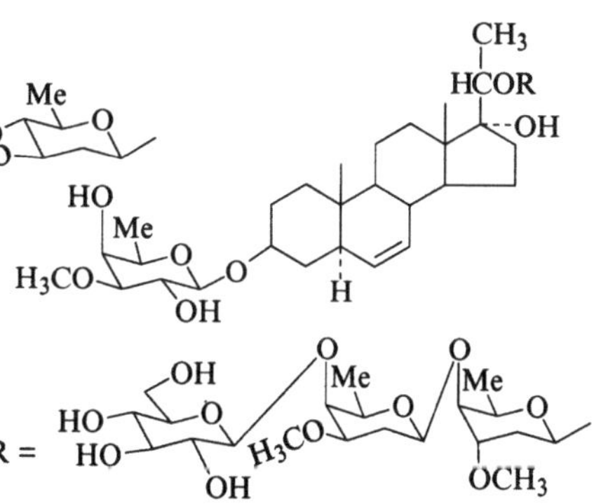

(**25**)

Table 2 (*continued*)

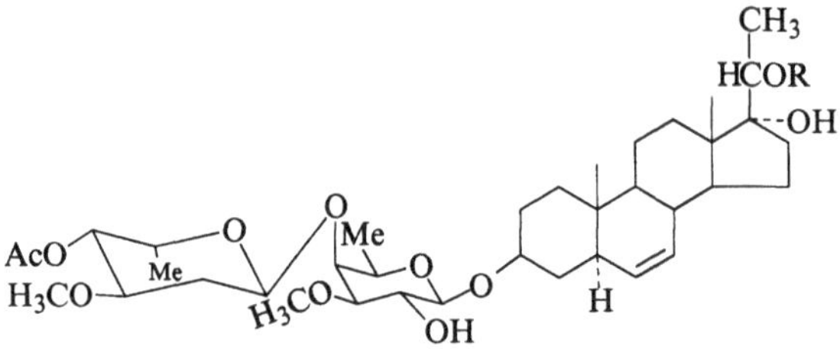

(**26**)

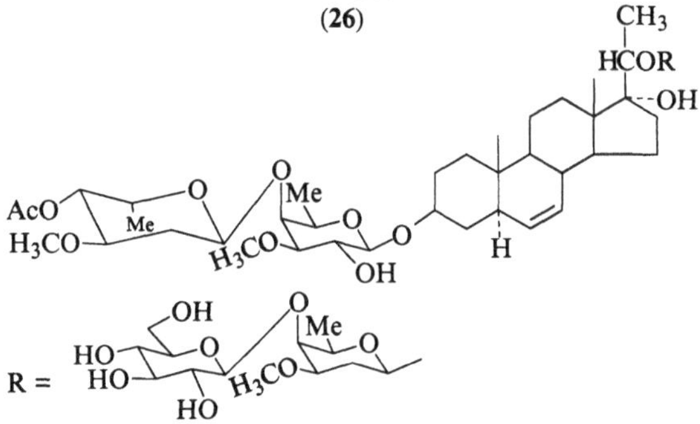

(**27**)

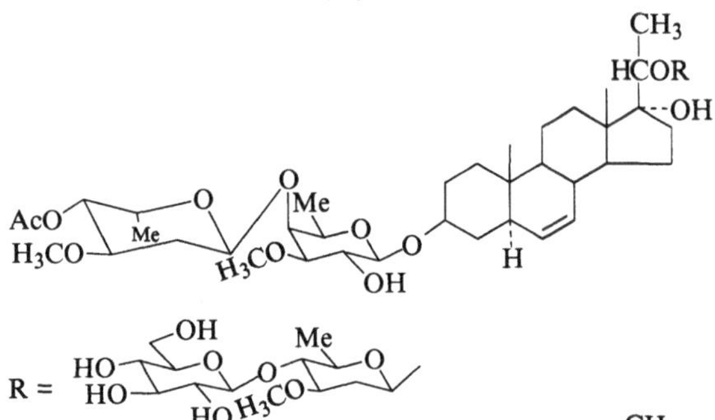

(**28**)

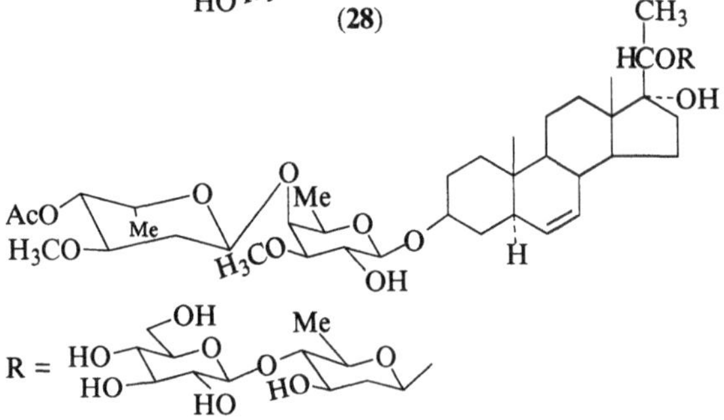

(**29**)

Table 2 (*continued*)

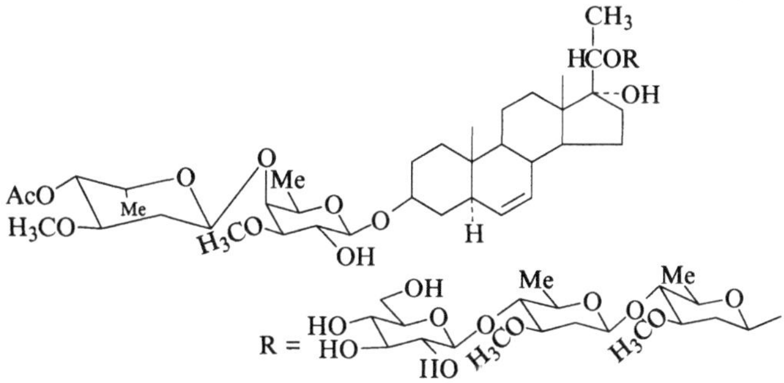

(**30**)

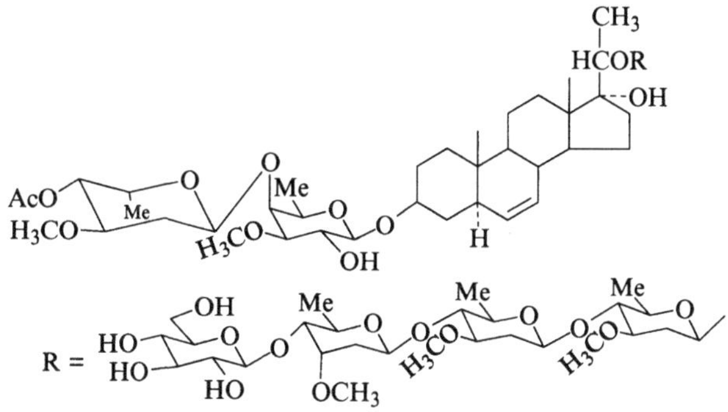

(**31**)

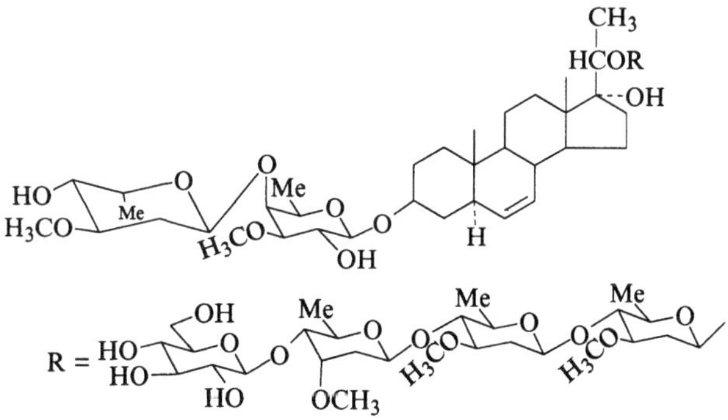

(**32**)

Table 2 (*continued*)

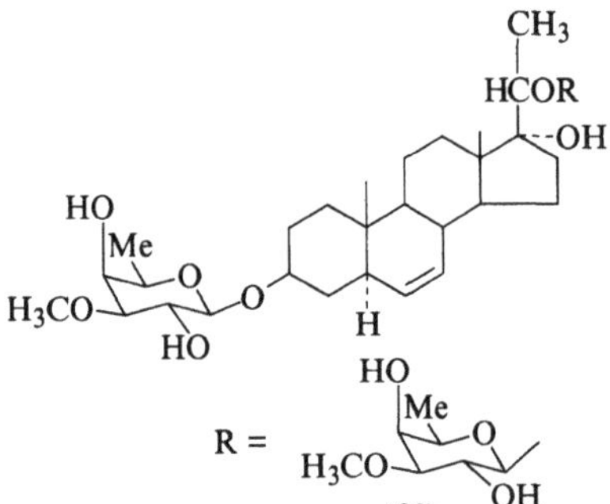

(33)

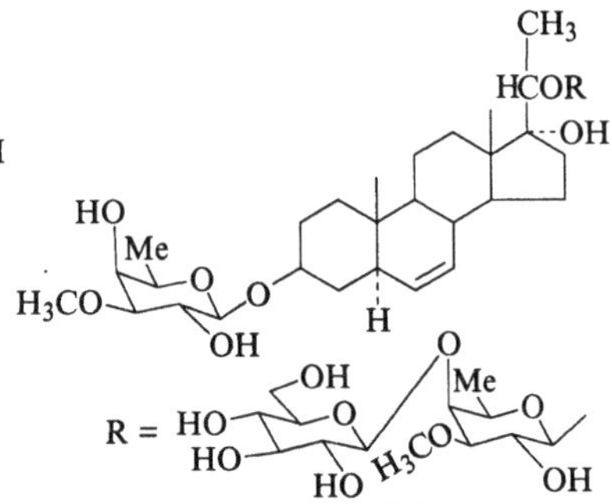

(34)

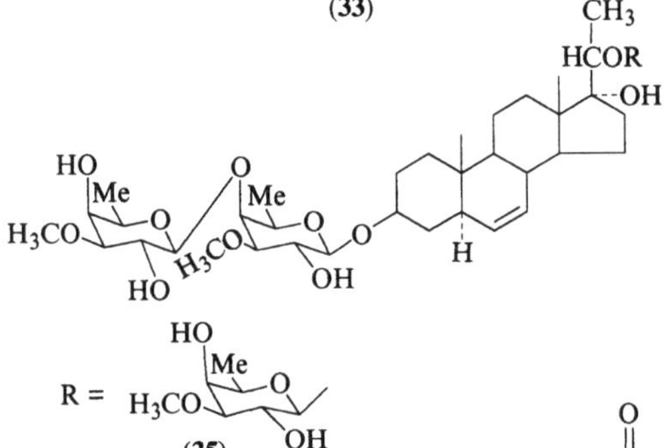

(35)

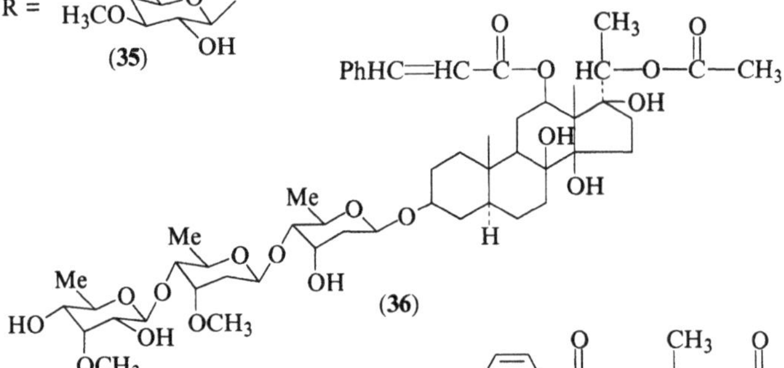

(36)

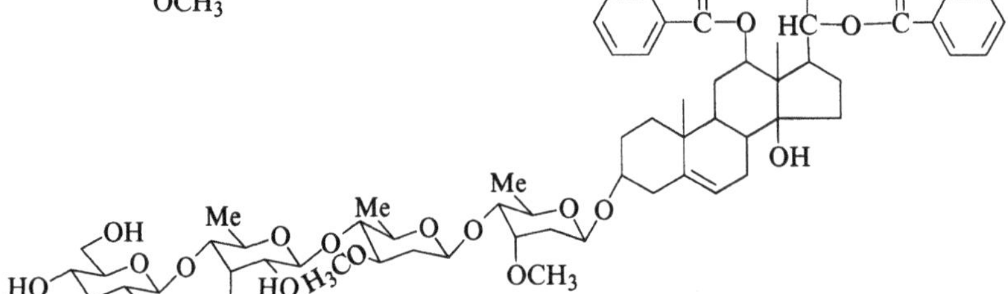

(37)

Table 2 (*continued*)

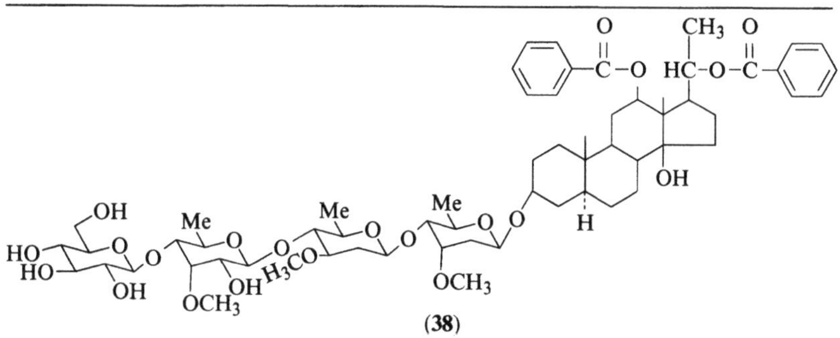

(**38**)

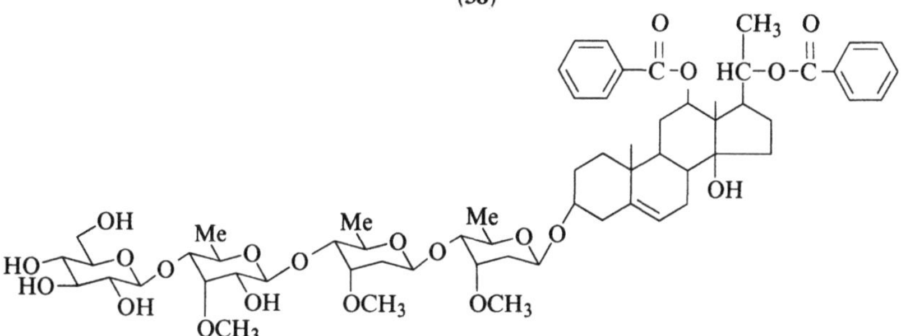

(**39**)

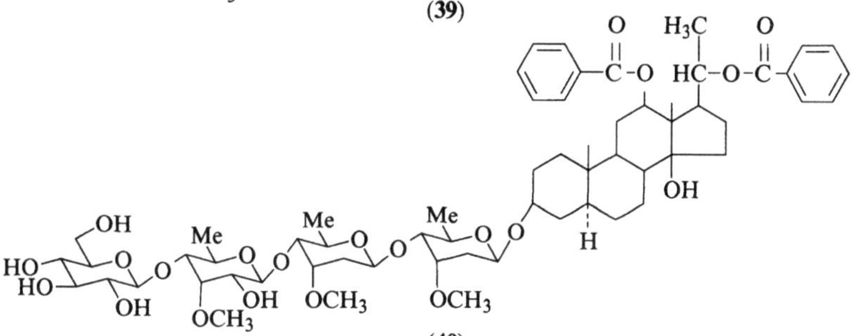

(**40**)

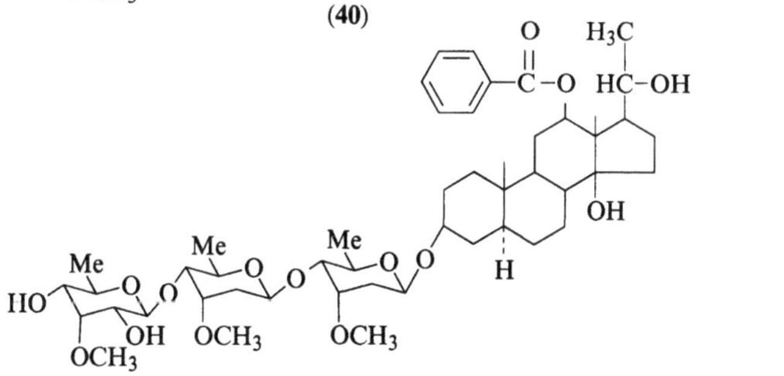

(**41**)

Table 2 (*continued*)

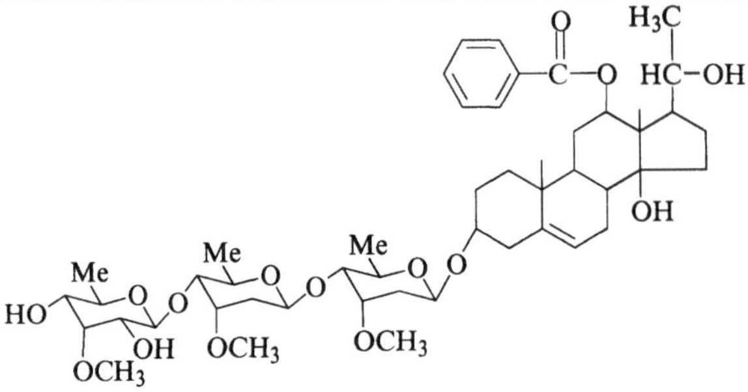

(**42**)

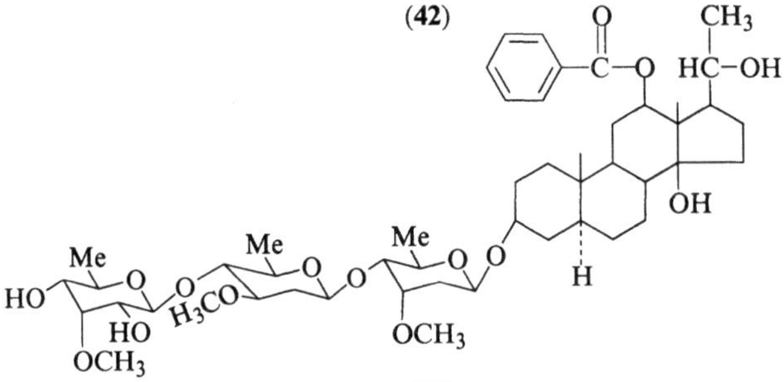

(**43**)

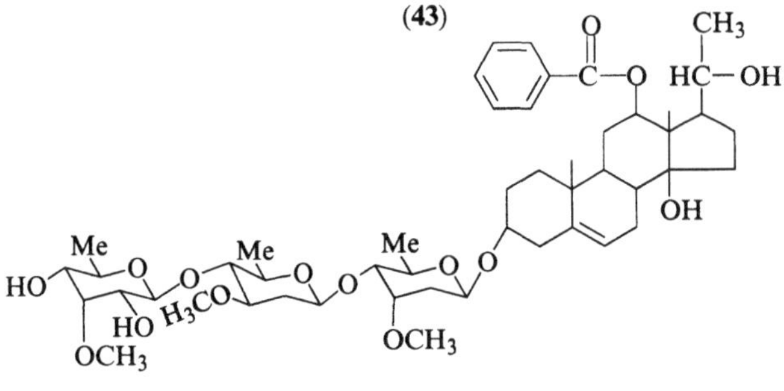

(**44**)

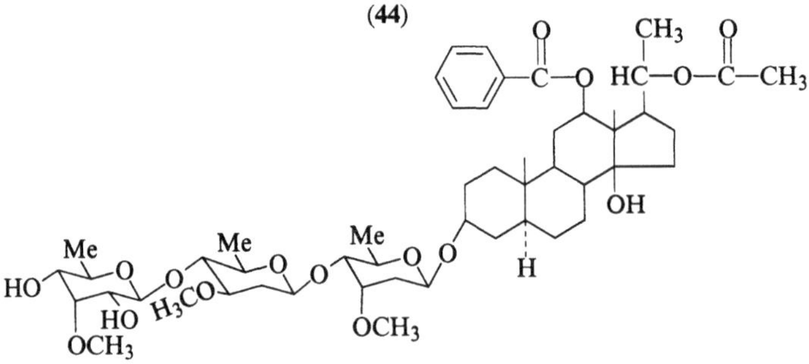

(**45**)

References, pp. 309–325

Table 2 (*continued*)

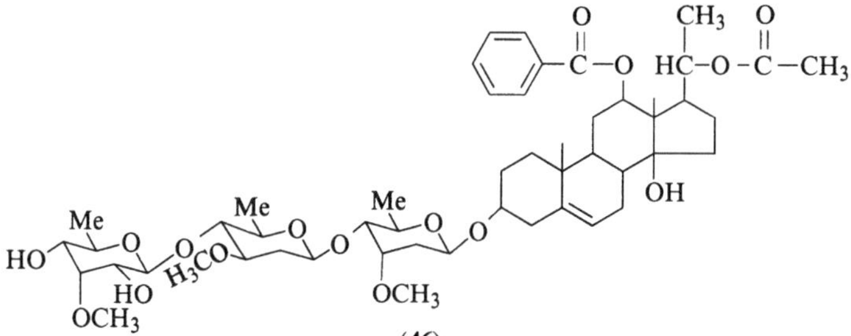

(**46**)

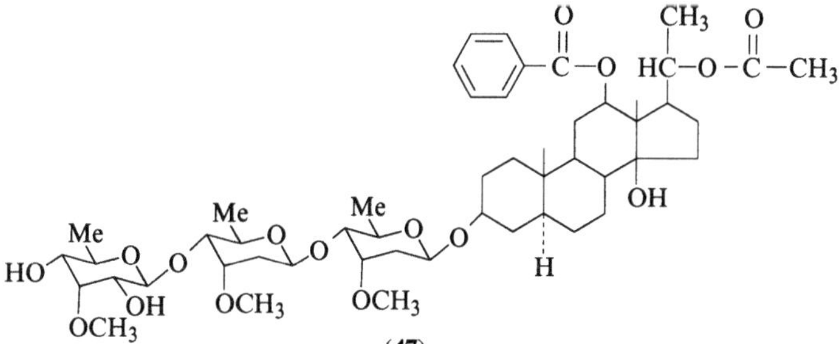

(**47**)

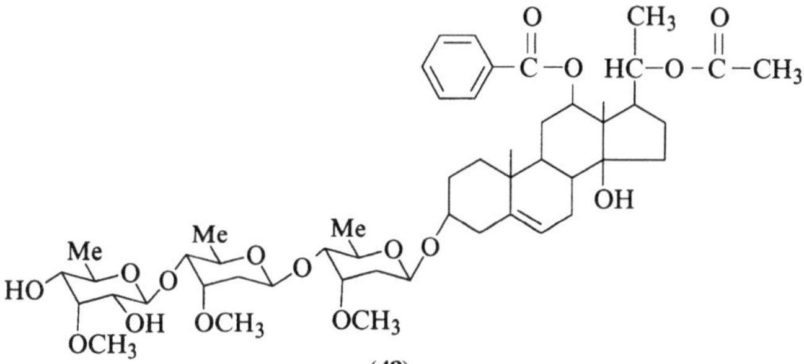

(**48**)

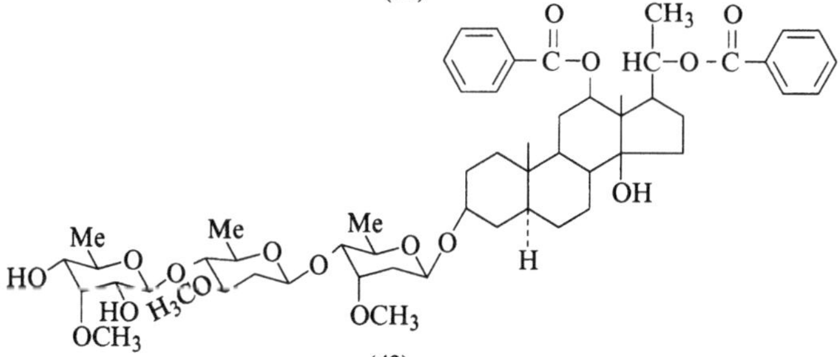

(**49**)

Table 2 (*continued*)

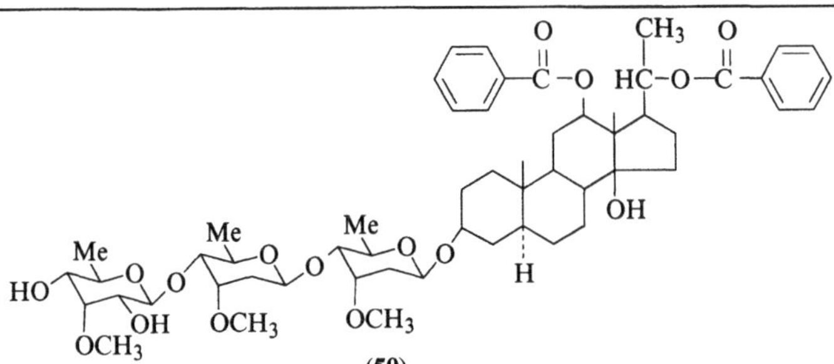

(**50**)

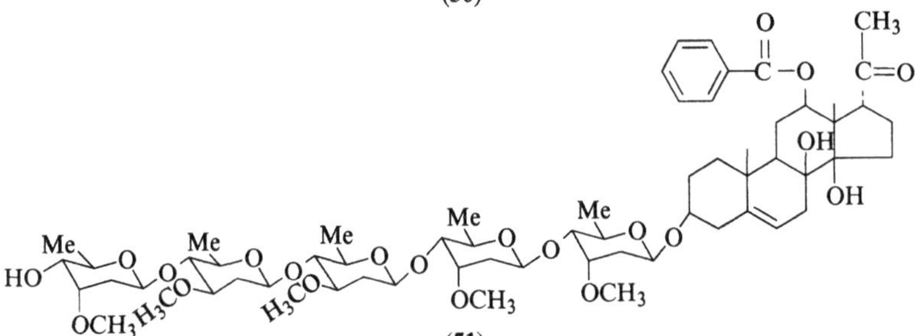

(**51**)

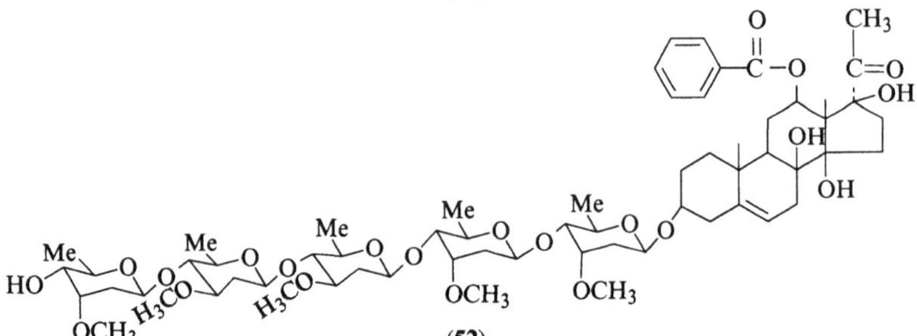

(**52**)

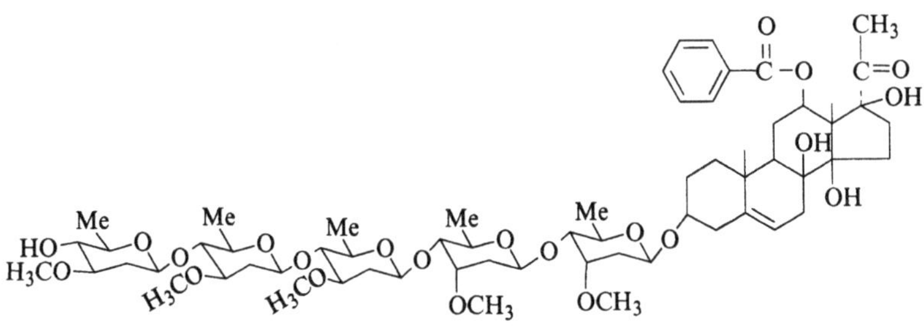

(**53**)

References, pp. 309–325

Table 2 (*continued*)

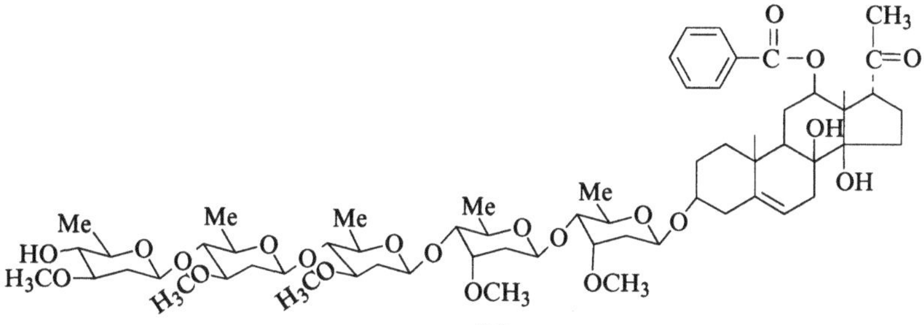

(**54**)

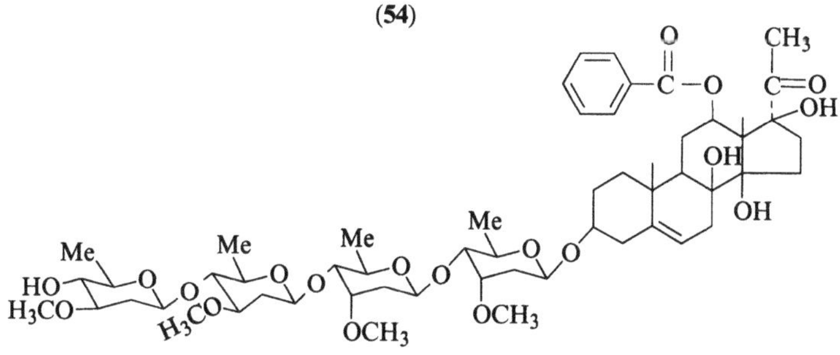

(**55**)

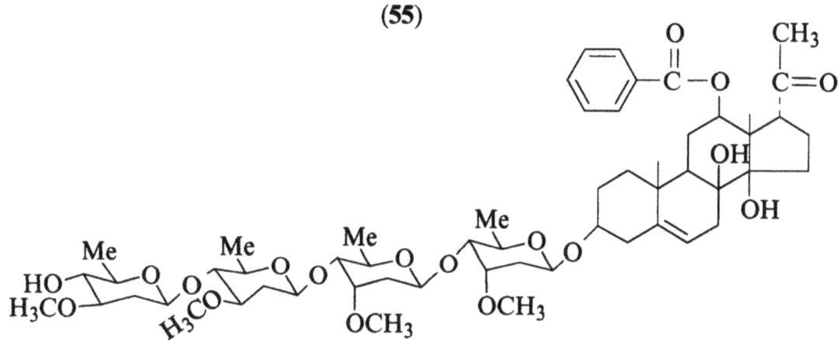

(**56**)

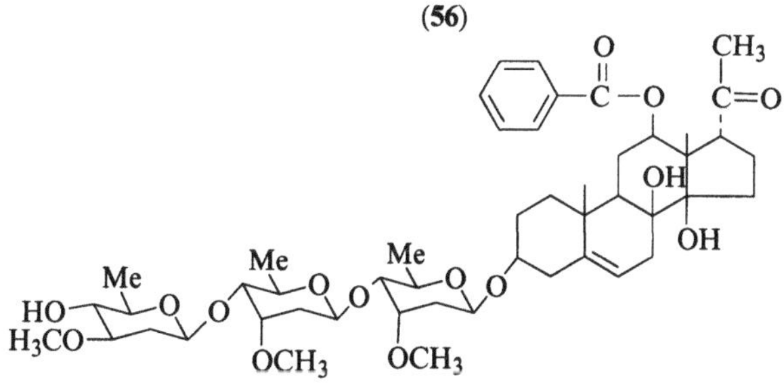

(**57**)

Table 2 (*continued*)

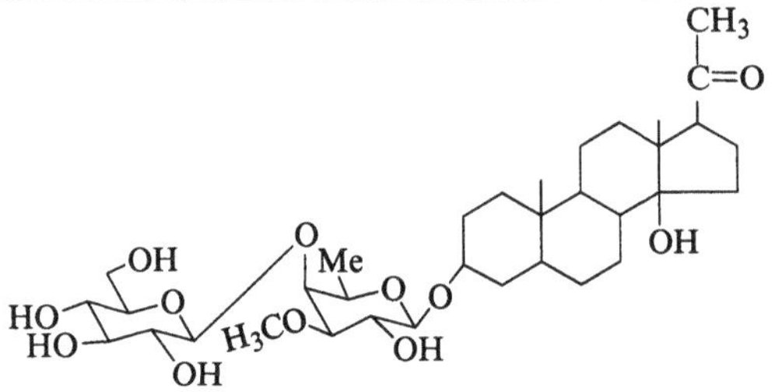

(**58**)

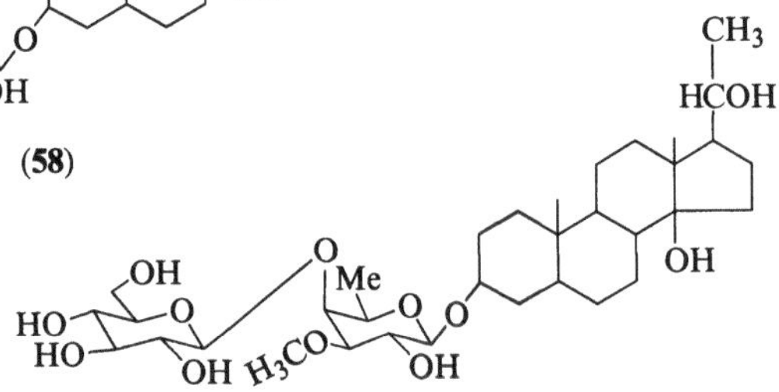

(**59**)

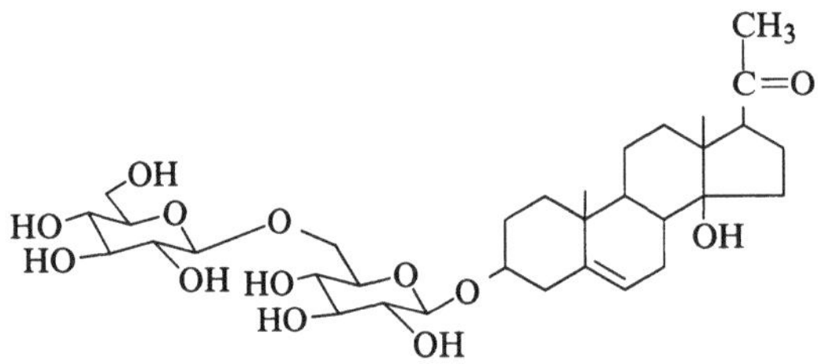

(**60**)

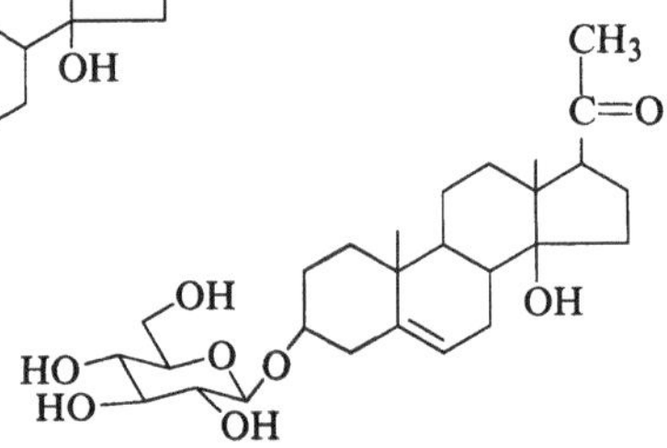

(**61**)

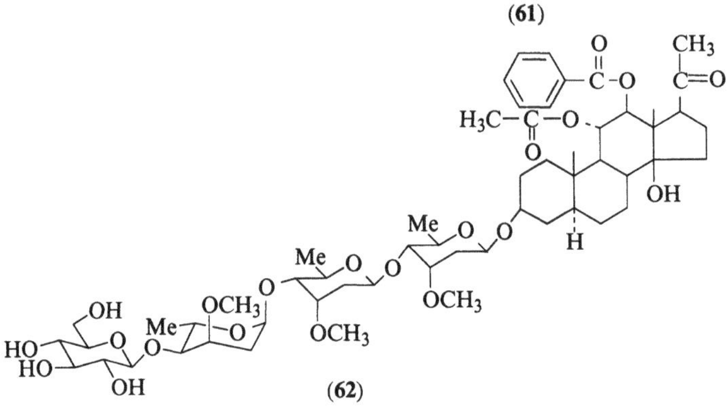

(**62**)

Table 2 (*continued*)

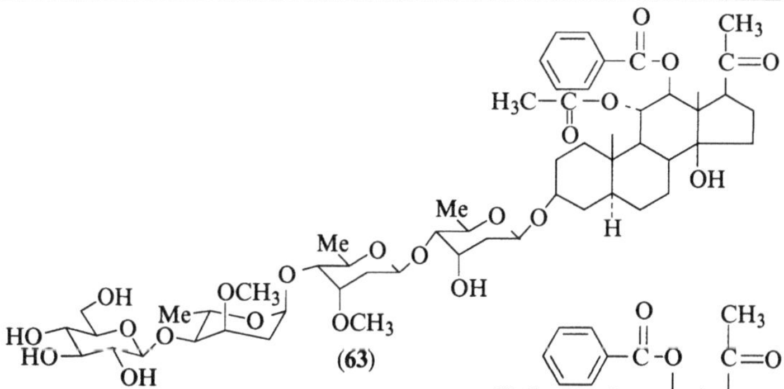

(**63**)

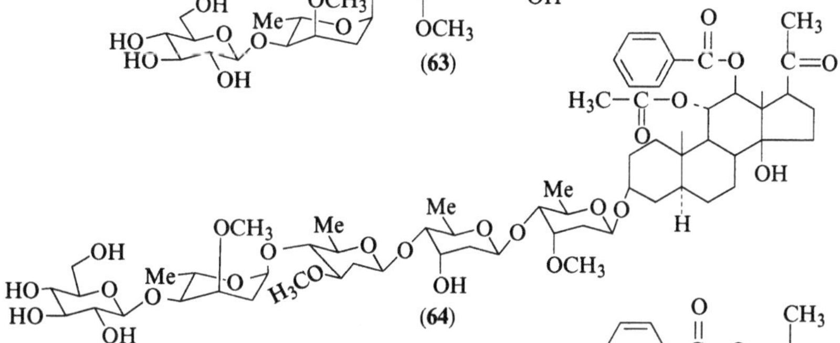

(**64**)

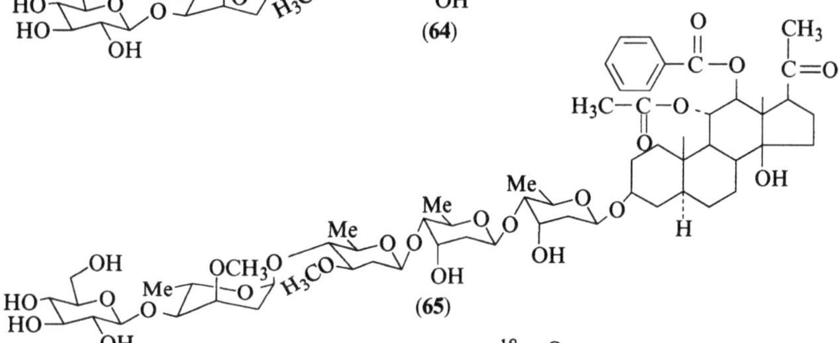

(**65**)

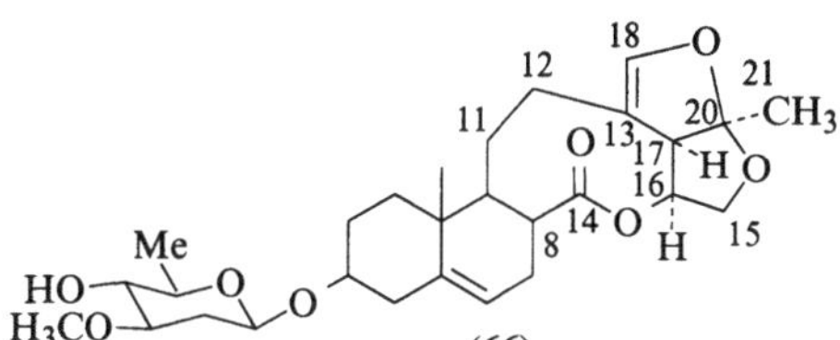

(**66**)

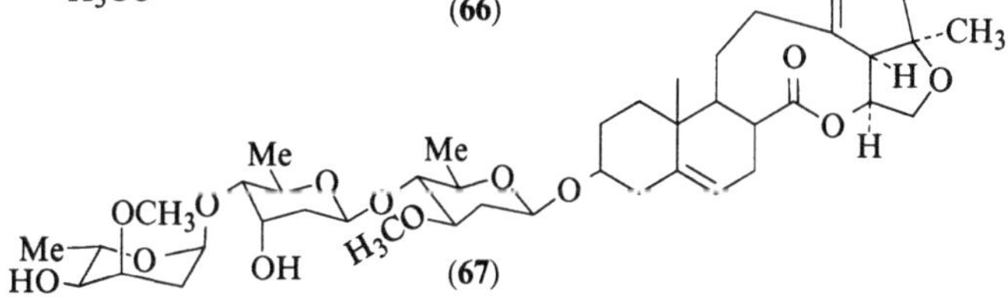

(**67**)

Table 2 (*continued*)

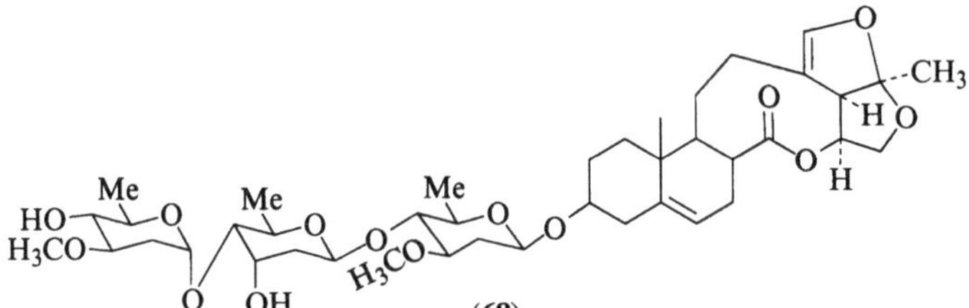

(**68**)

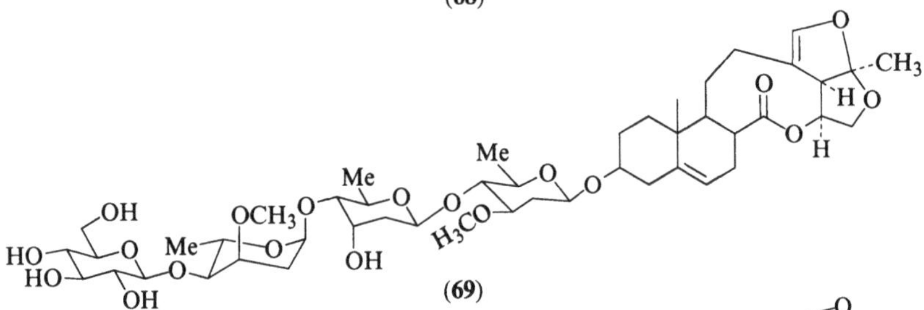

(**69**)

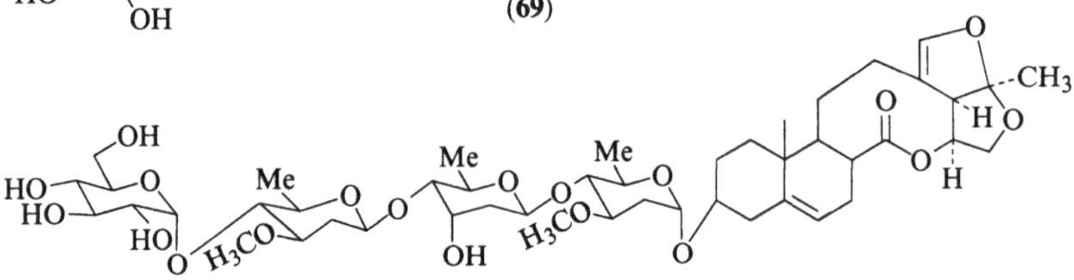

(**70**)

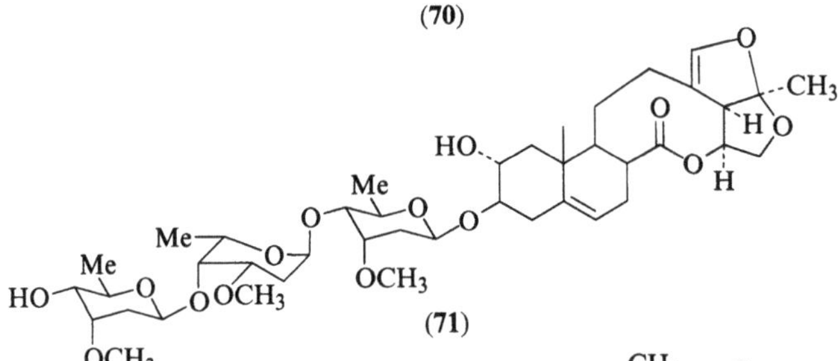

(**71**)

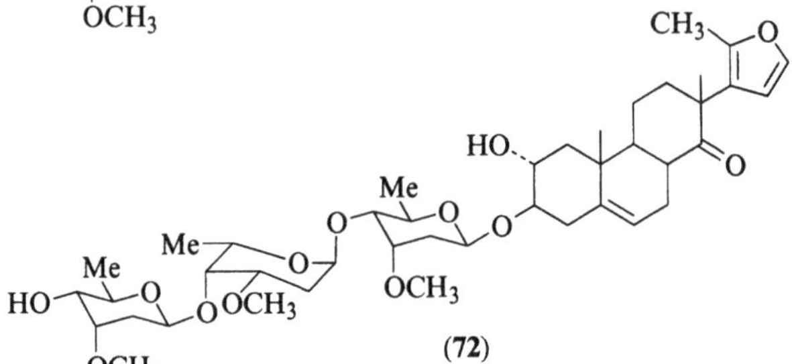

(**72**)

Table 2 (*continued*)

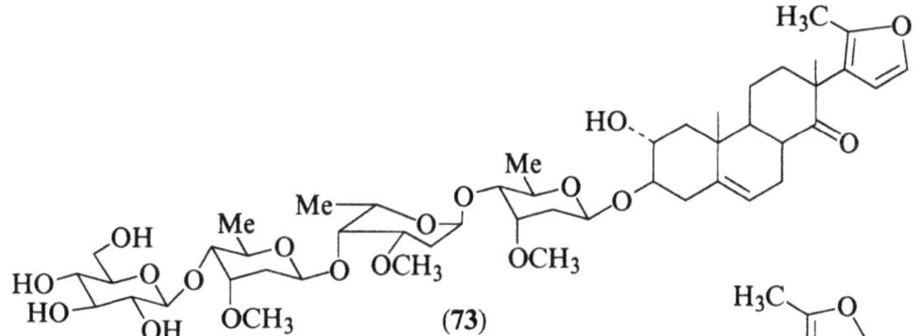

(**73**)

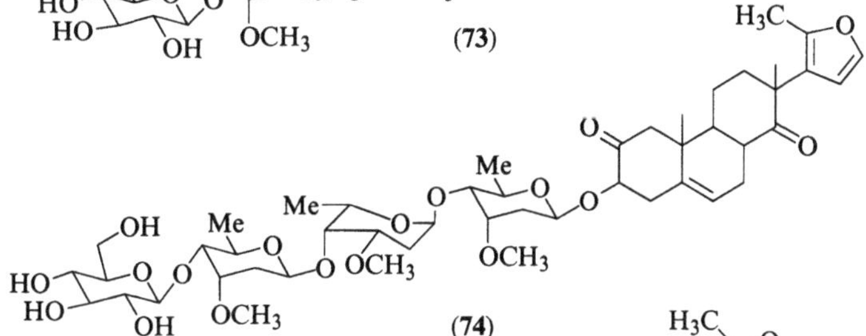

(**74**)

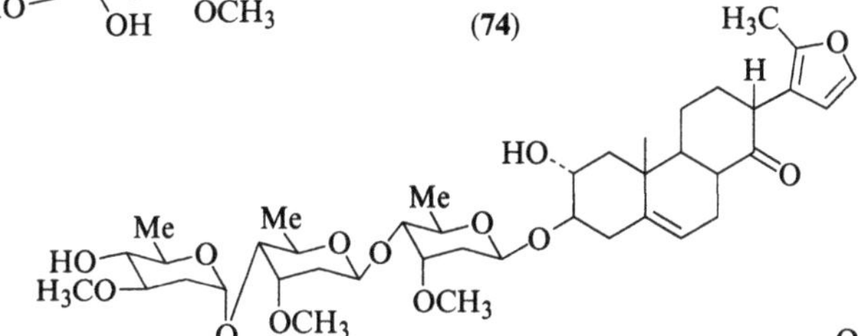

(**75**)

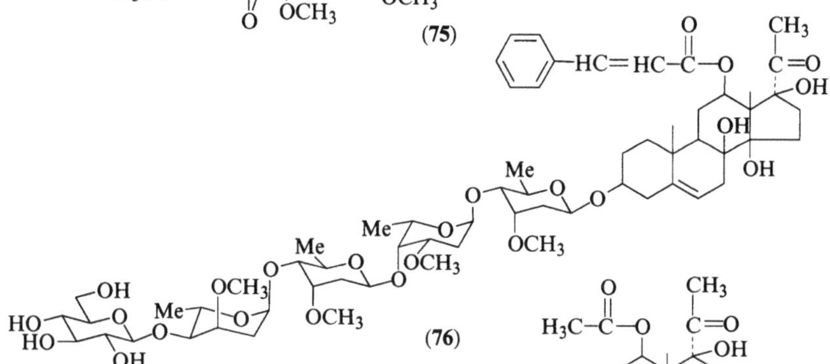

(**76**)

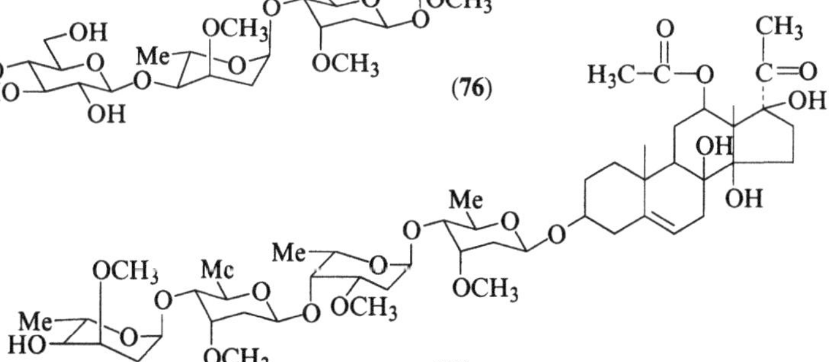

(**77**)

Table 2 (*continued*)

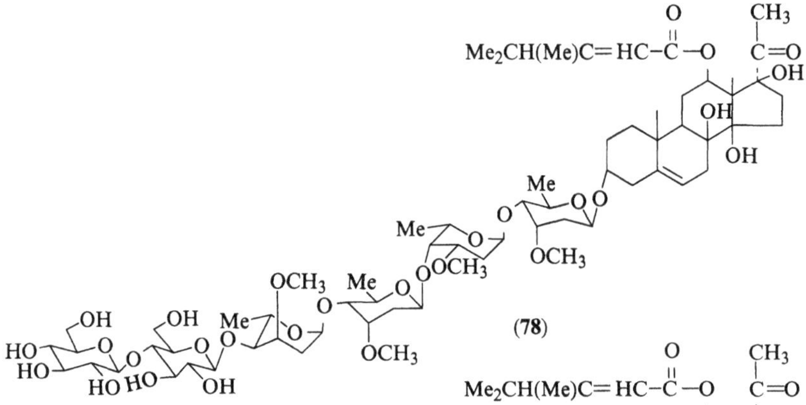

(**78**)

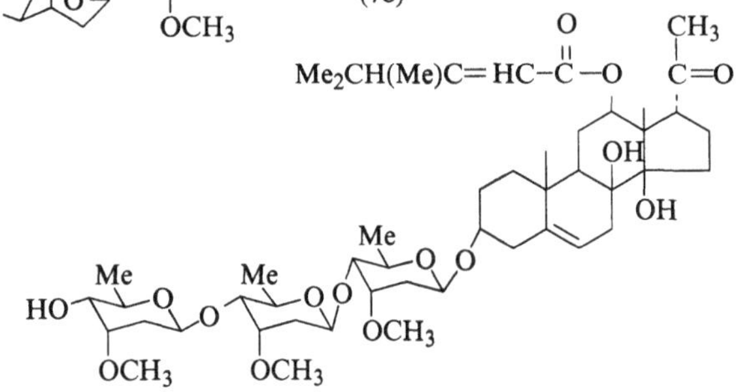

(**79**)

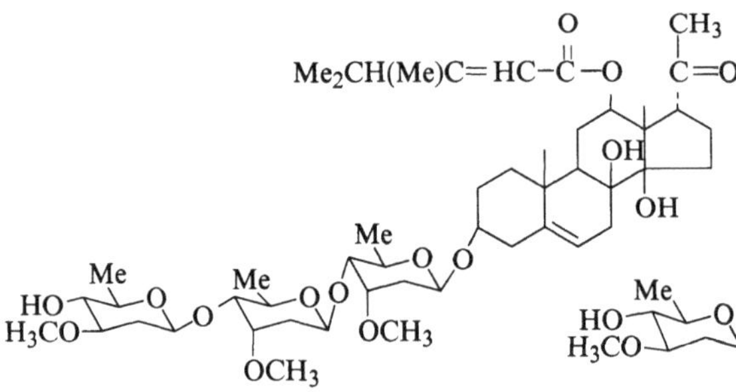

(**80**)

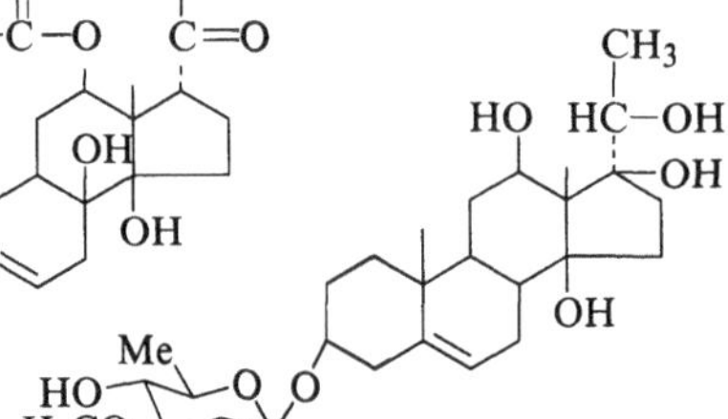

(**81**)

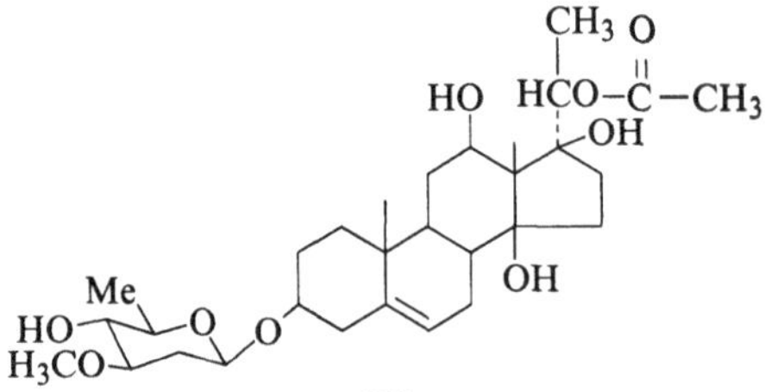

(**82**)

References, pp. 309–325

Table 2 (*continued*)

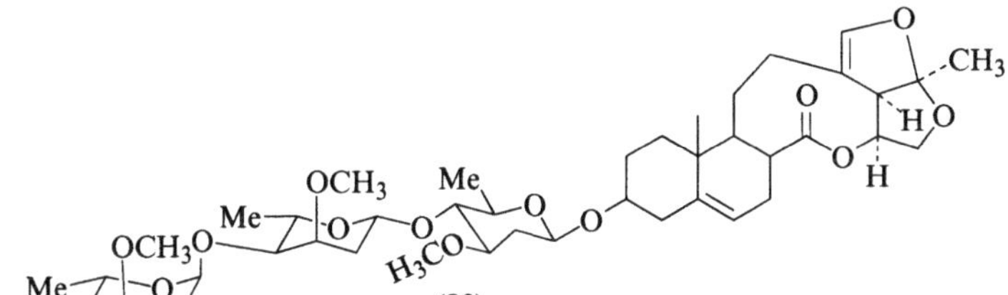

(**83**)

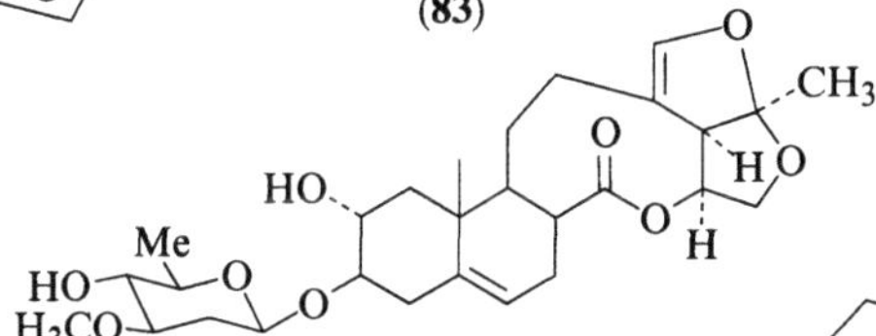

(**84**)

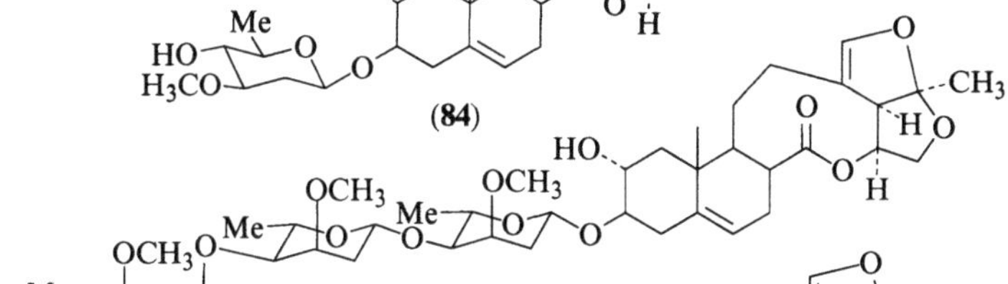

(**85**)

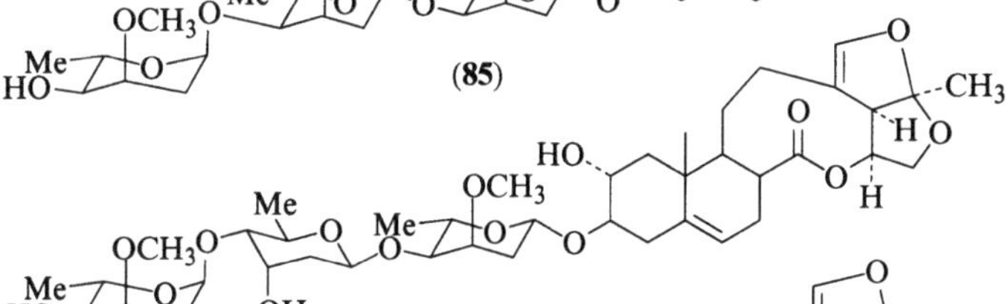

(**86**)

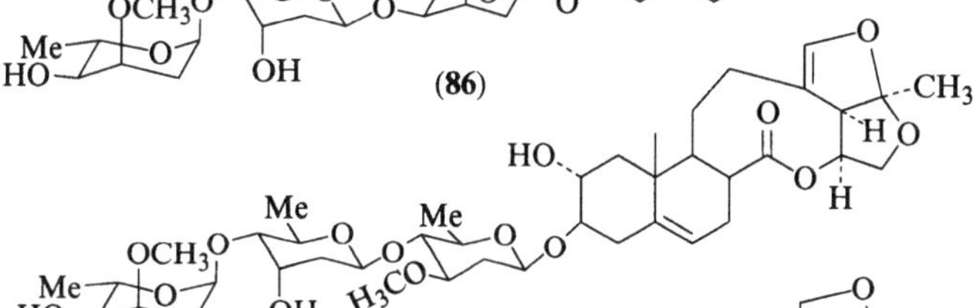

(**87**)

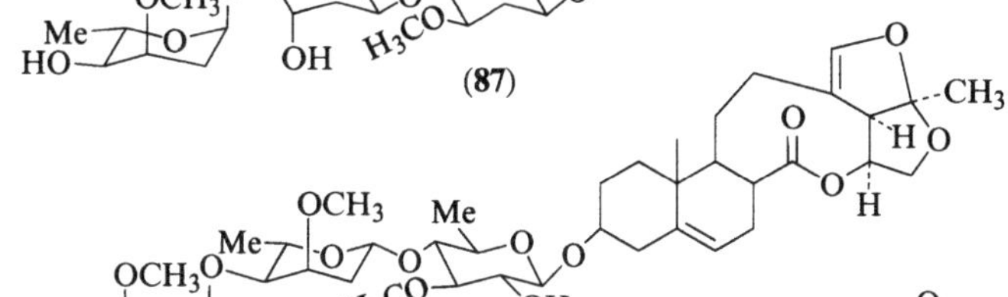

(**88**)

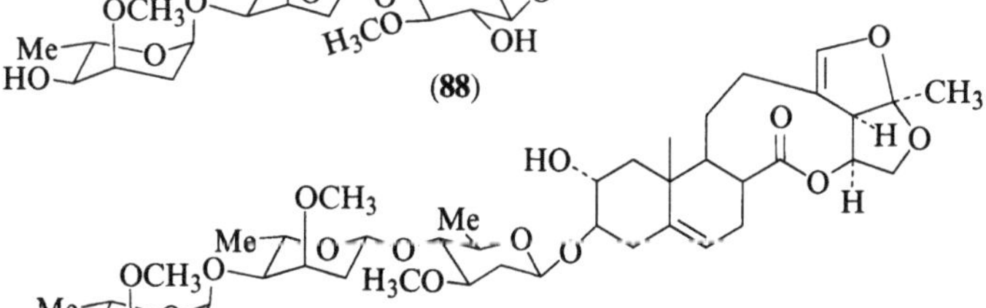

(**89**)

Table 2 (*continued*)

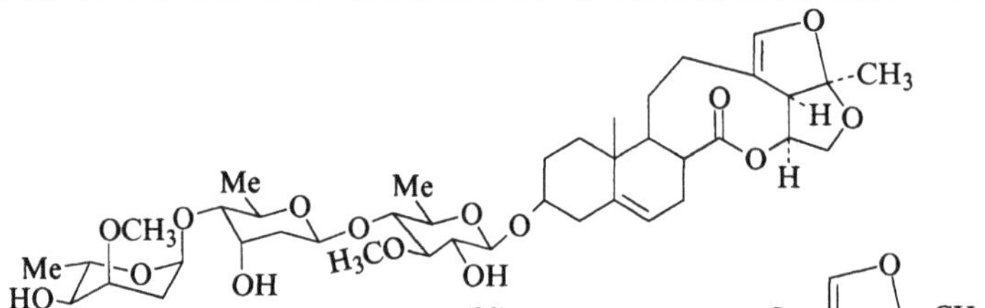

(**90**)

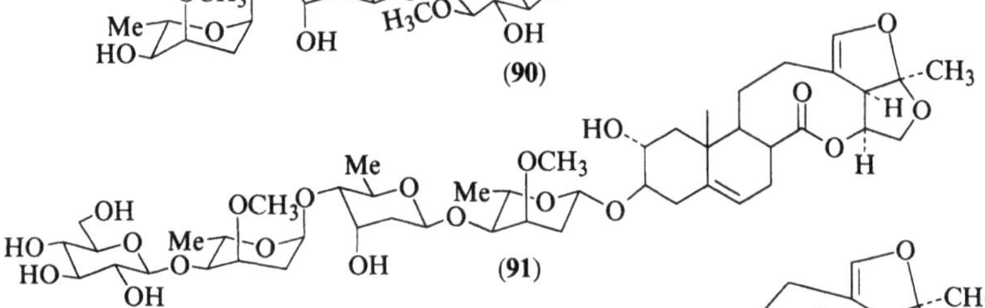

(**91**)

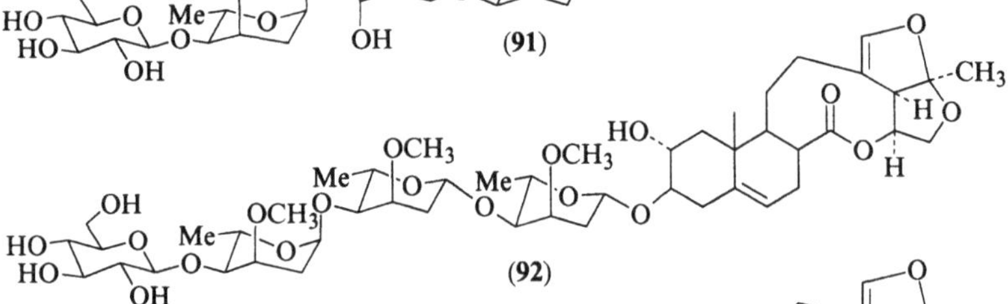

(**92**)

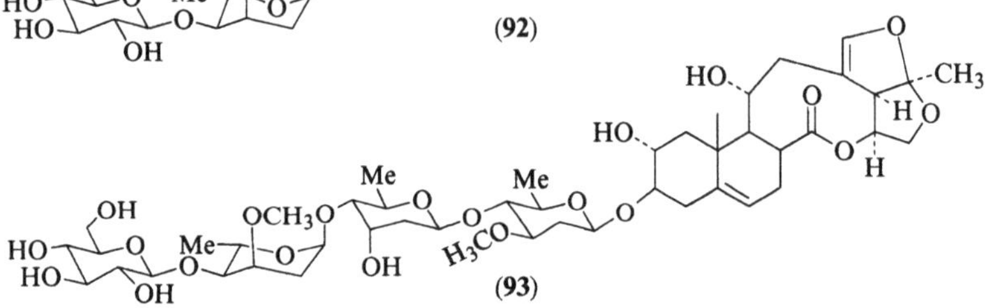

(**93**)

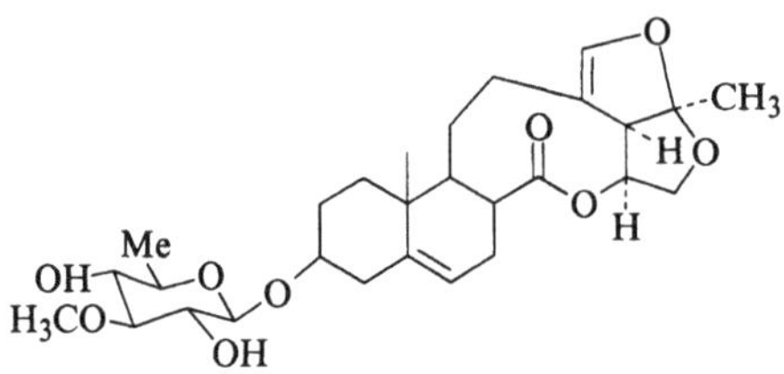

(**94**)

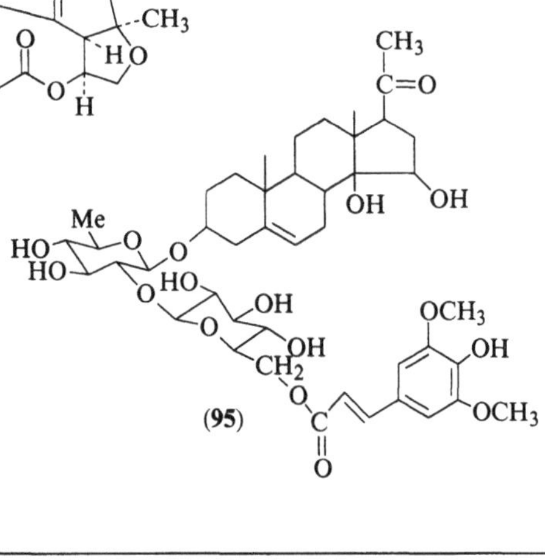

(**95**)

Table 2 (*continued*)

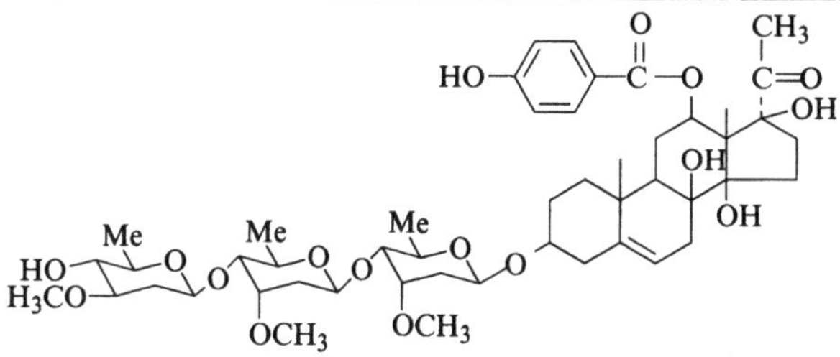

(**96**)

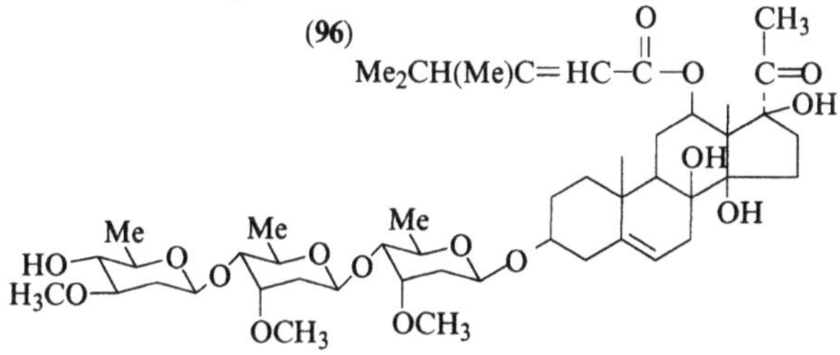

(**97**)

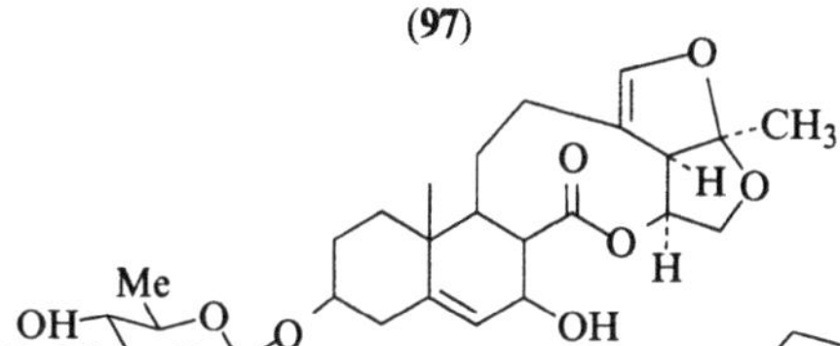

(**98**)

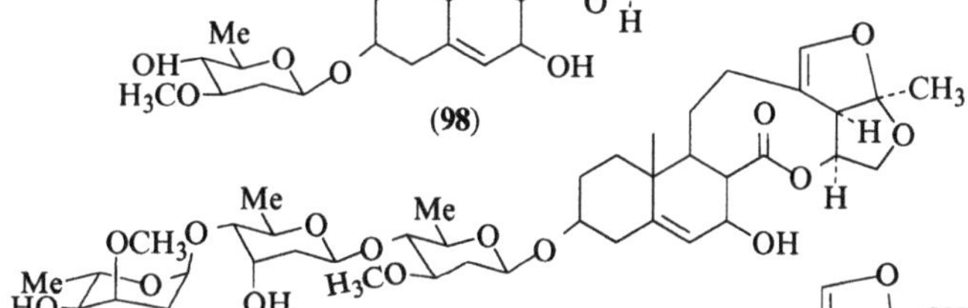

(**99**)

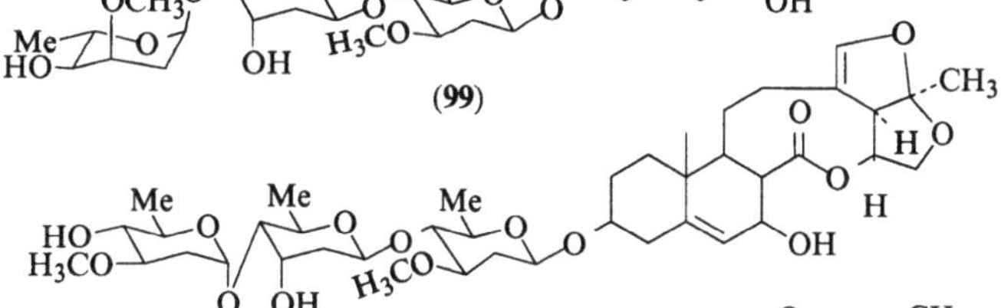

(**100**)

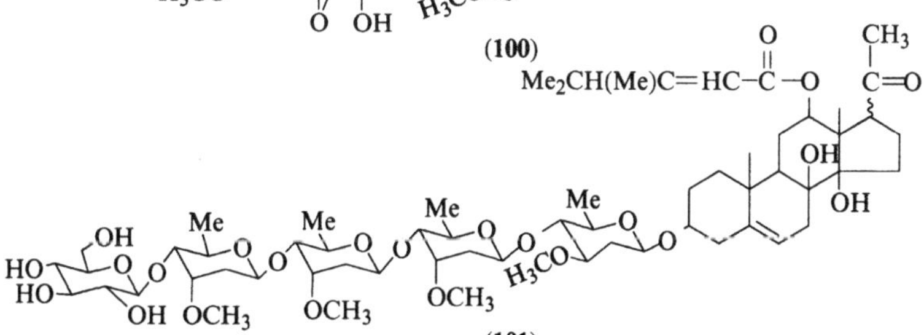

(**101**)

Table 2 (*continued*)

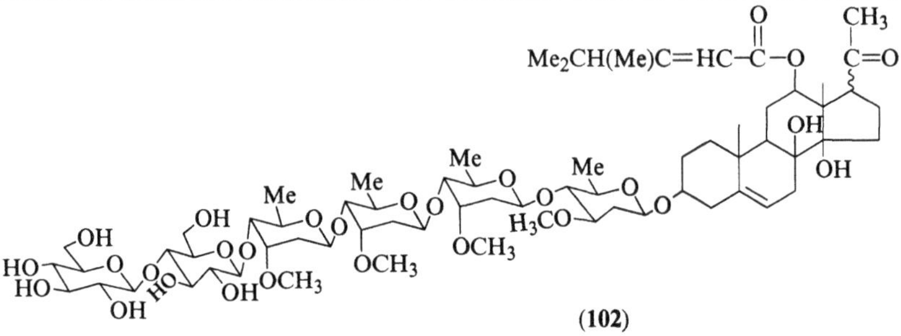

(**102**)

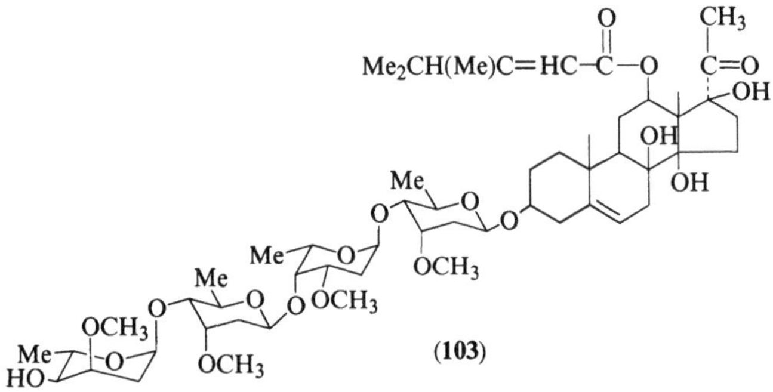

(**103**)

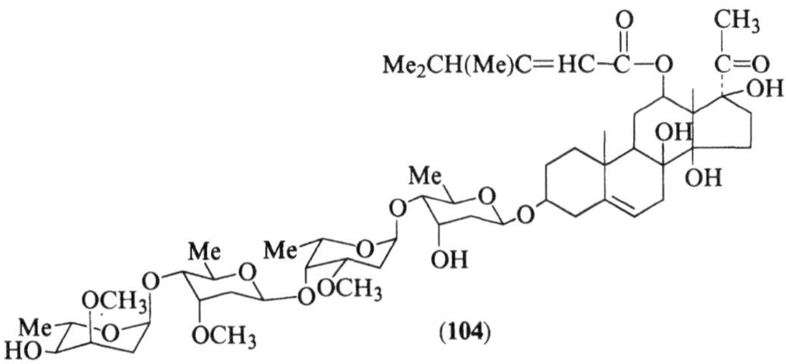

(**104**)

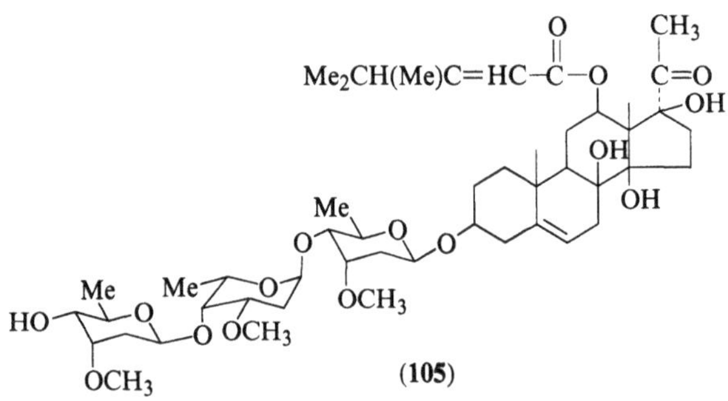

(**105**)

Table 2 (*continued*)

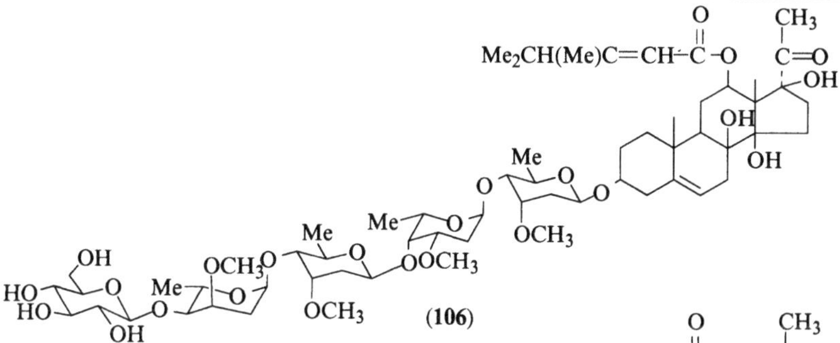

(**106**)

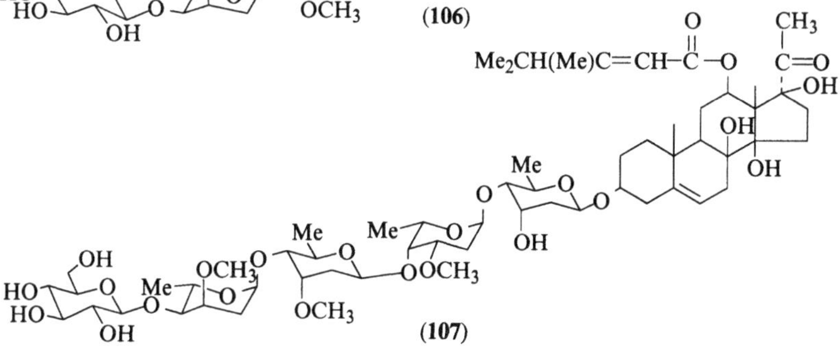

(**107**)

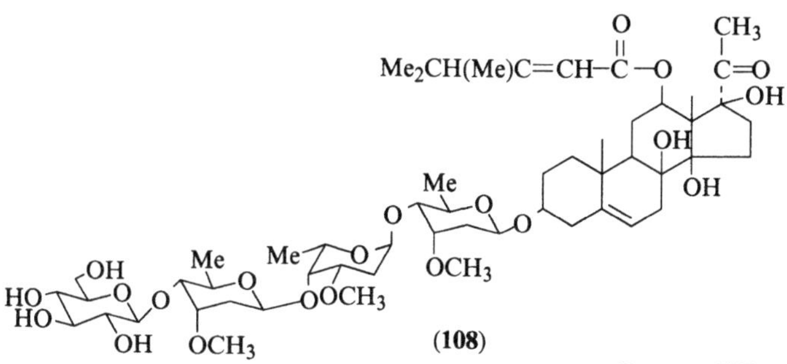

(**108**)

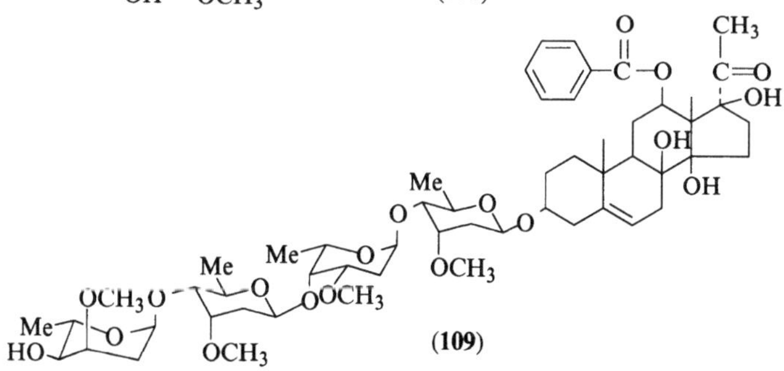

(**109**)

Table 2 (*continued*)

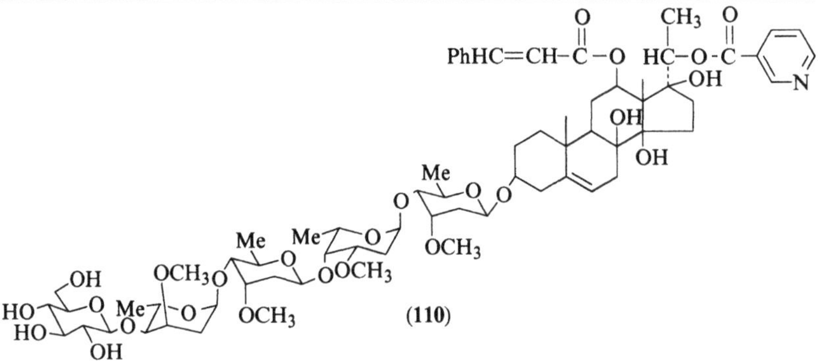

(**110**)

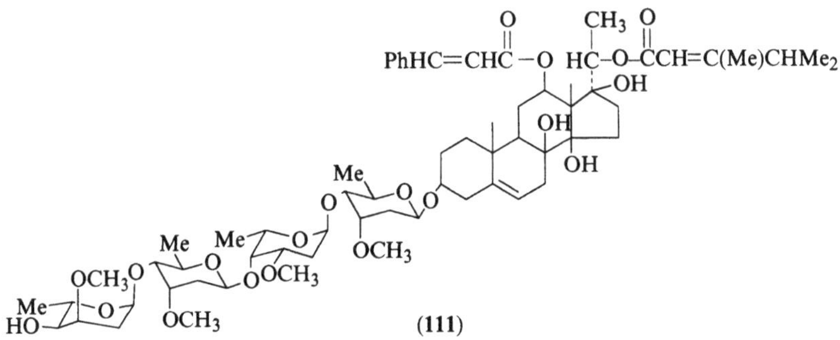

(**111**)

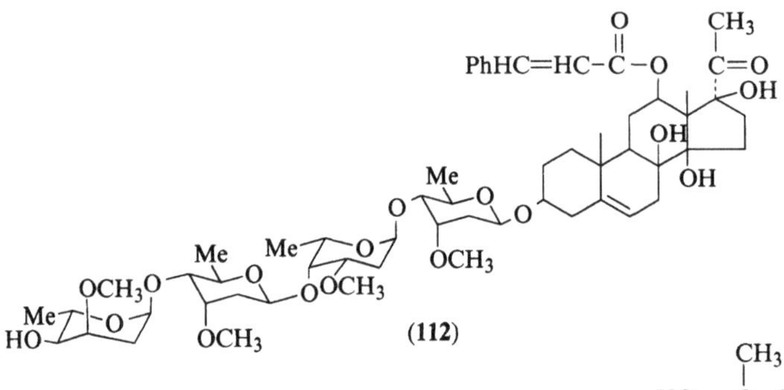

(**112**)

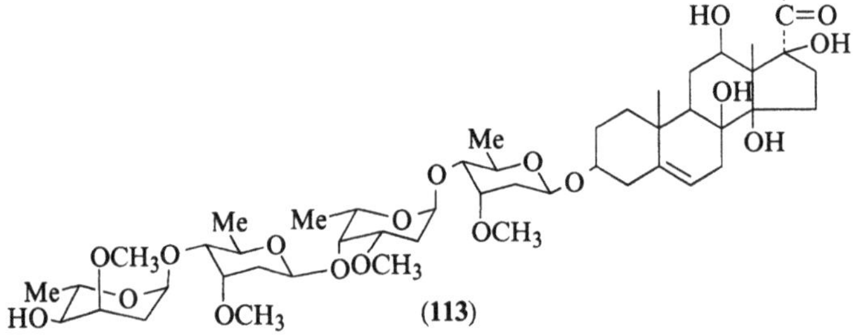

(**113**)

Table 2 (*continued*)

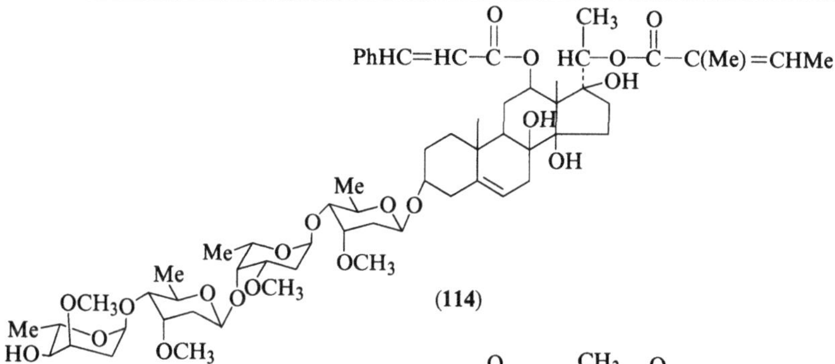

(**114**)

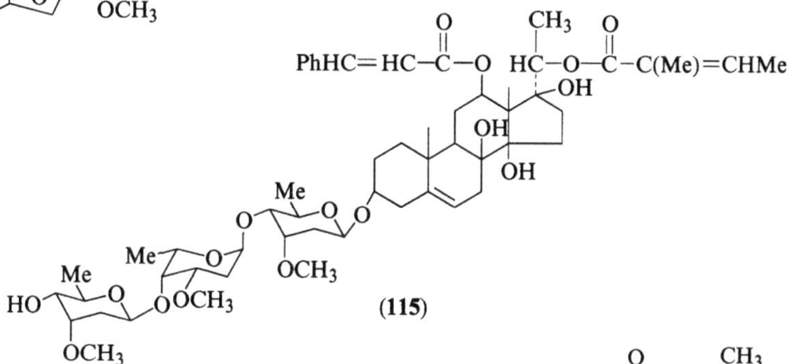

(**115**)

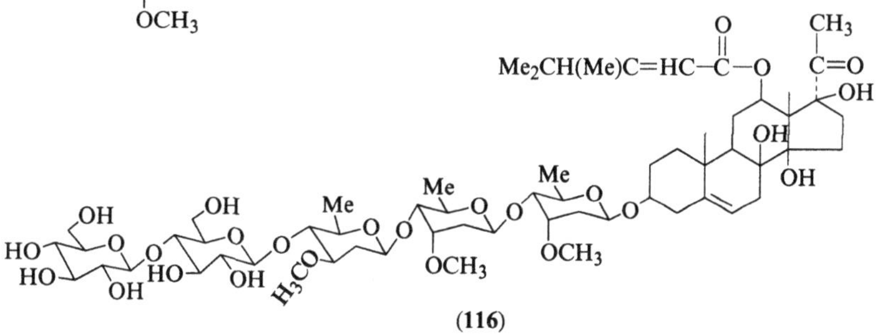

(**116**)

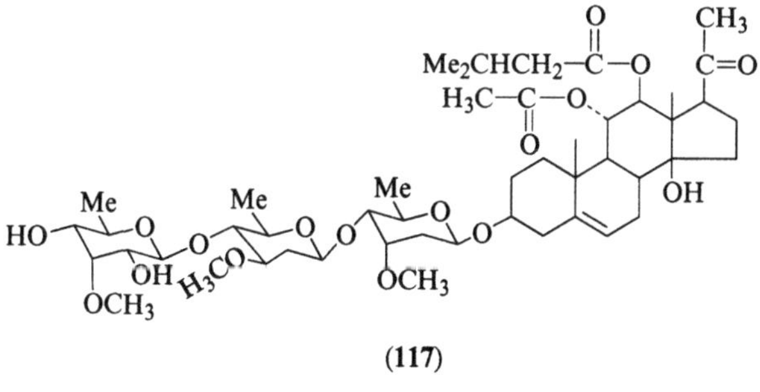

(**117**)

Table 2 (*continued*)

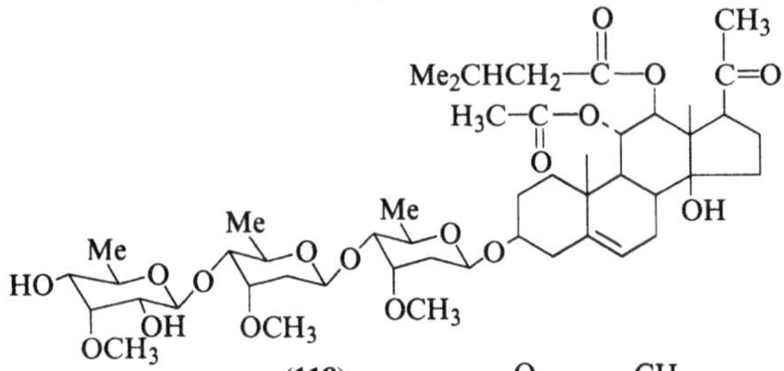

(**118**)

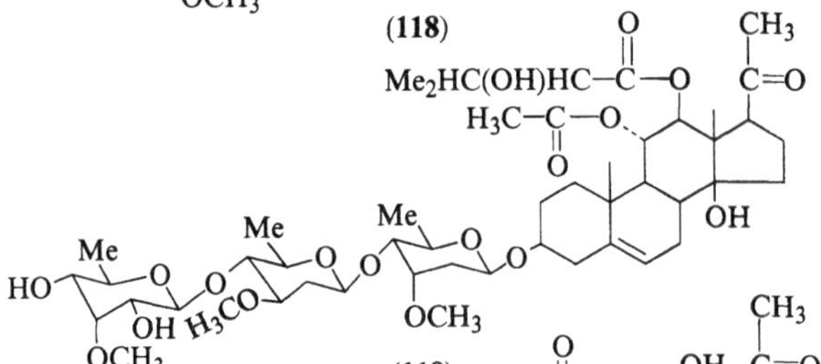

(**119**)

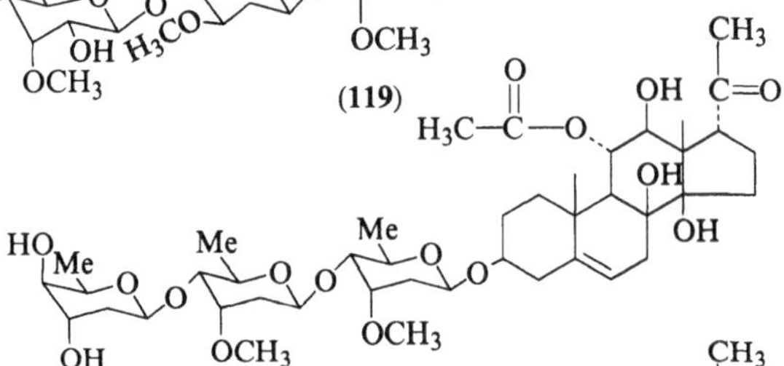

(**120**)

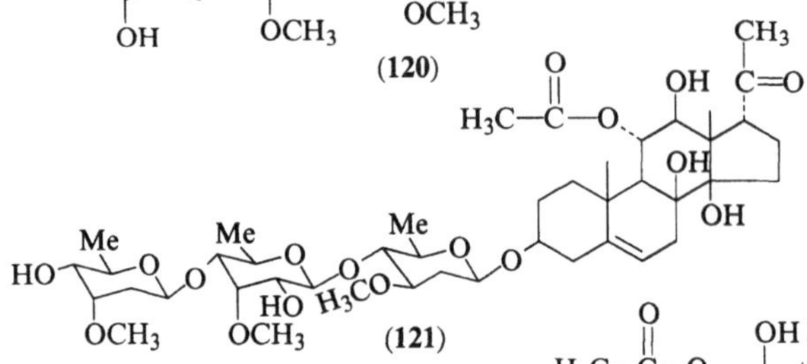

(**121**)

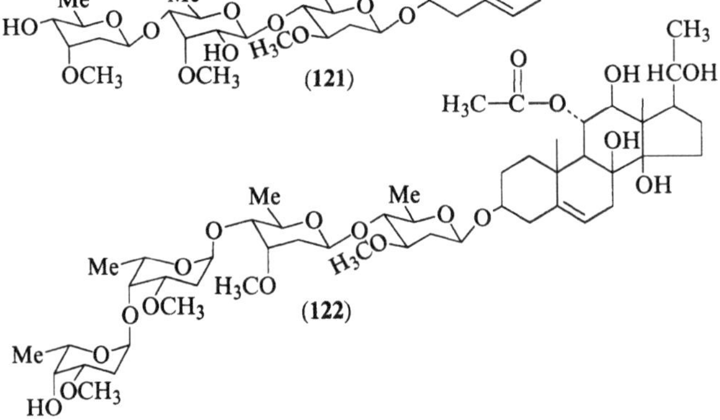

(**122**)

Table 2 (*continued*)

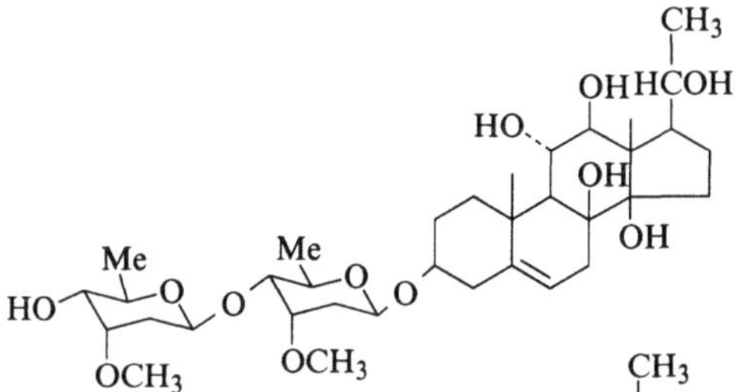

(**123**)

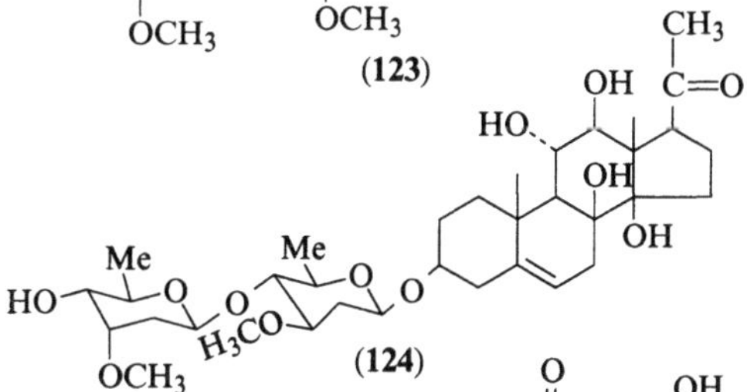

(**124**)

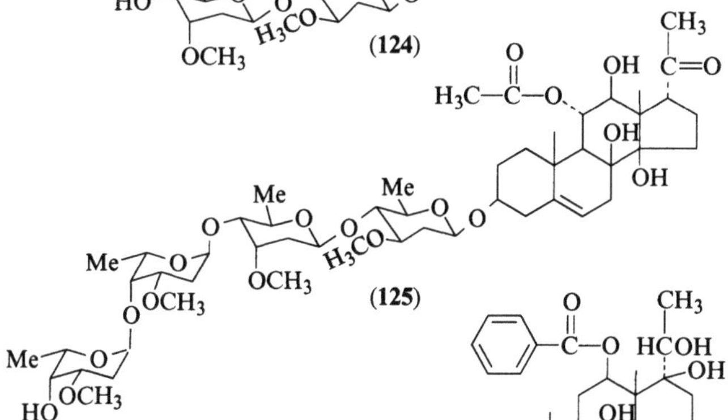

(**125**)

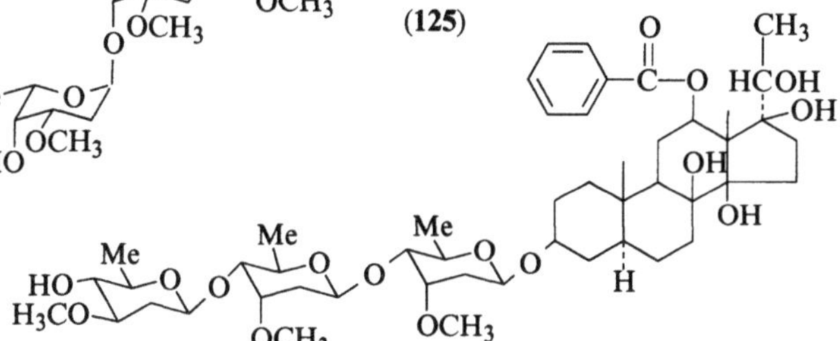

(**126**)

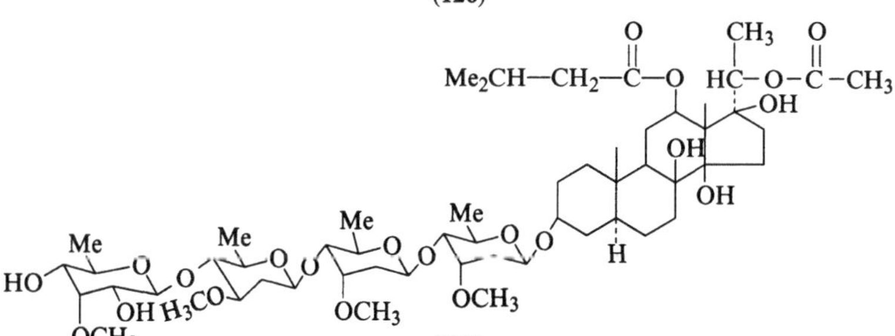

(**127**)

Table 2 (*continued*)

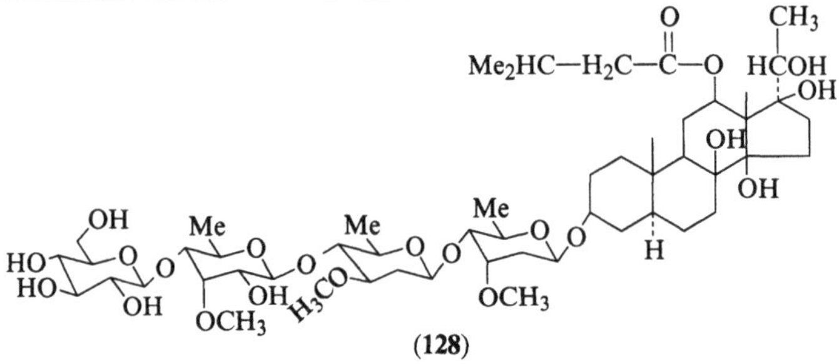

(**128**)

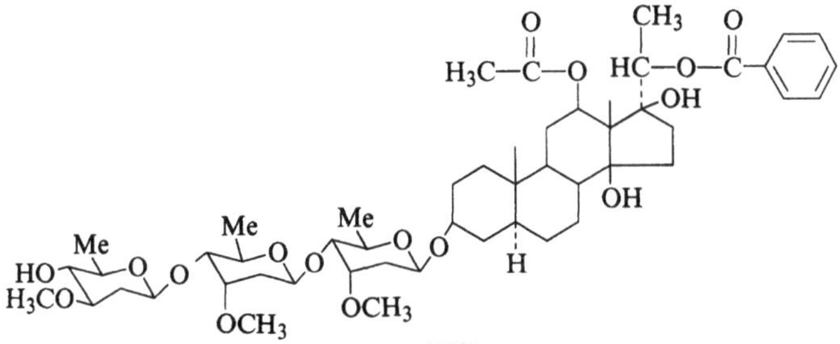

(**129**)

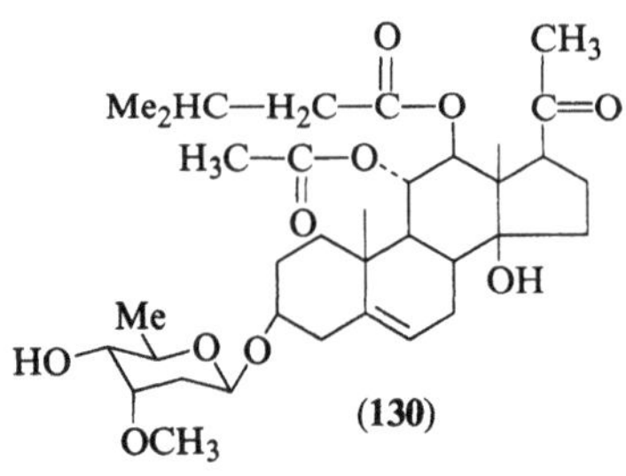

(**130**)

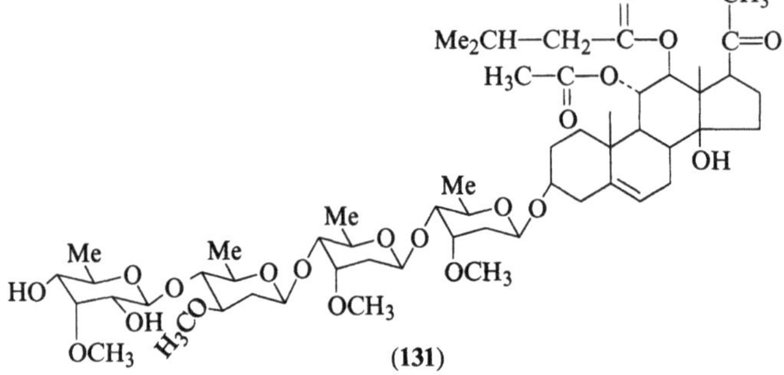

(**131**)

References, pp. 309–325

Table 2 (*continued*)

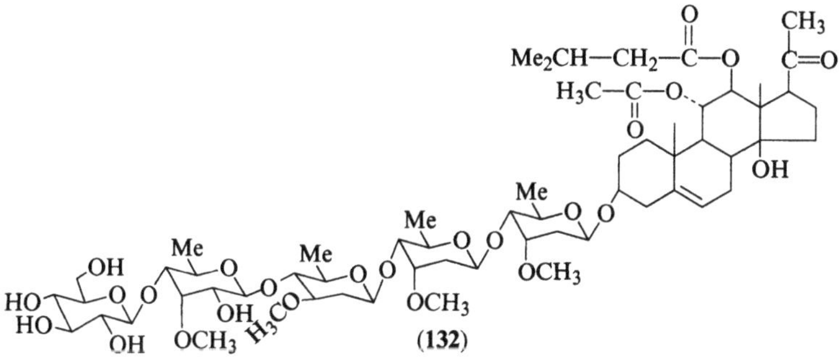

(**132**)

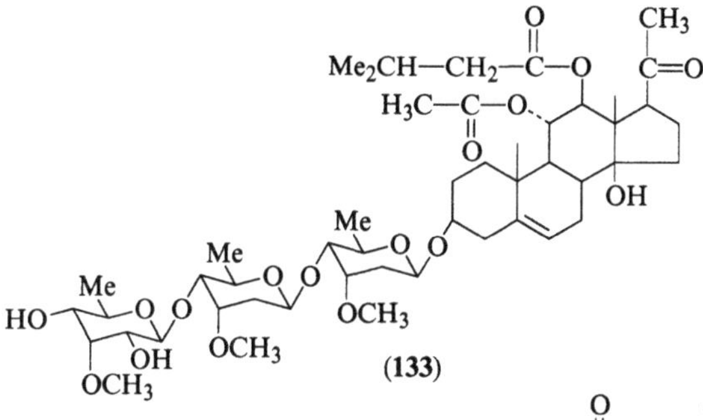

(**133**)

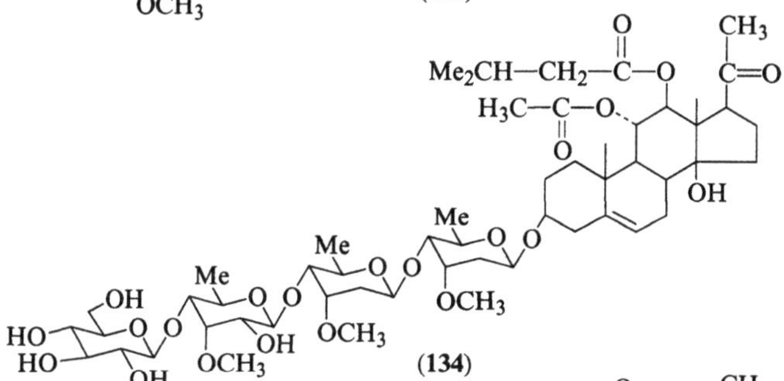

(**134**)

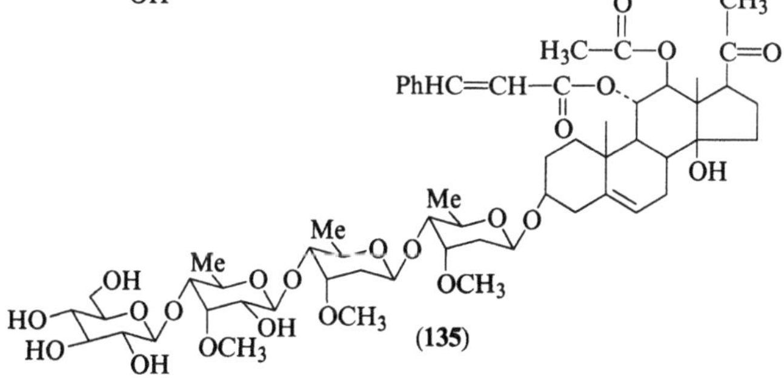

(**135**)

Table 2 (*continued*)

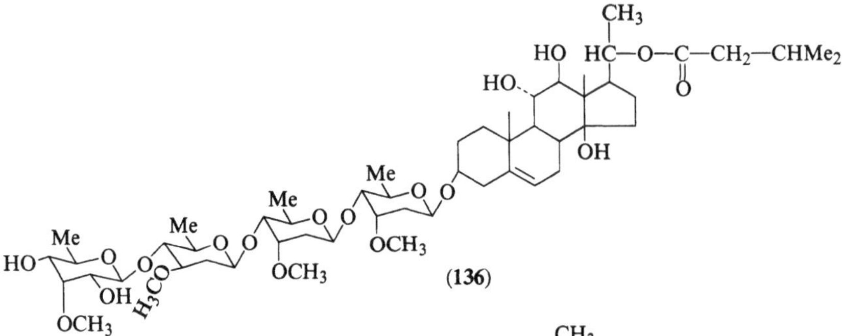

(**136**)

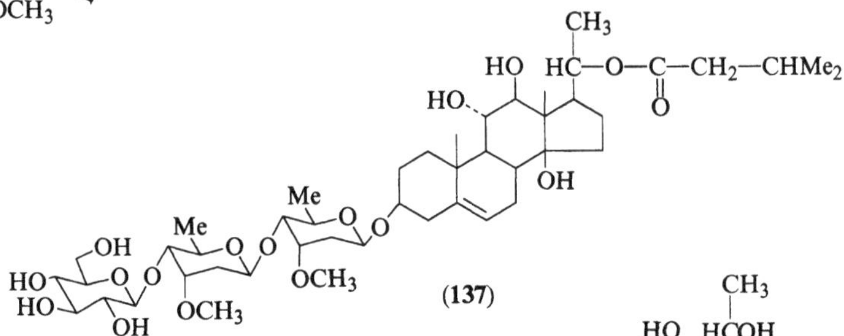

(**137**)

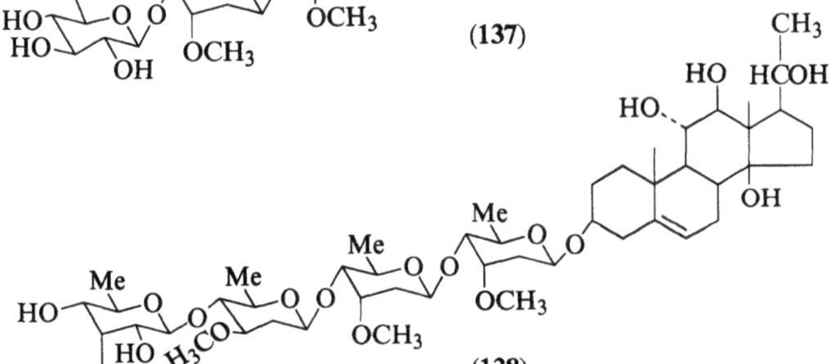

(**138**)

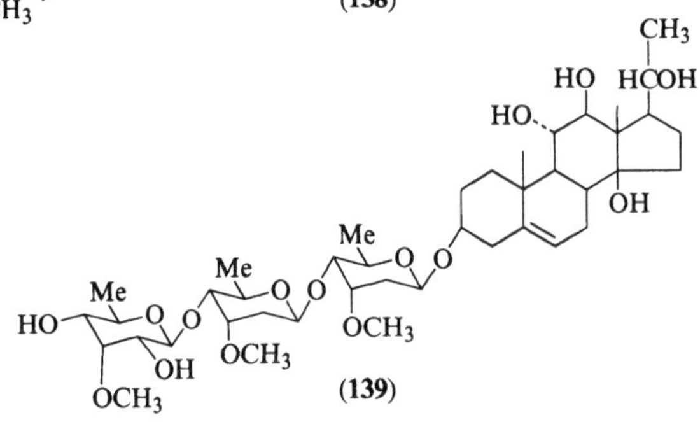

(**139**)

Table 2 (*continued*)

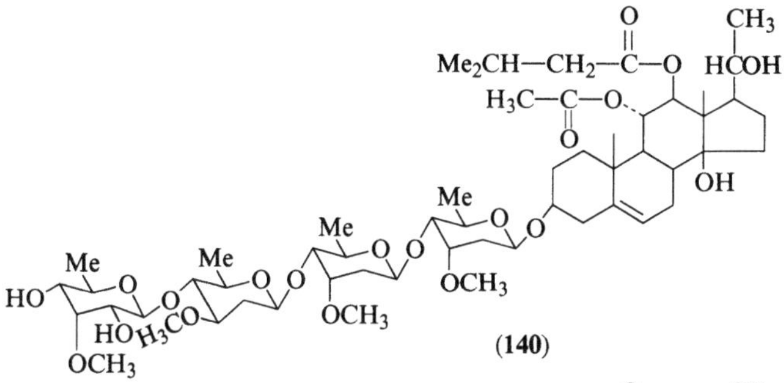

(**140**)

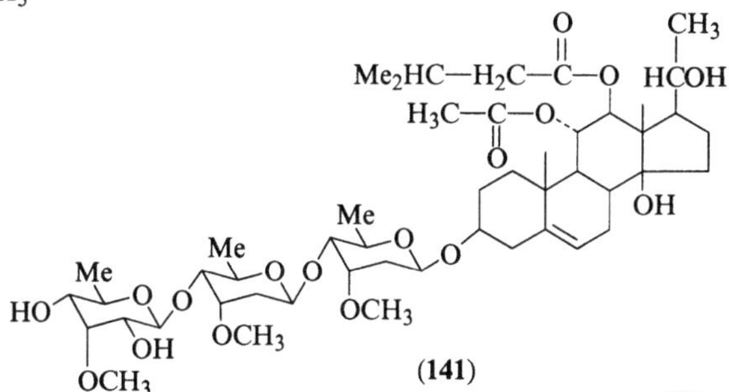

(**141**)

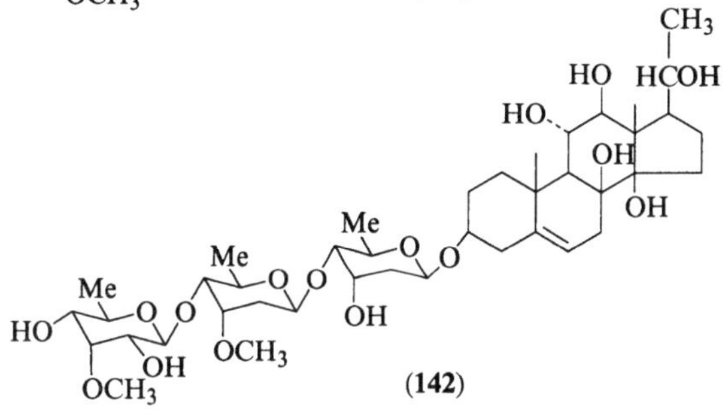

(**142**)

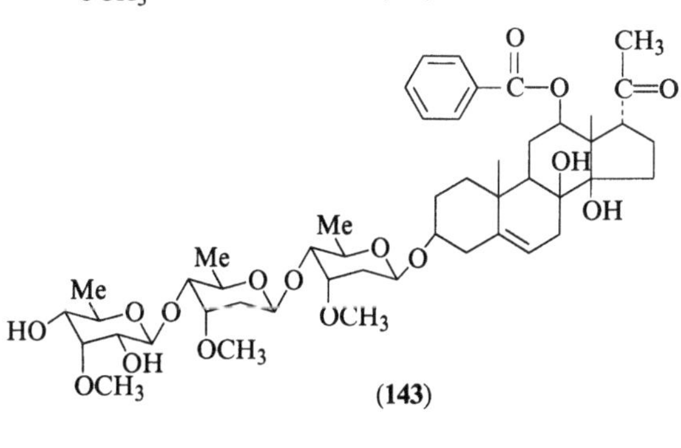

(**143**)

Table 2 (*continued*)

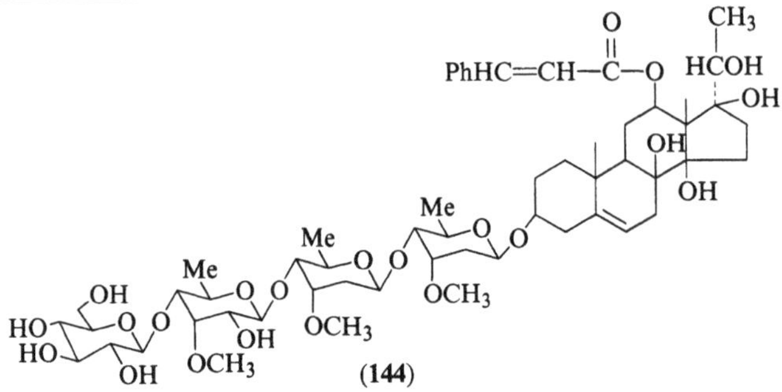

(**144**)

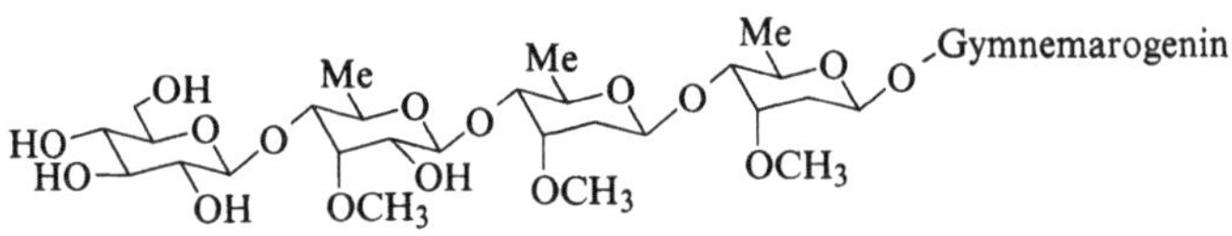

(**145**)

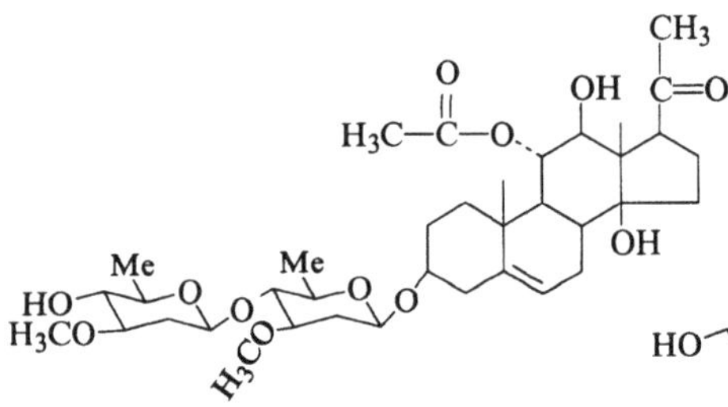

(**146**)

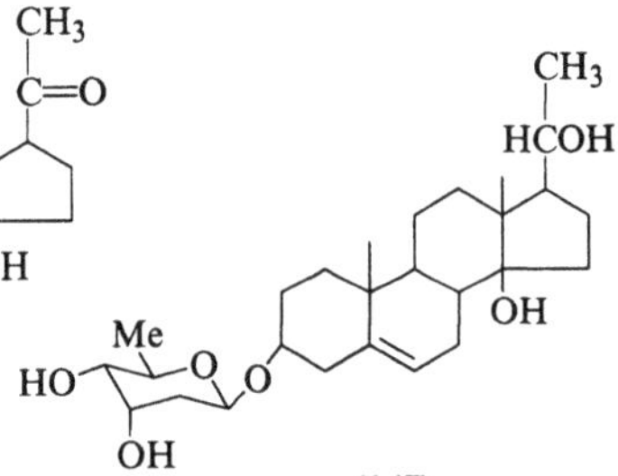

(**147**)

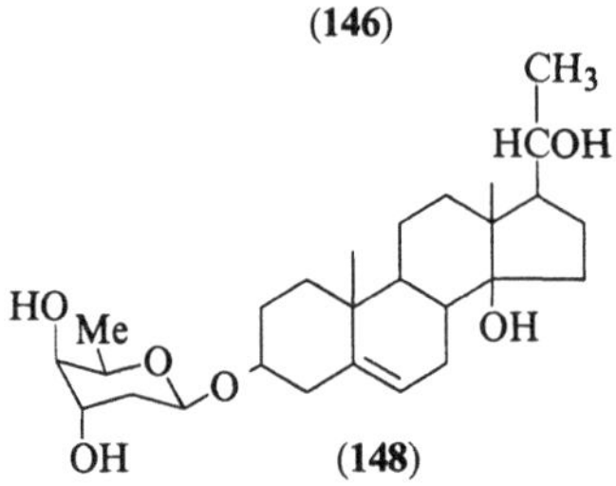

(**148**)

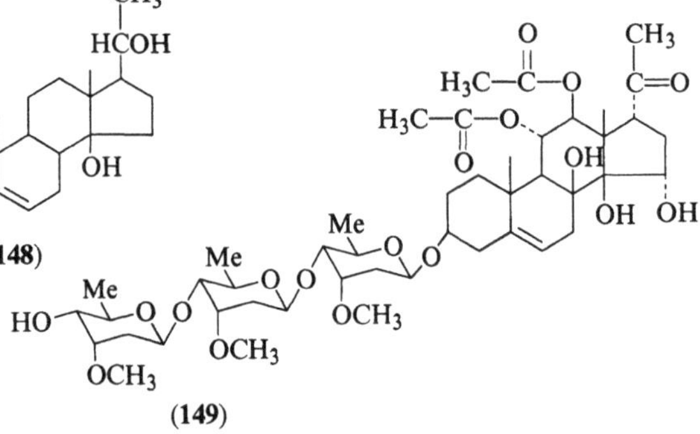

(**149**)

Table 2 (*continued*)

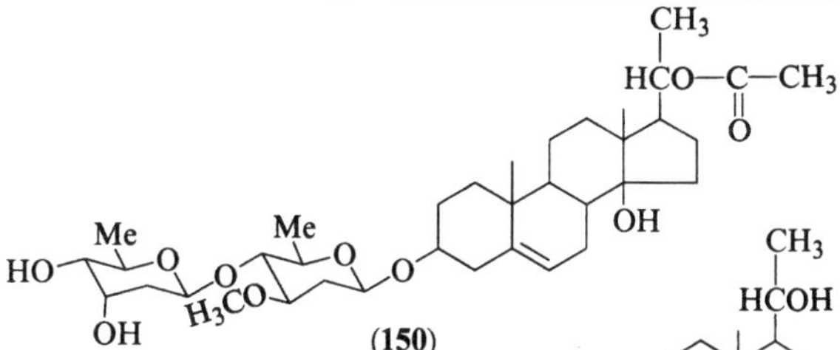

(**150**)

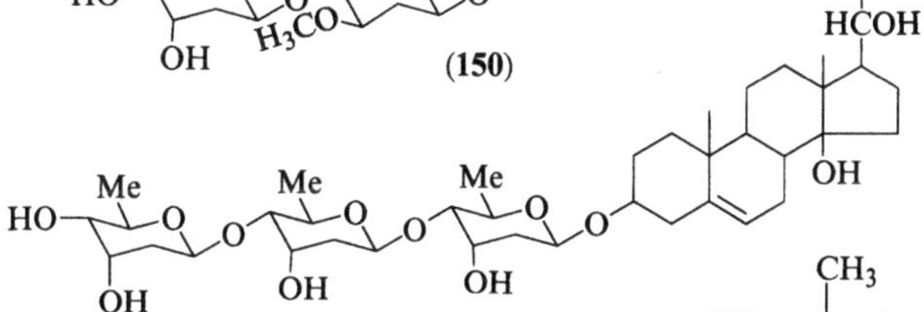

(**151**)

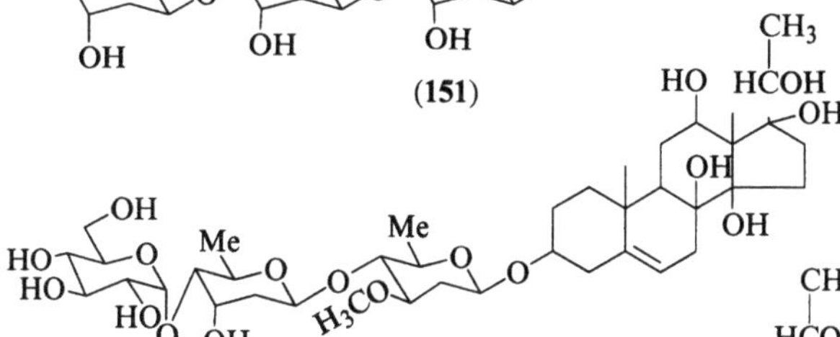

(**152**)

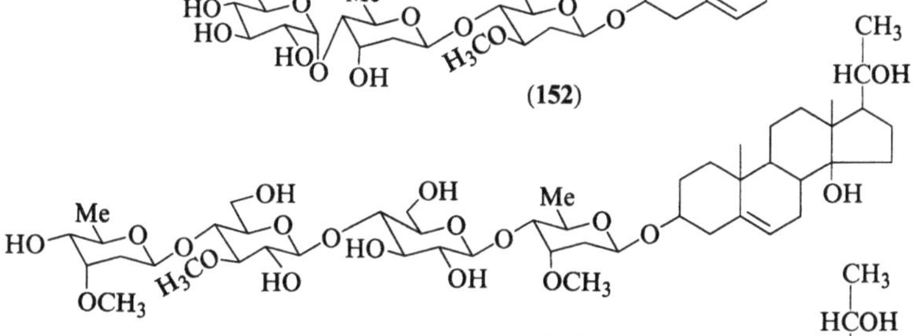

(**153**)

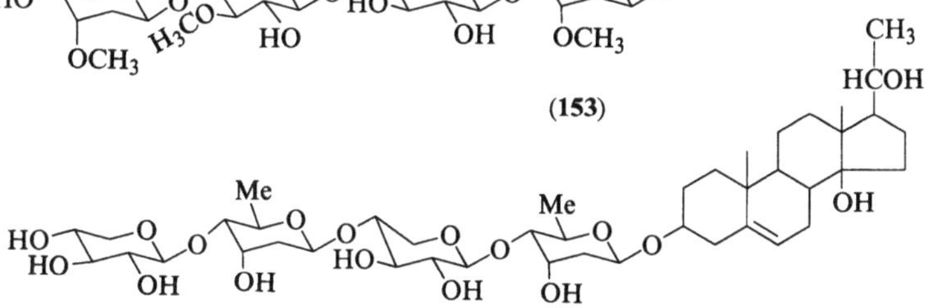

(**154**)

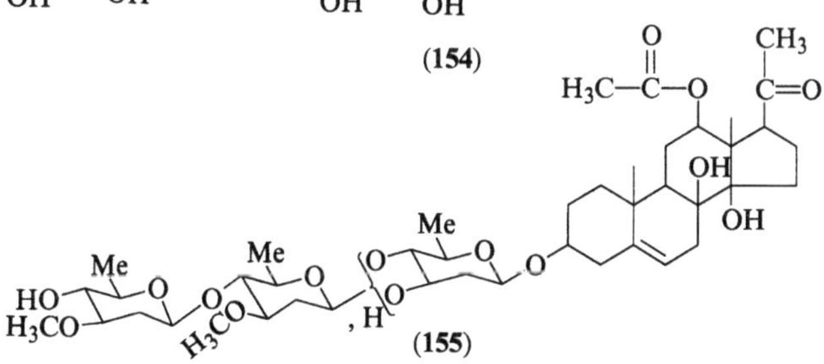

(**155**)

Table 2 (*continued*)

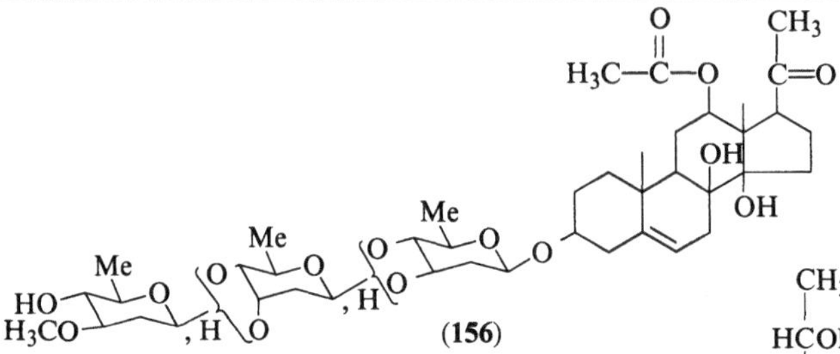

(**156**)

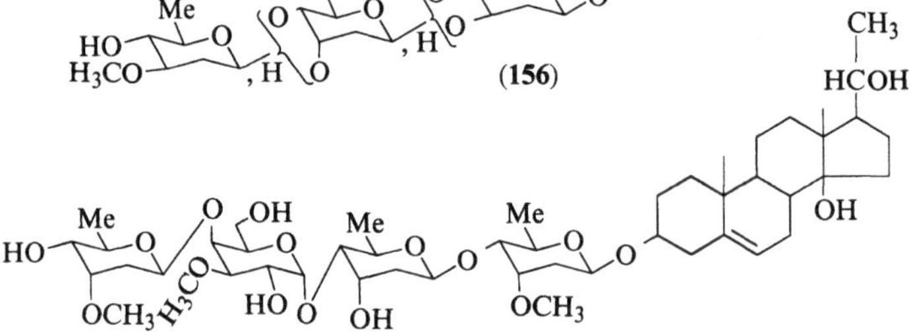

(**157**)

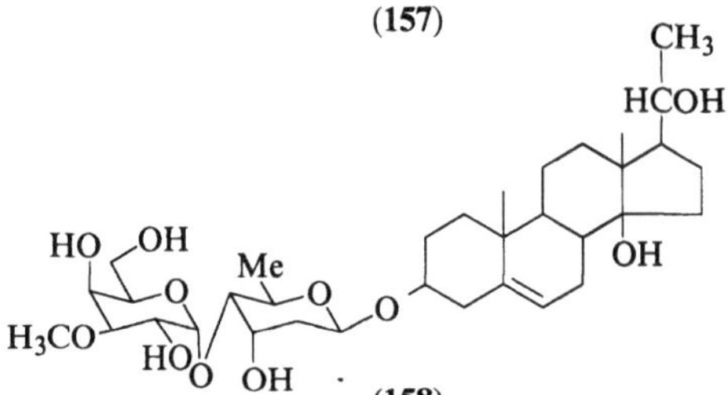

(**158**)

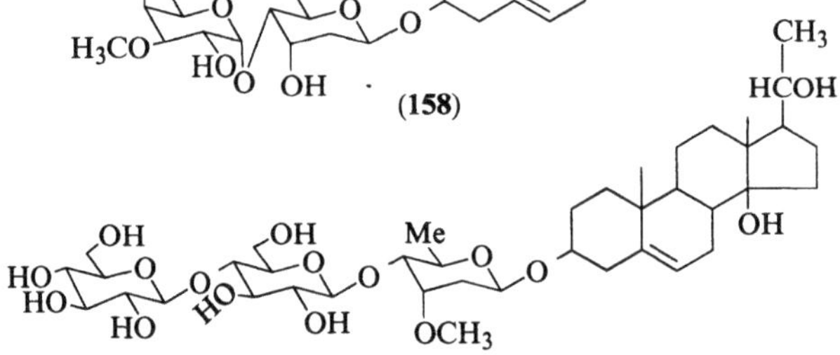

(**159**)

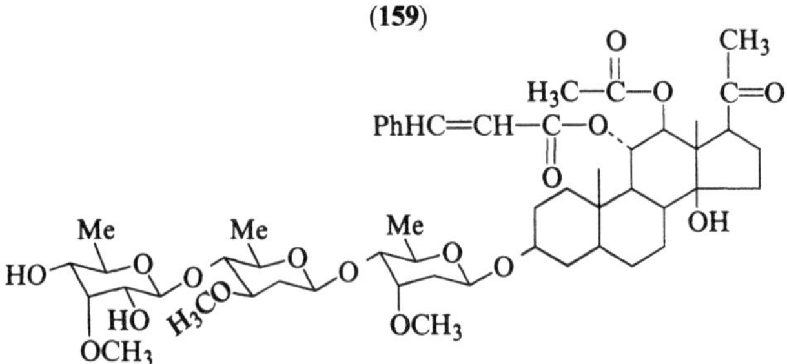

(**160**)

Table 2 (*continued*)

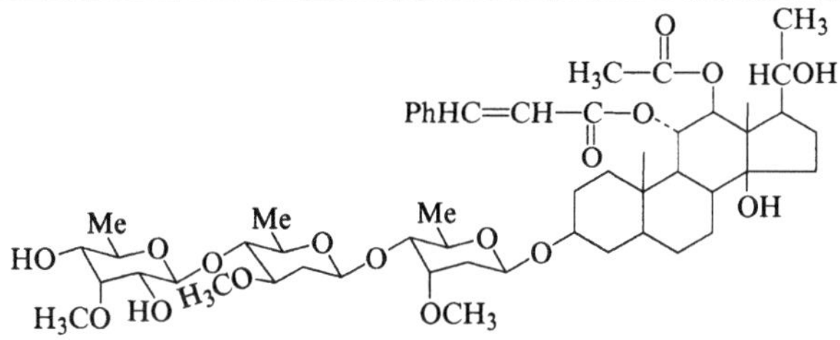

(**161**)

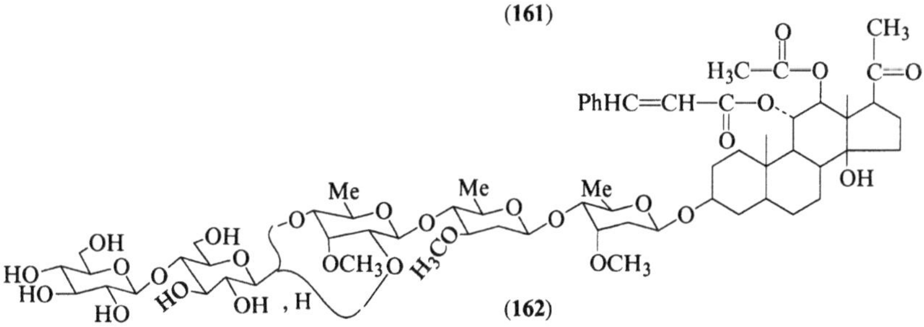

(**162**)

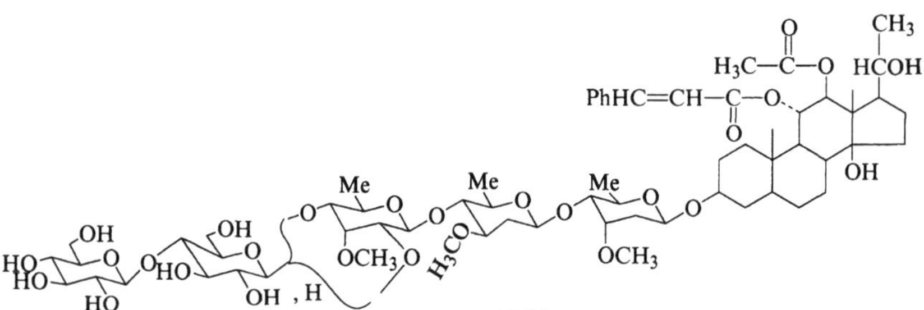

(**163**)

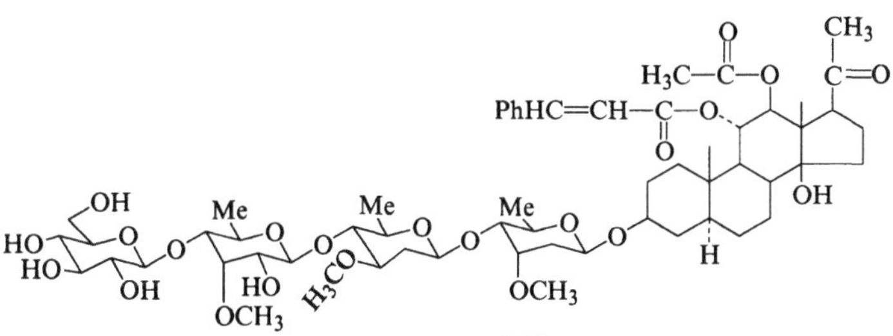

(**164**)

Table 2 (*continued*)

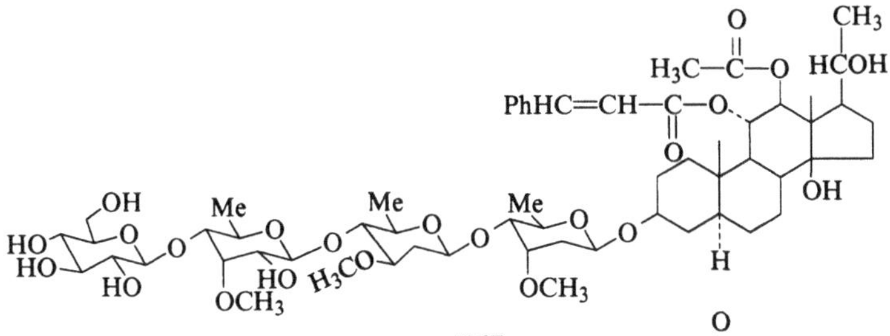

(**165**)

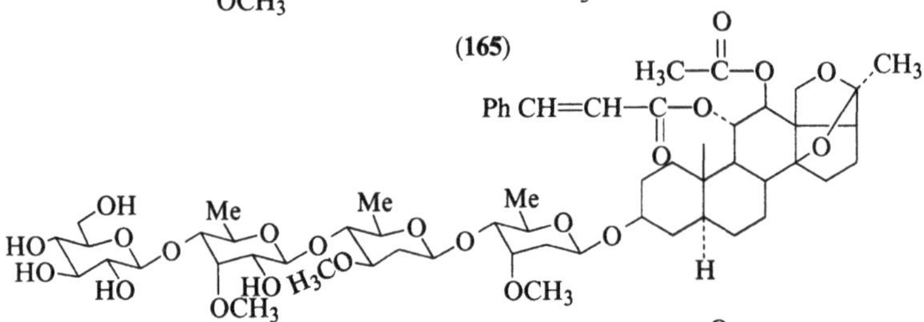

(**166**)

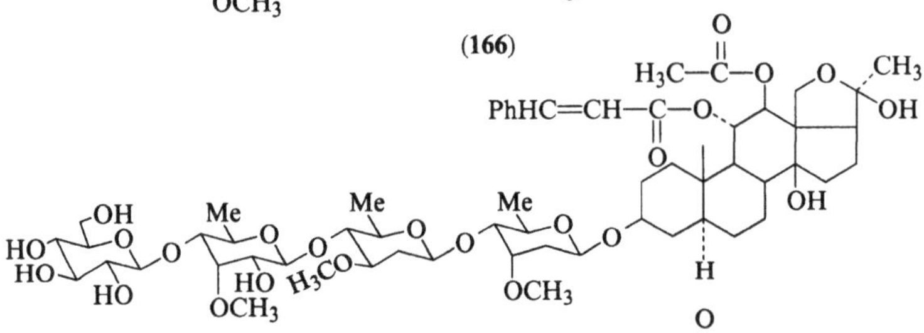

(**167**)

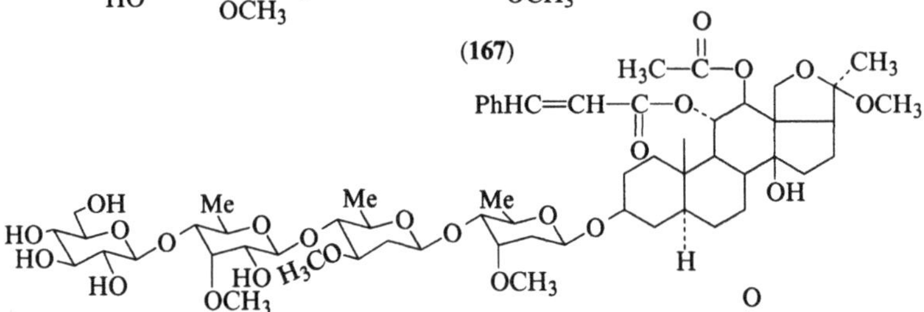

(**168**)

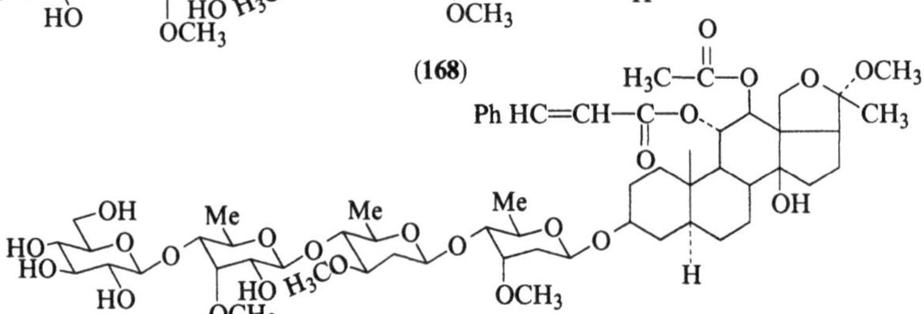

(**169**)

Table 2 (*continued*)

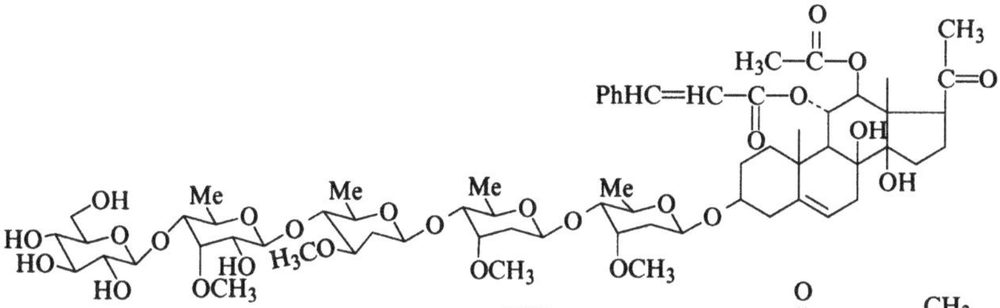

(170)

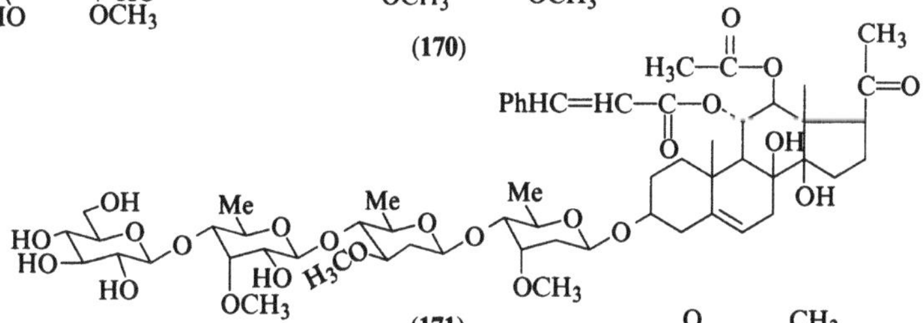

(171)

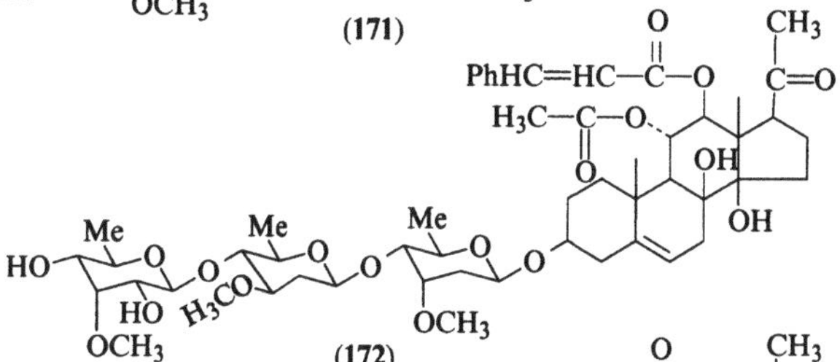

(172)

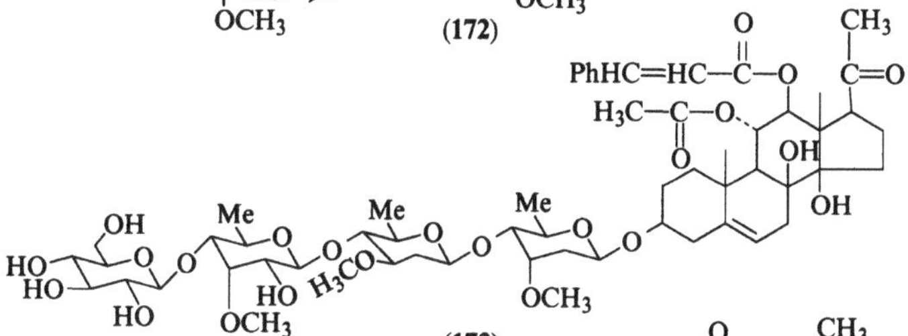

(173)

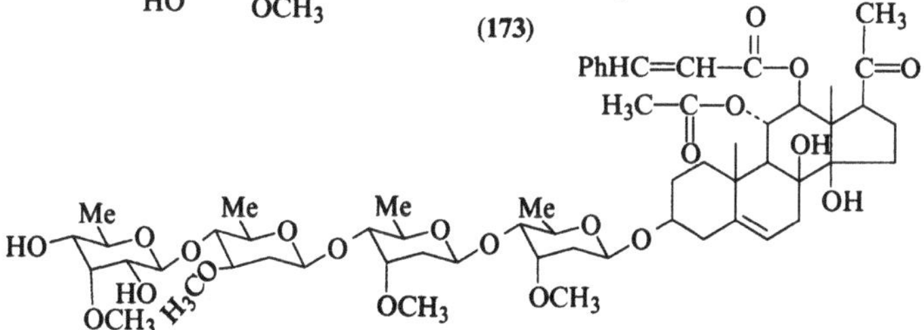

(174)

Table 2 (*continued*)

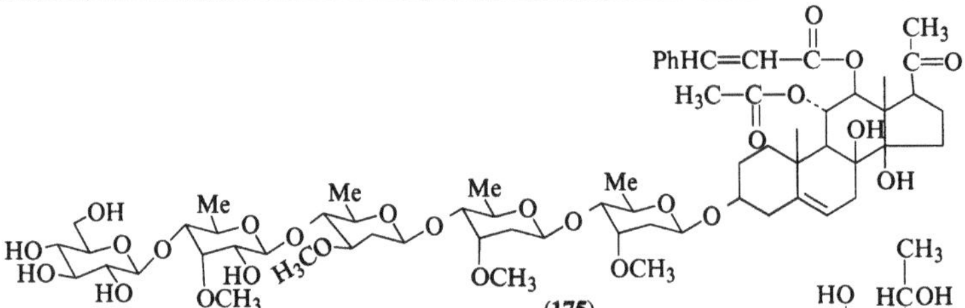

(175)

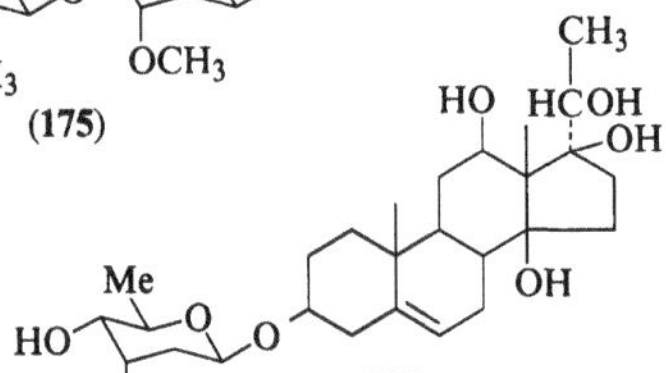

(177)

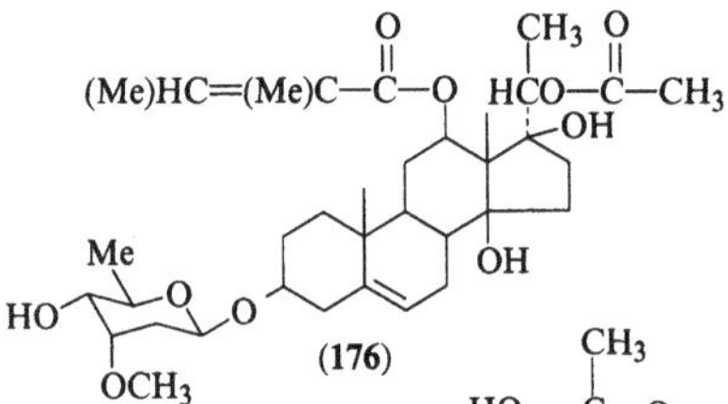

(176)

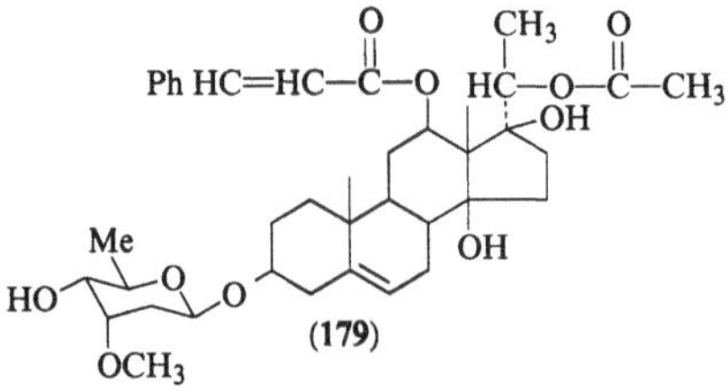

(179)

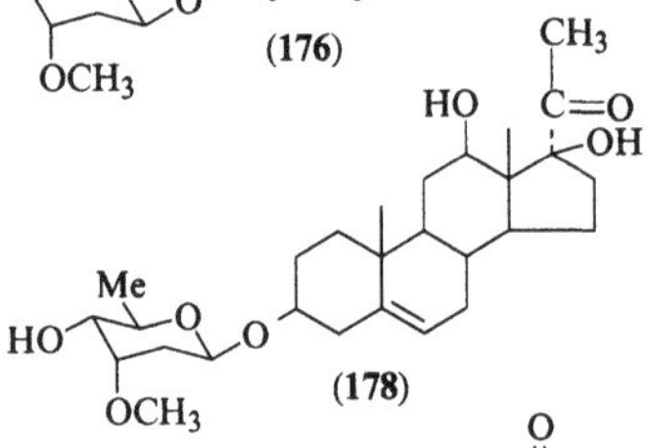

(178)

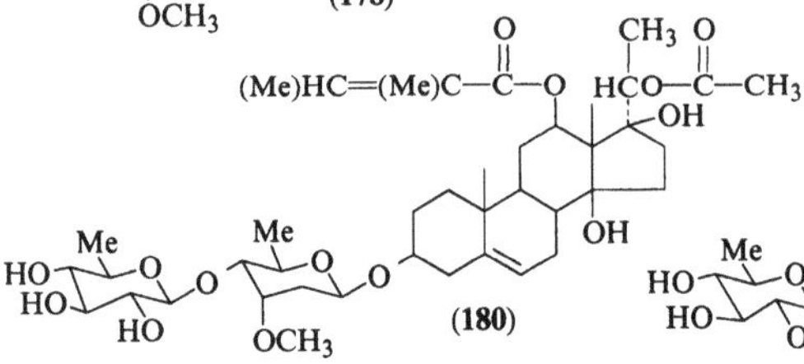

(180)

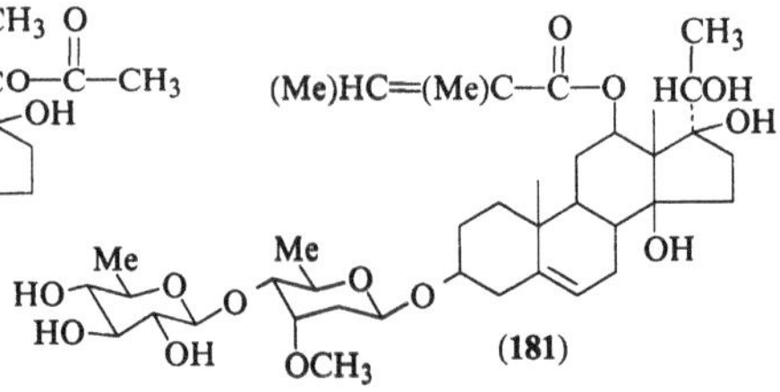

(181)

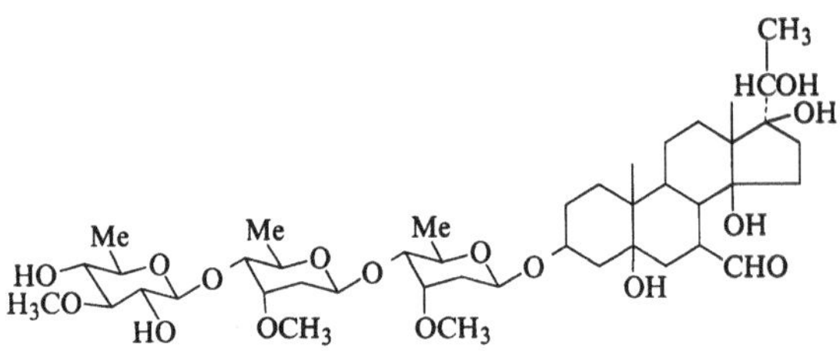

(182)

Table 2 (*continued*)

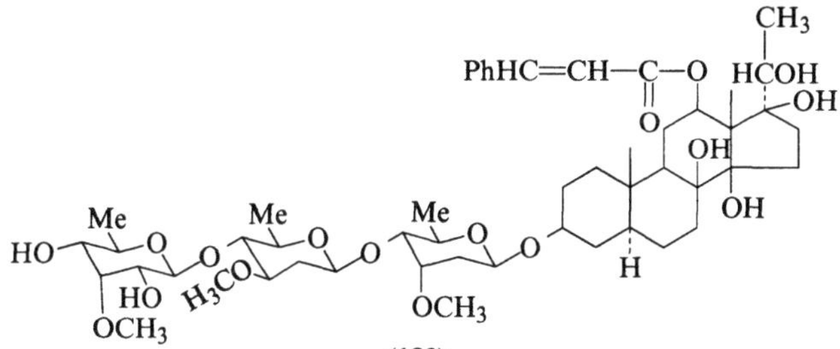

(**183**)

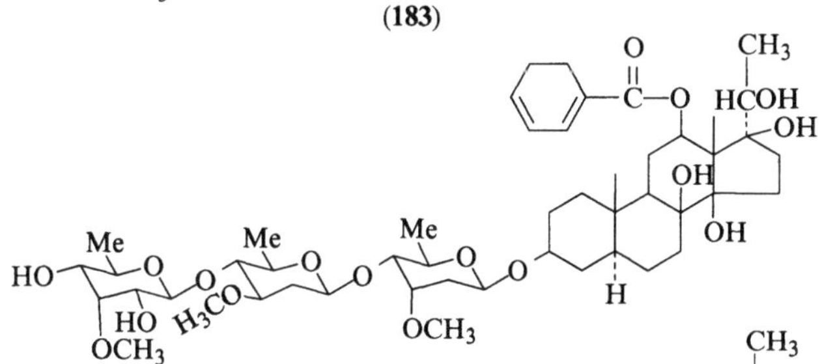

(**184**)

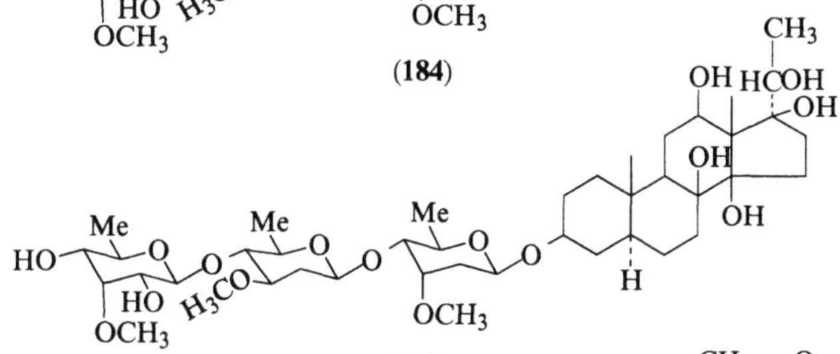

(**185**)

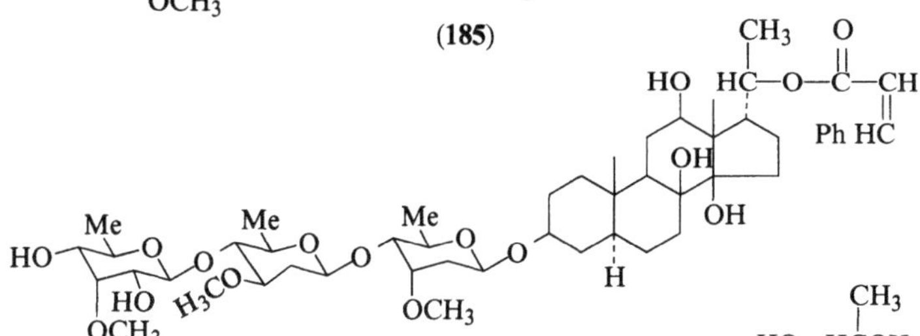

(**186**)

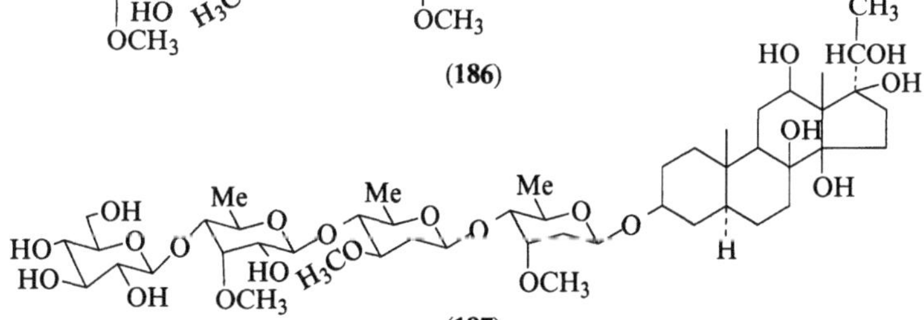

(**187**)

Table 2 (*continued*)

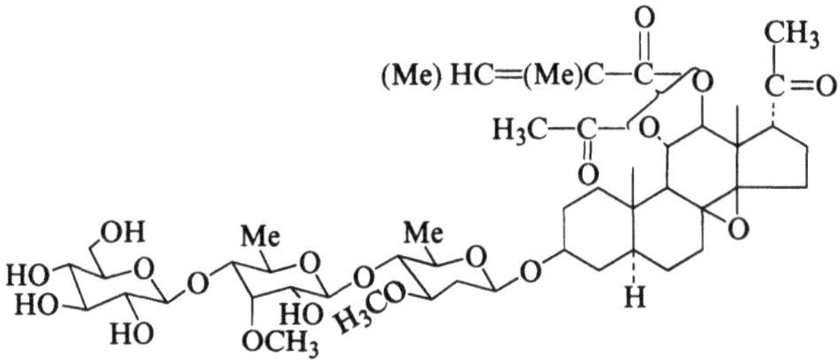

(**188**)

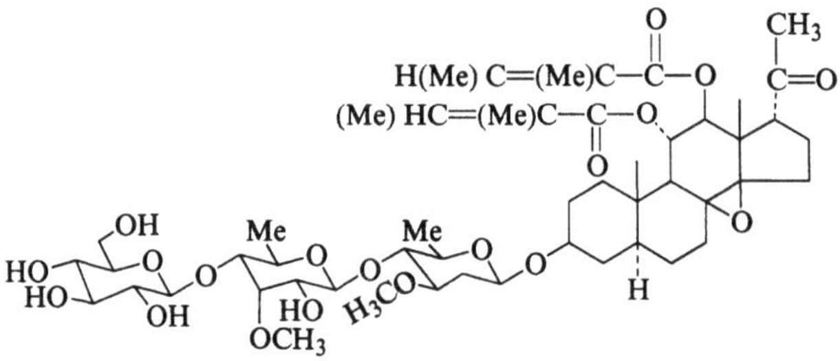

(**189**)

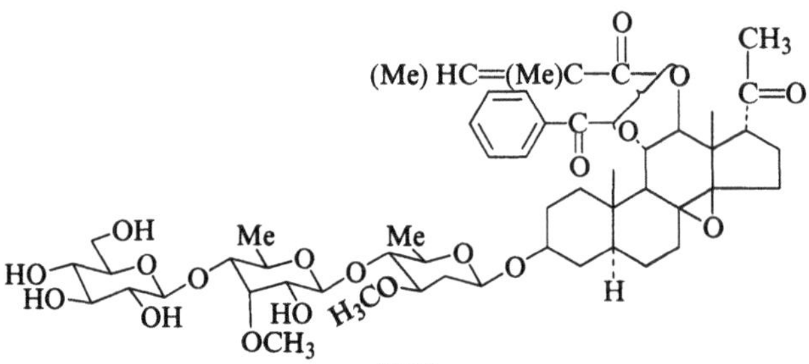

(**190**)

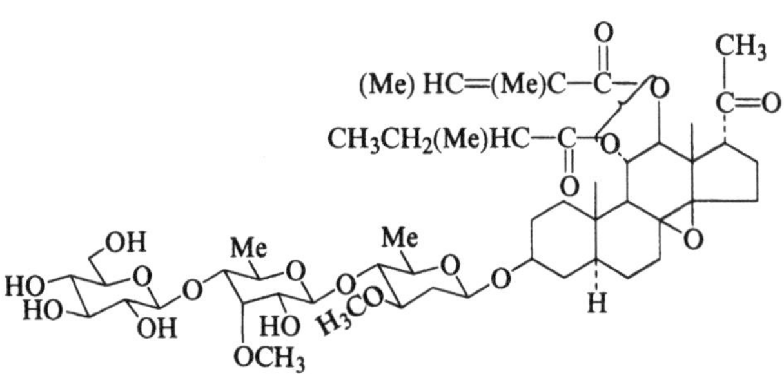

(**191**)

Table 2 (*continued*)

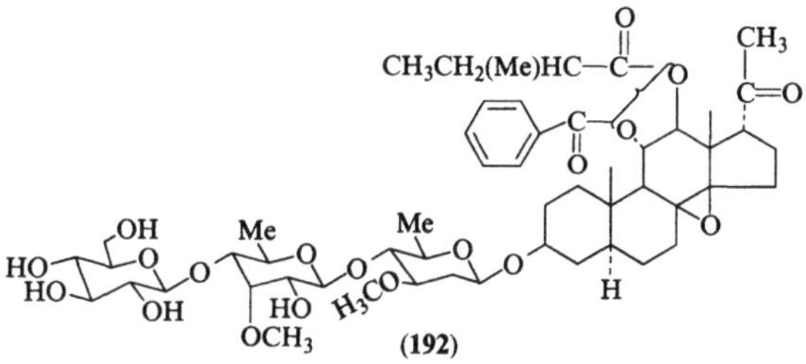

(192)

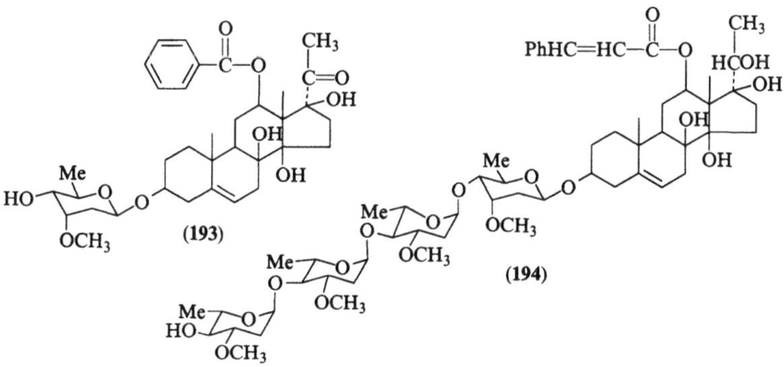

(193)

(194)

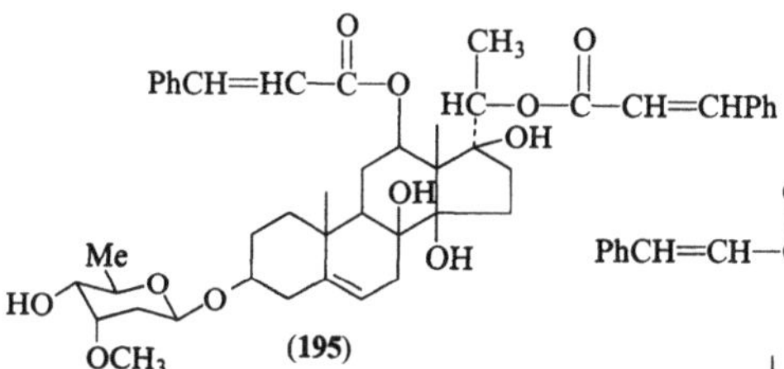

(195)

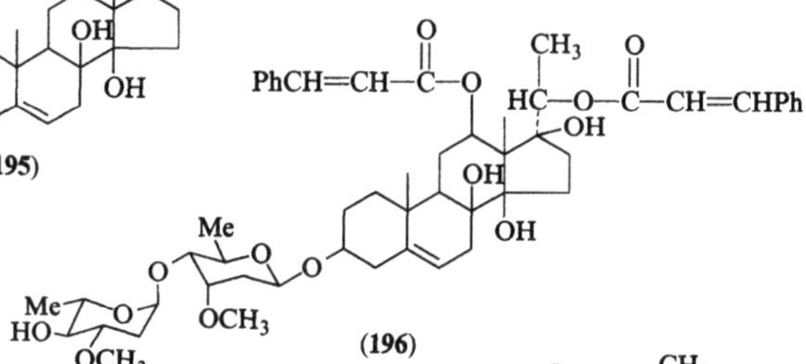

(196)

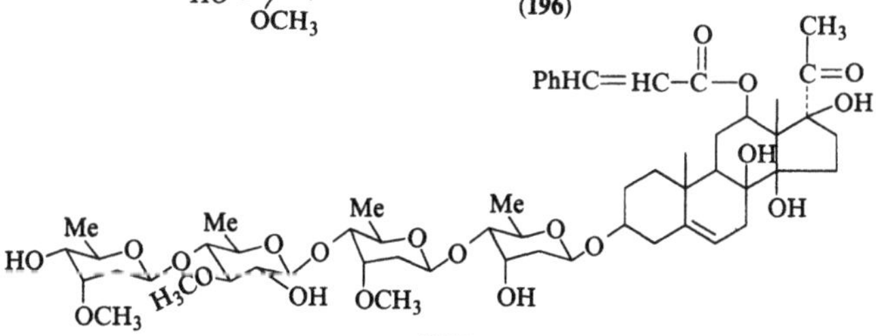

(197)

Table 2 (*continued*)

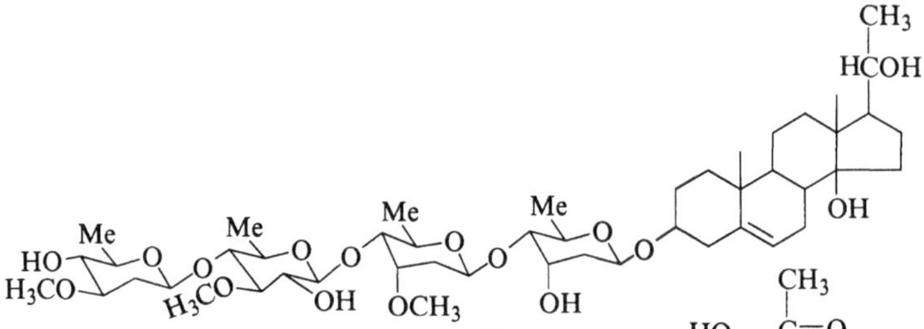

(198)

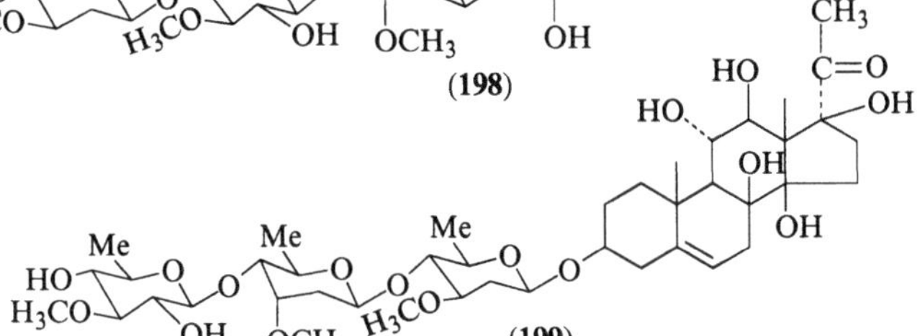

(199)

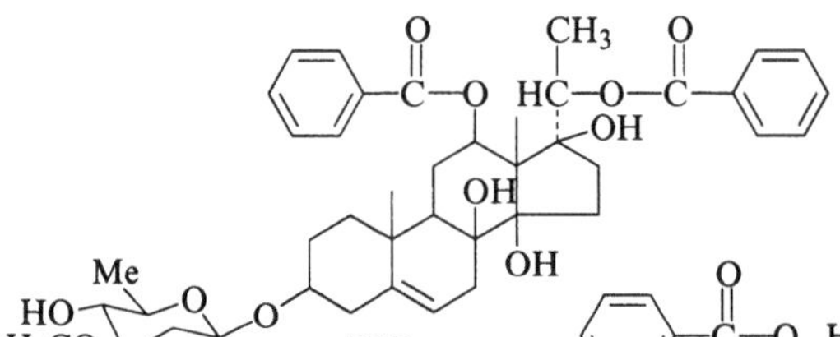

(200)

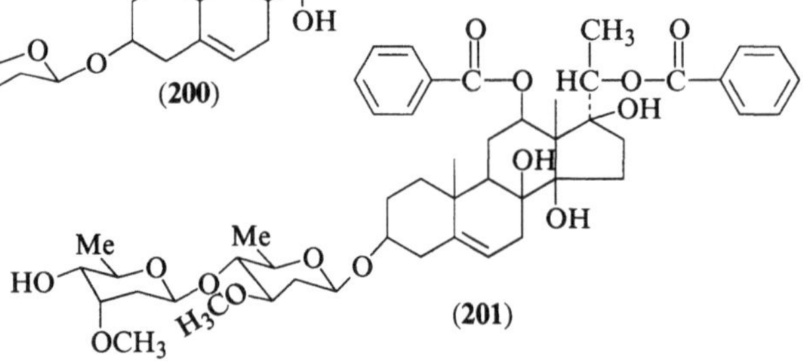

(201)

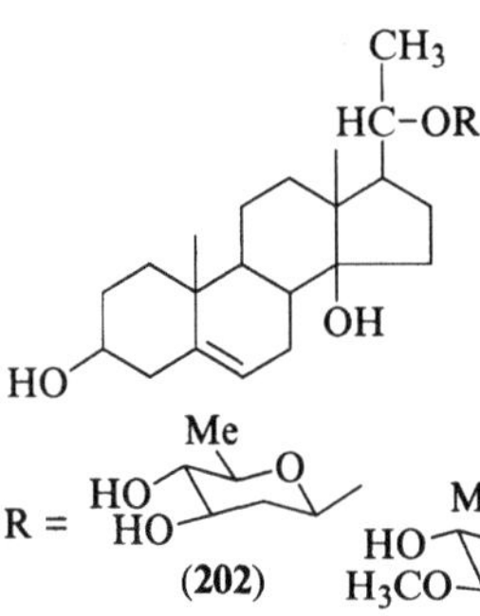

(202)

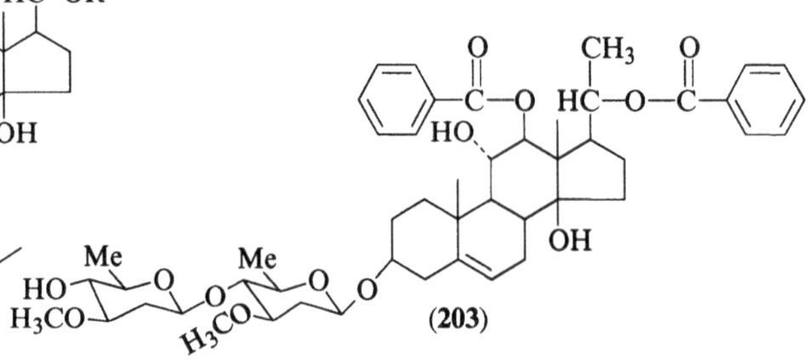

(203)

References, pp. 309–325

Table 2 (*continued*)

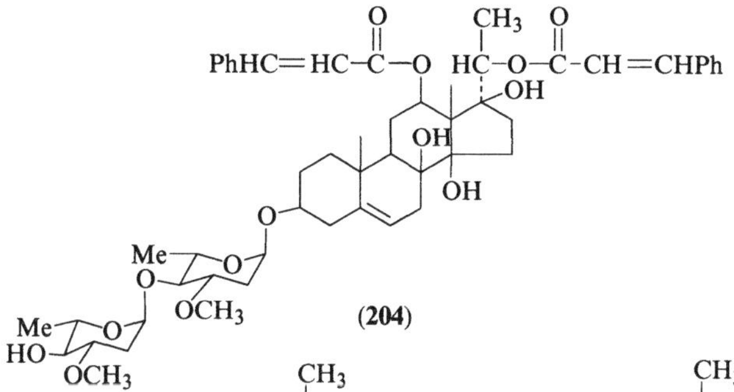

(**204**)

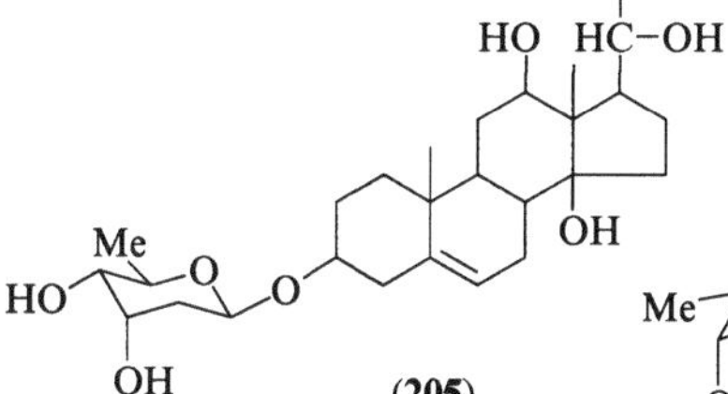

(**205**)

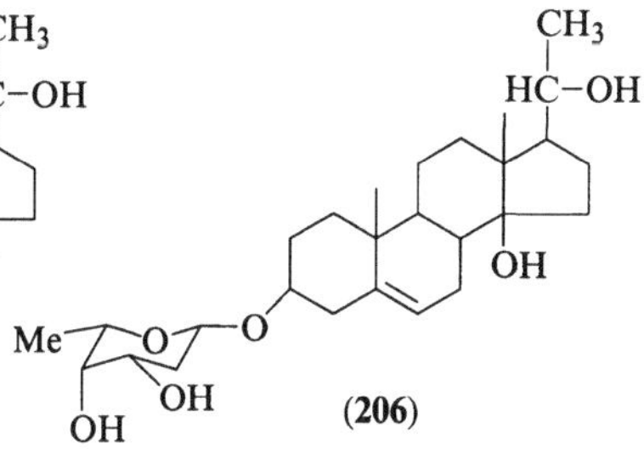

(**206**)

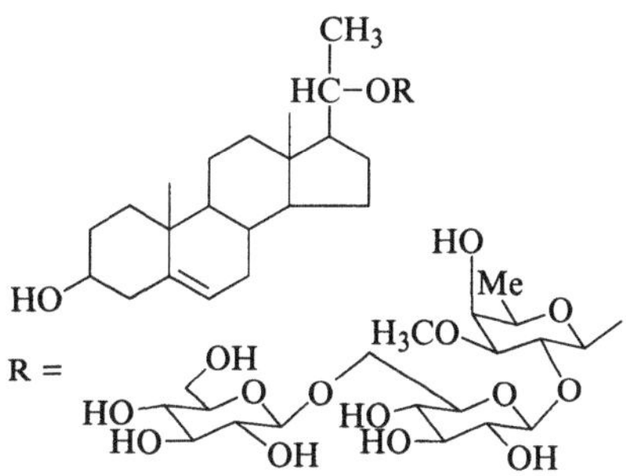

(**207**)

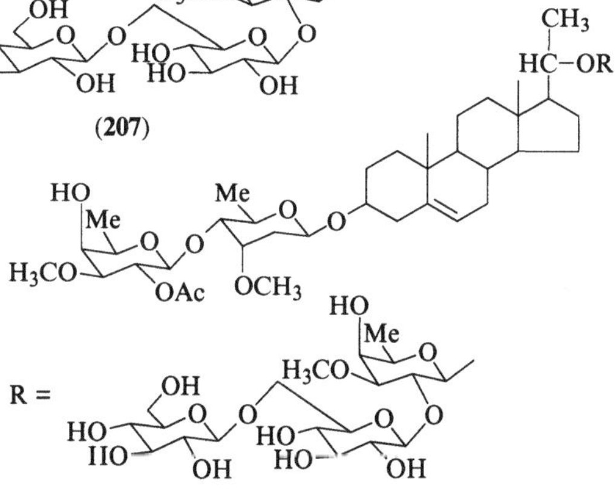

(**208**)

Table 2 (*continued*)

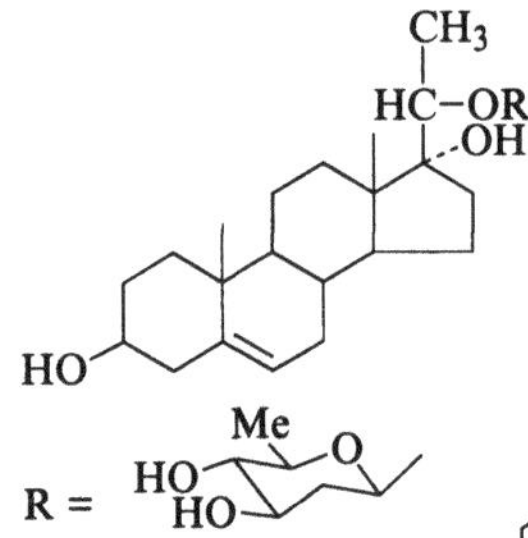

(**209**)

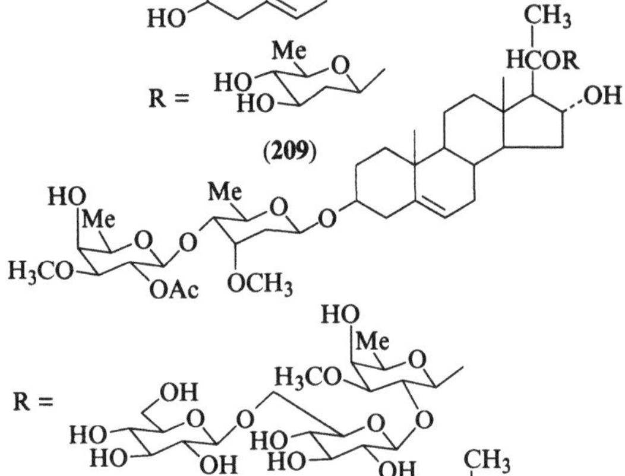

(**210**)

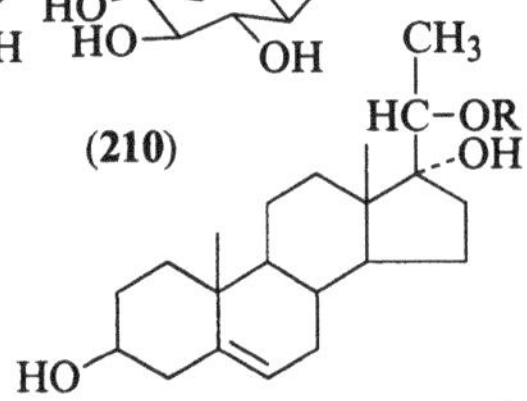

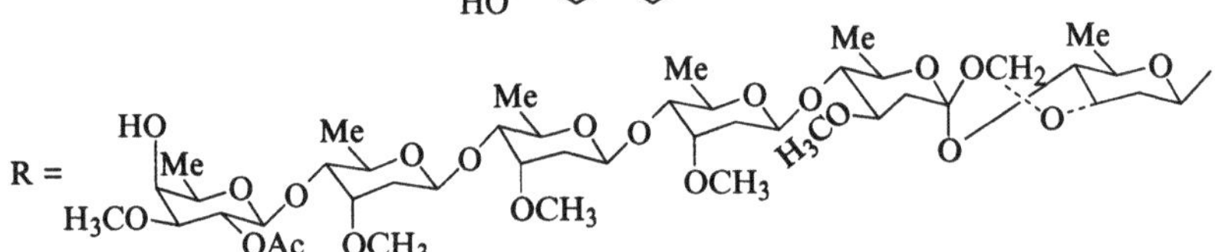

(**211**)

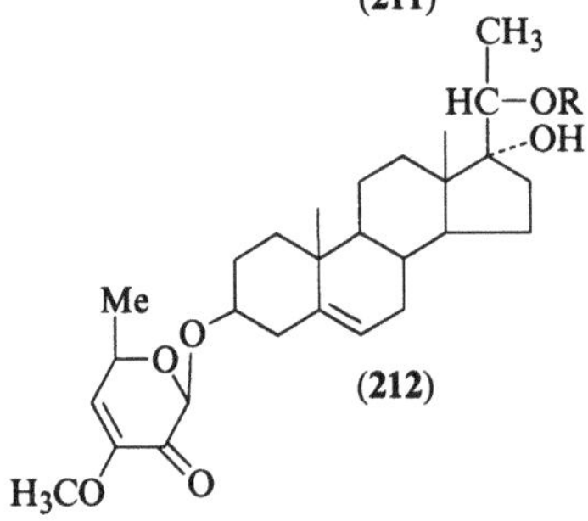

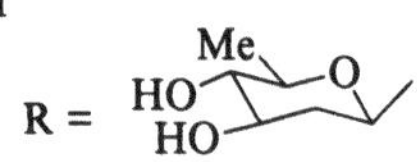

(**212**)

Table 2 (*continued*)

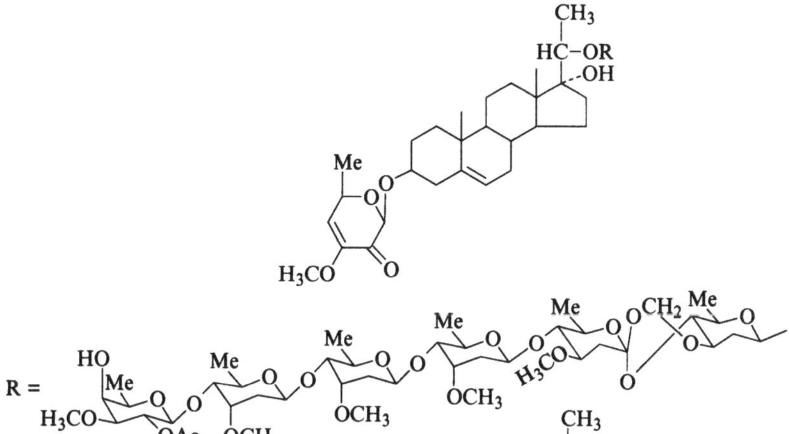

(**213**)

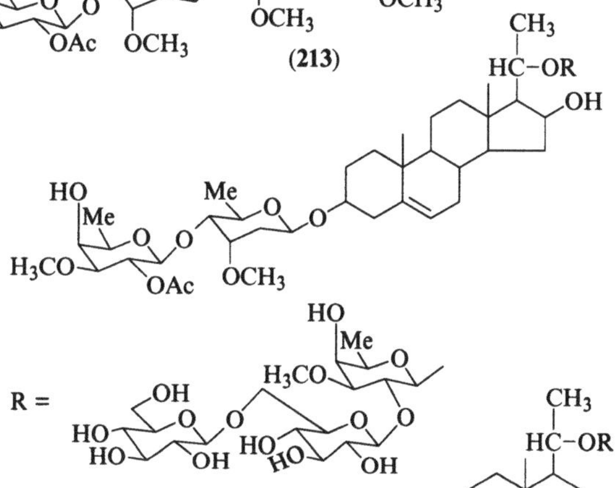

(**214**)

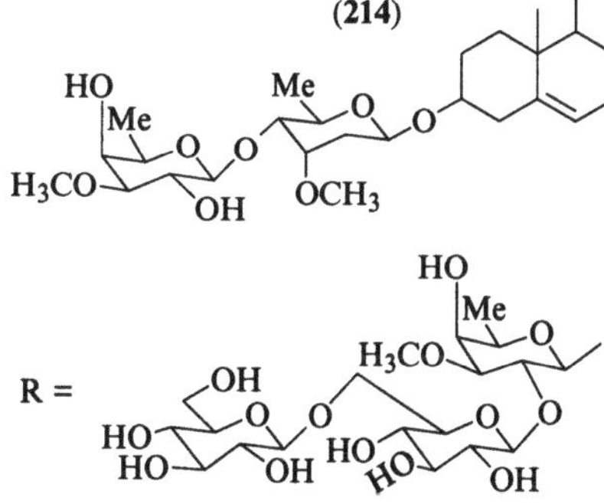

(**215**)

Table 2 (*continued*)

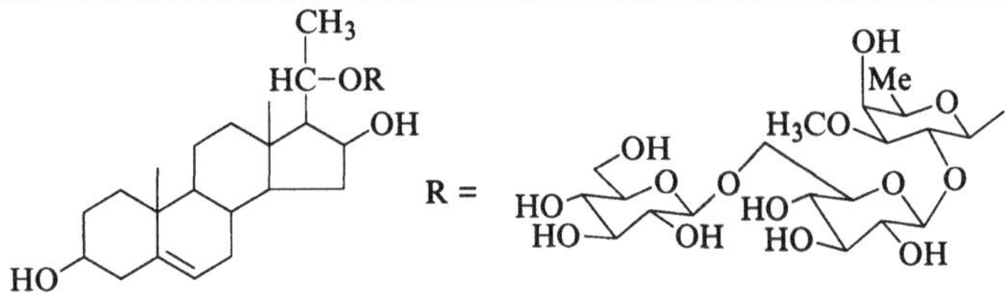

(**216**)

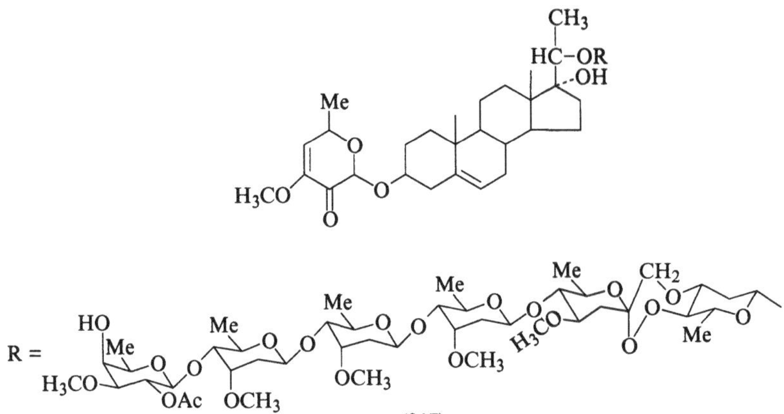

(**217**)

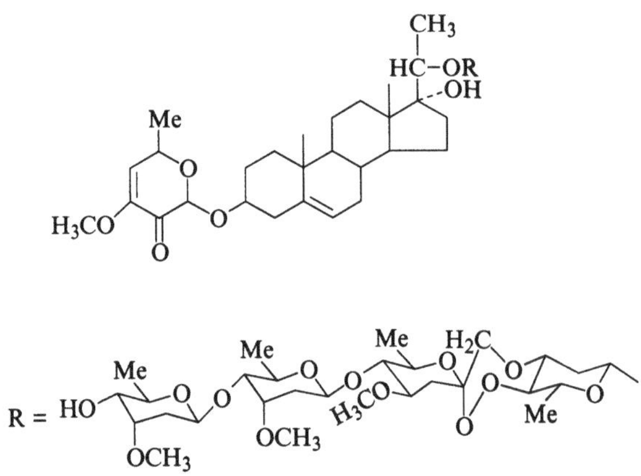

(**218**)

References, pp. 309–325

Table 2 (*continued*)

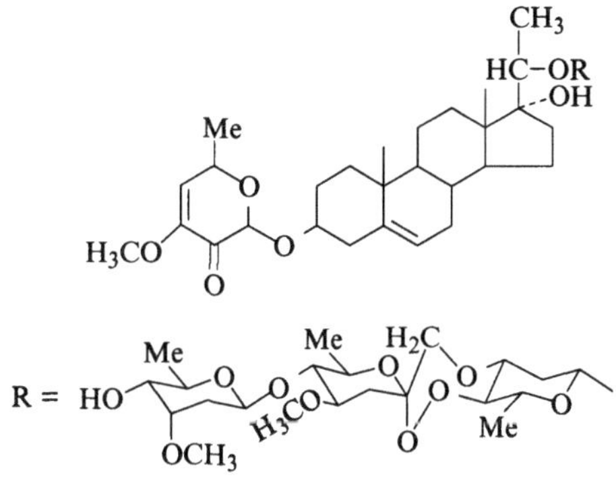

(**219**)

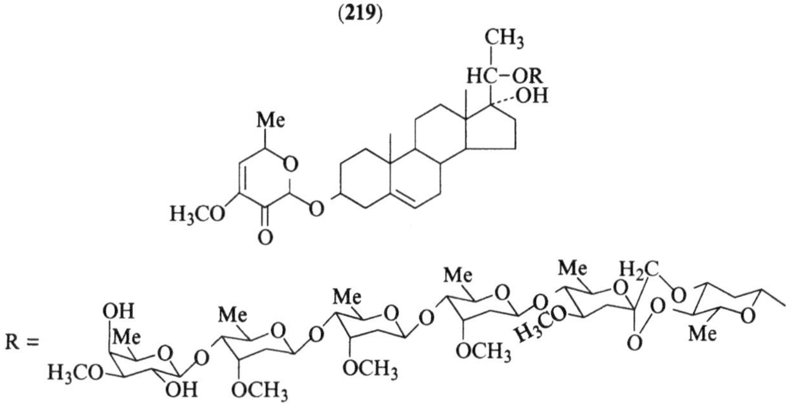

(**220**)

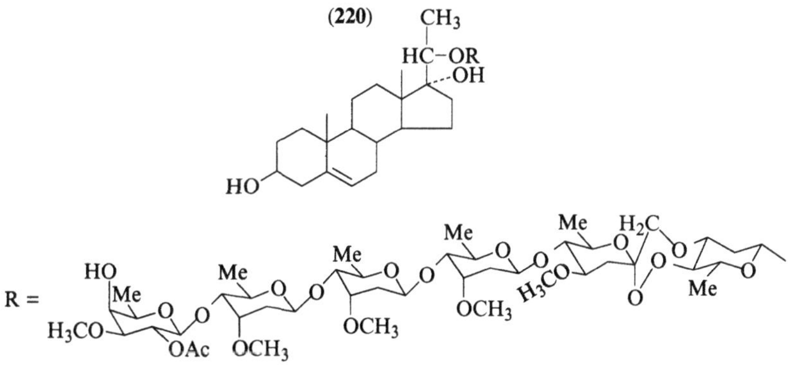

(**221**)

Table 2 (*continued*)

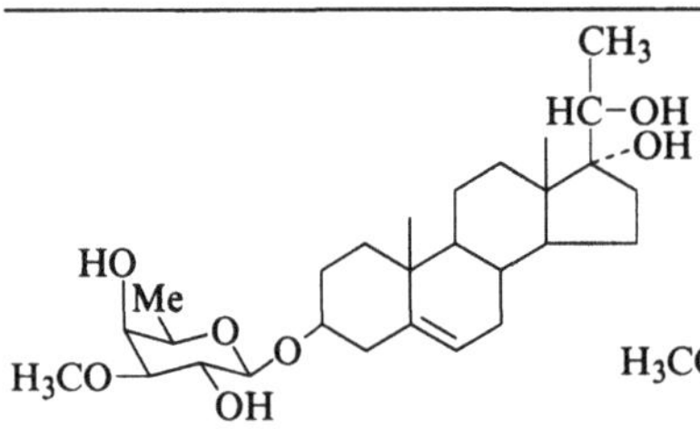

(**222**)

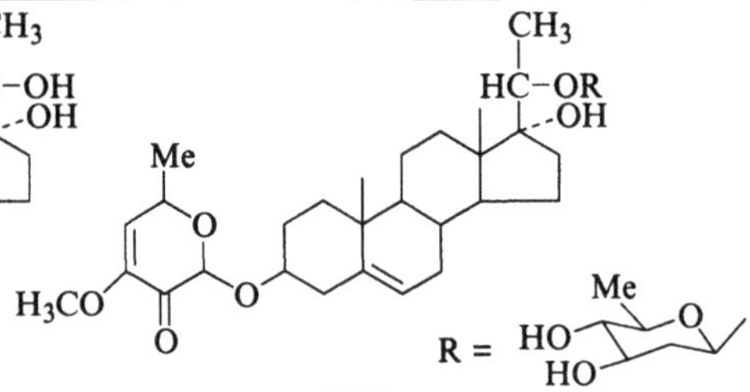

(**223**)

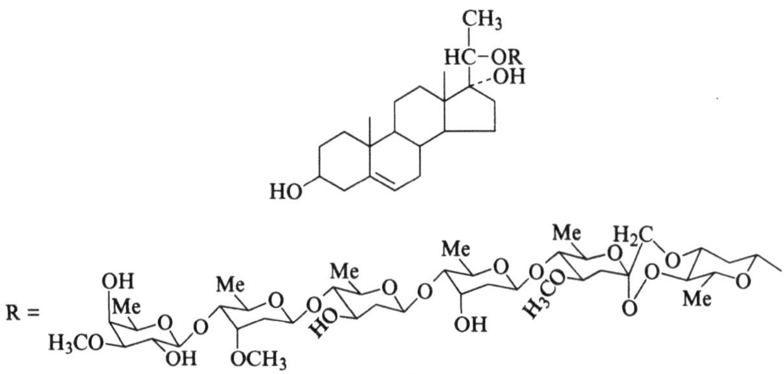

(**224**)

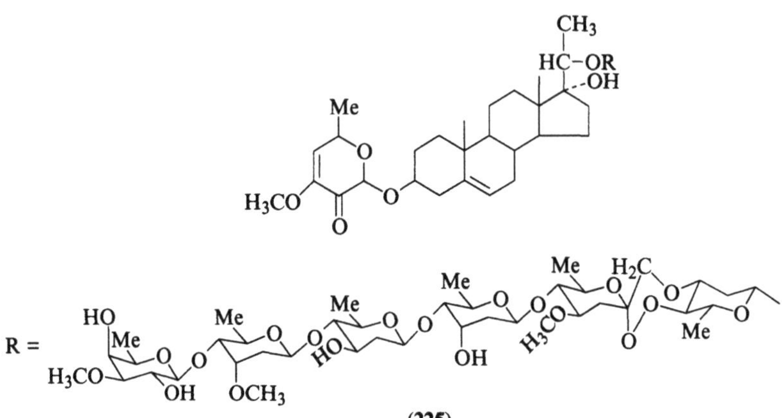

(**225**)

Table 2 (*continued*)

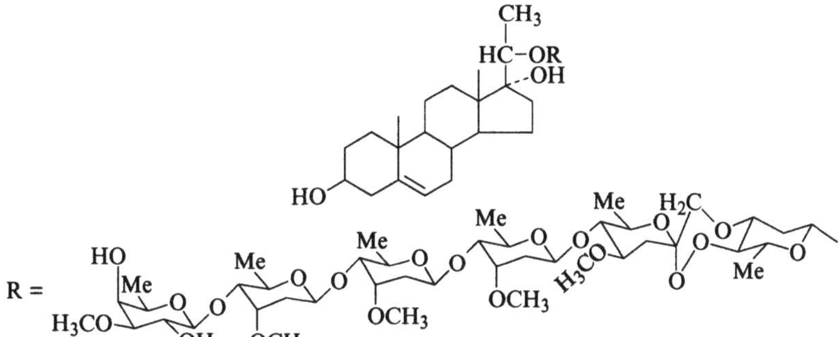

(**226**)

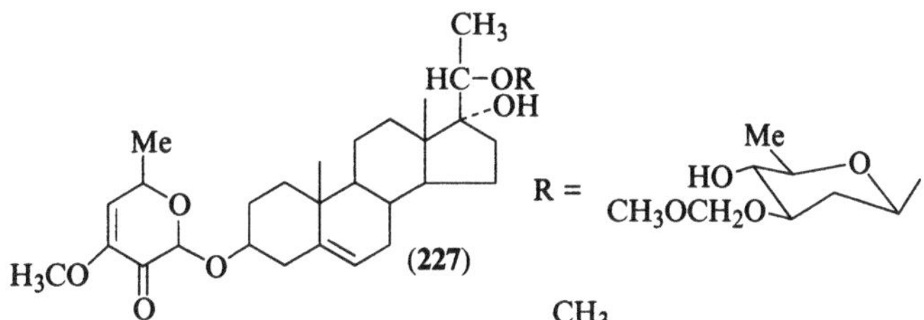

(**227**)

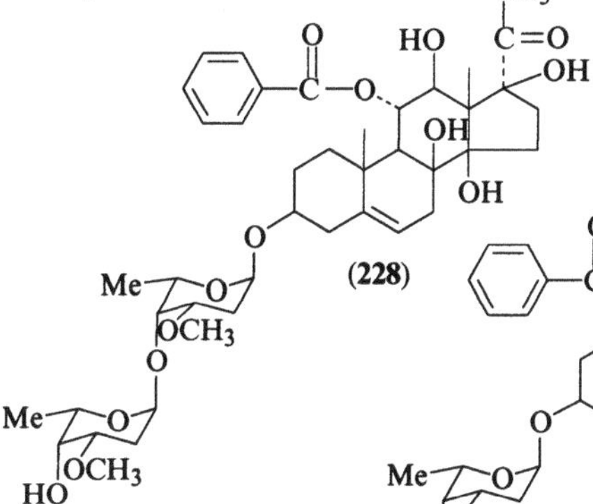

(**228**)

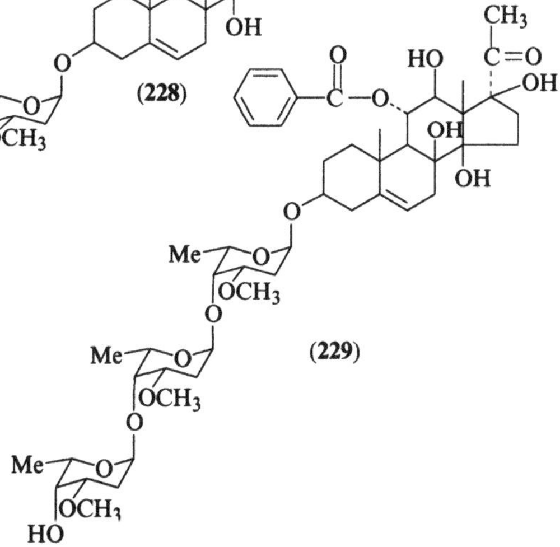

(**229**)

Table 2 (*continued*)

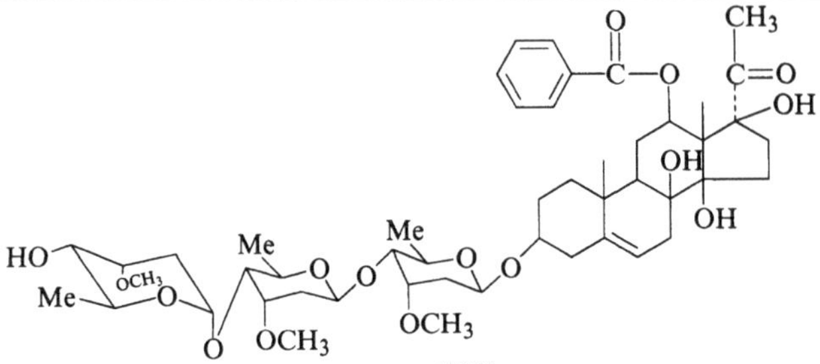

(**230**)

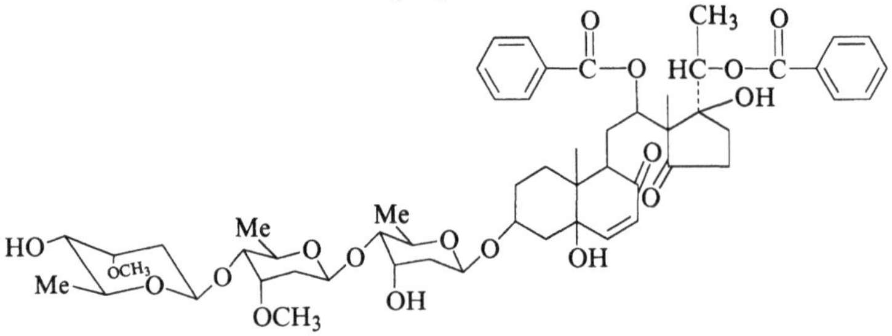

(**231**)

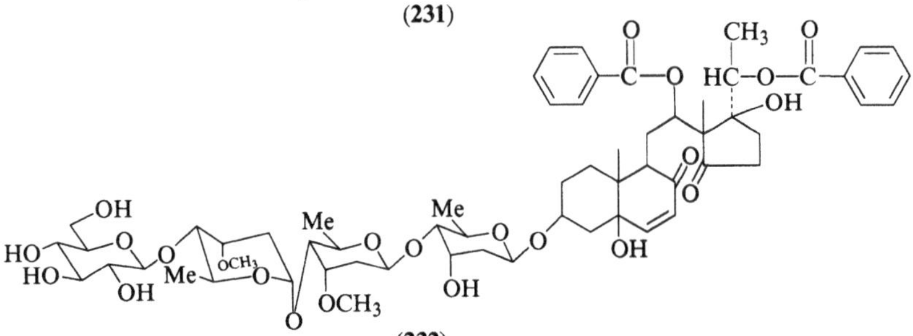

(**232**)

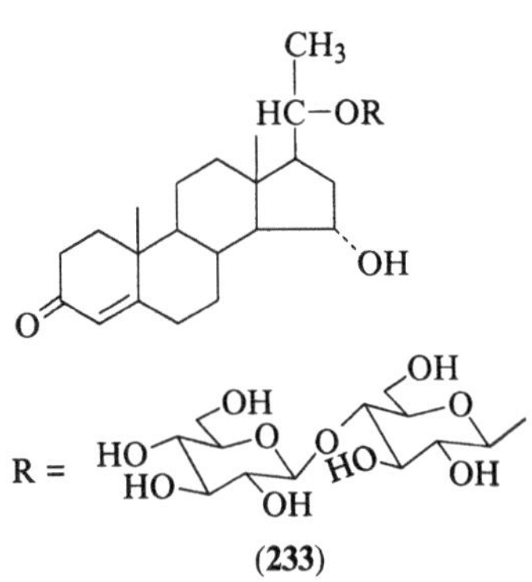

(**233**)

Table 2 (*continued*)

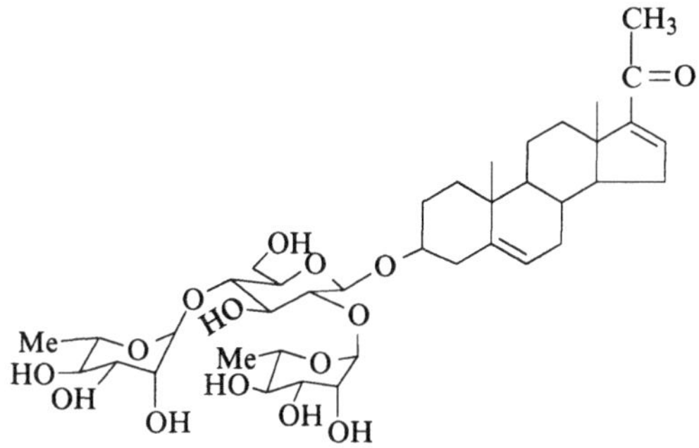

(**234**)

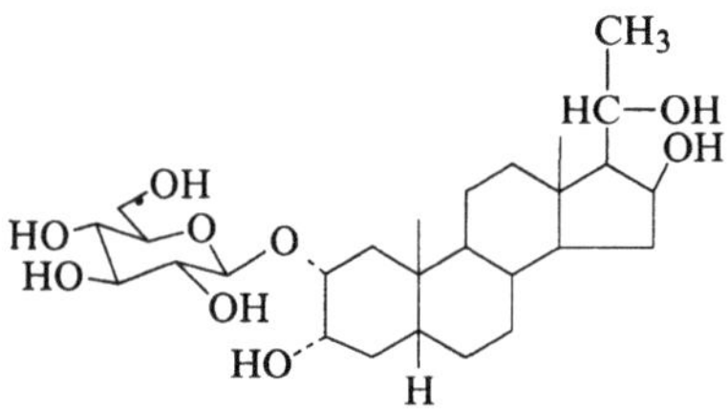

(**235**)

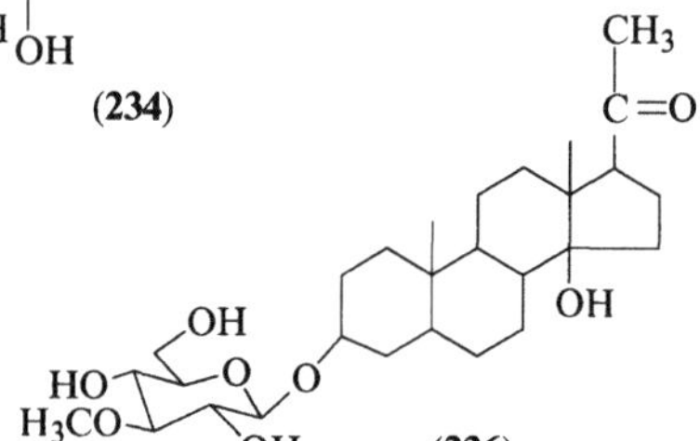

(**236**)

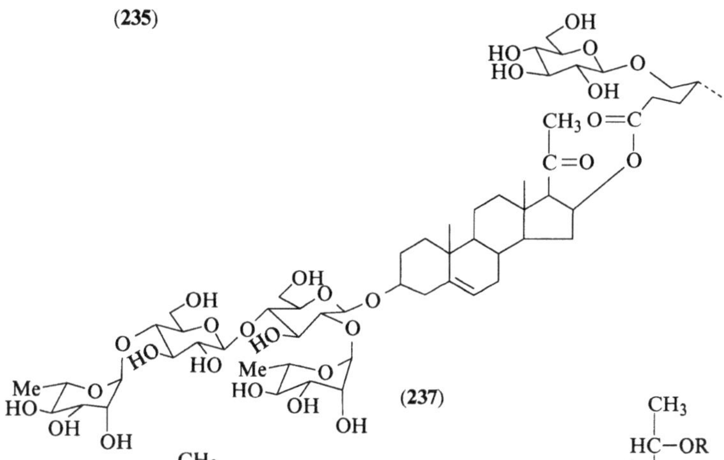

(**237**)

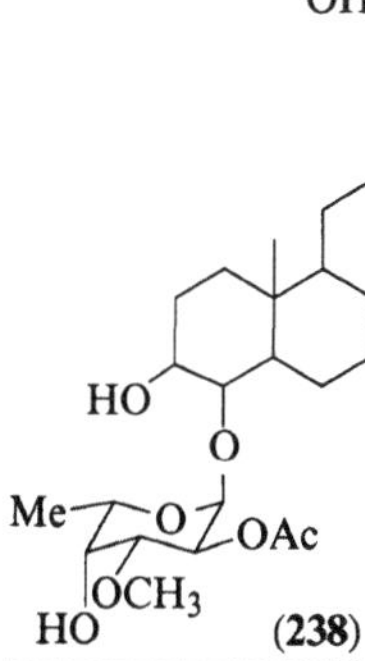

(**238**)

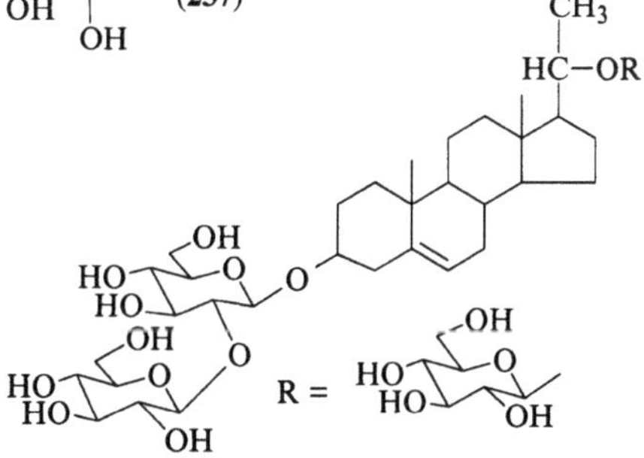

(**239**)

Table 2 (*continued*)

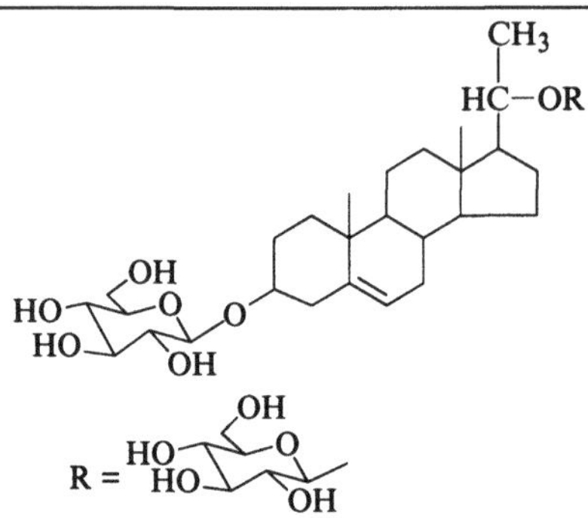

(**240**)

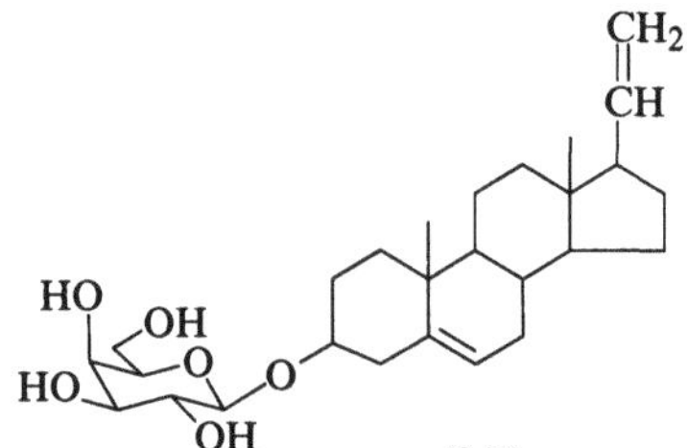

(**241**)

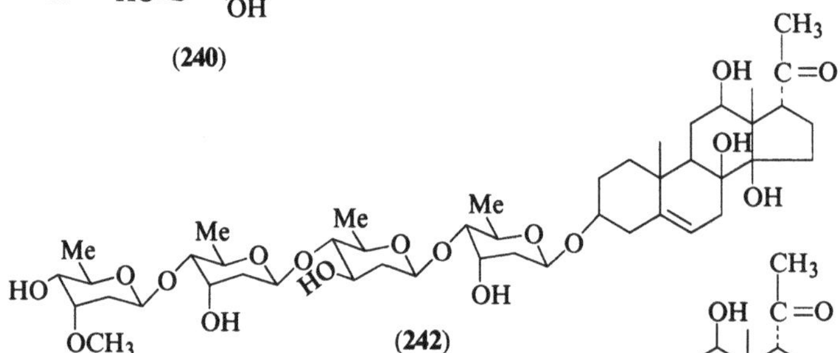

(**242**)

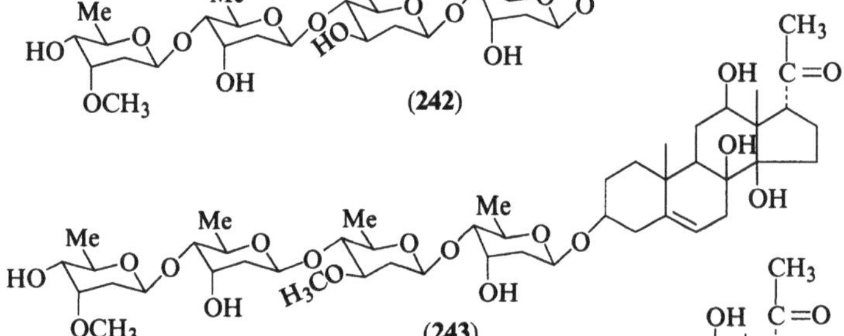

(**243**)

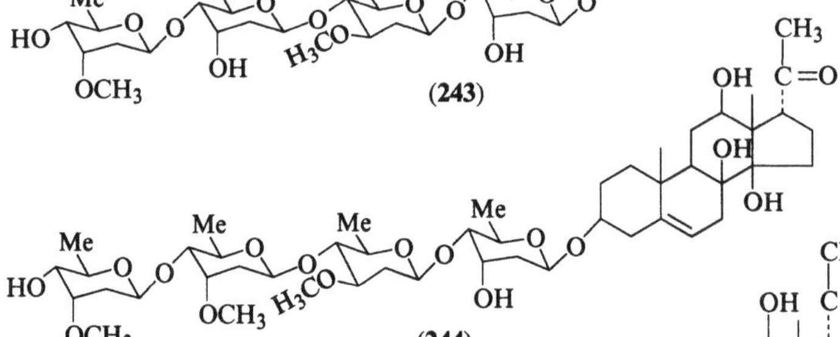

(**244**)

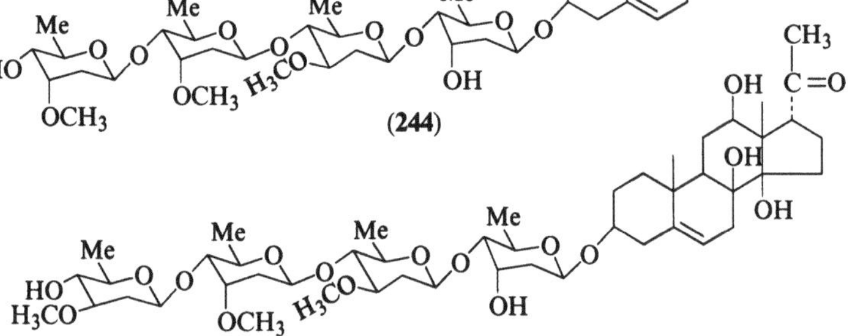

(**245**)

Table 2 (*continued*)

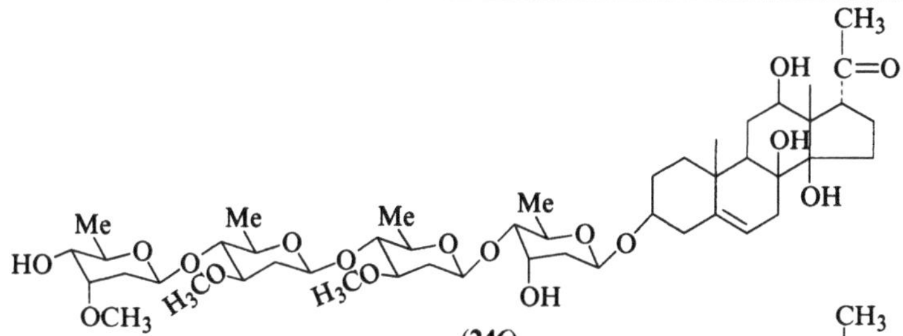

(**246**)

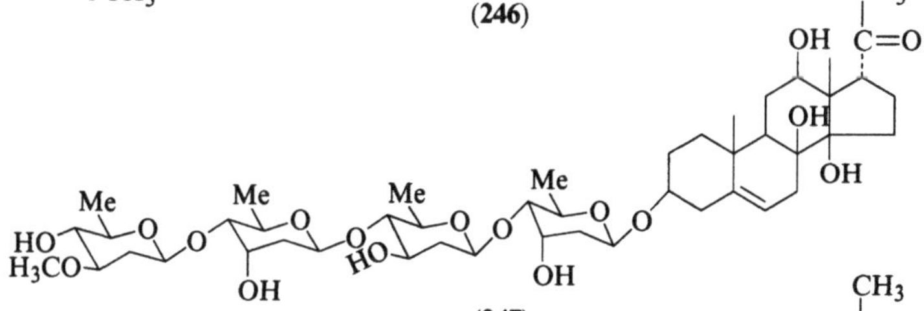

(**247**)

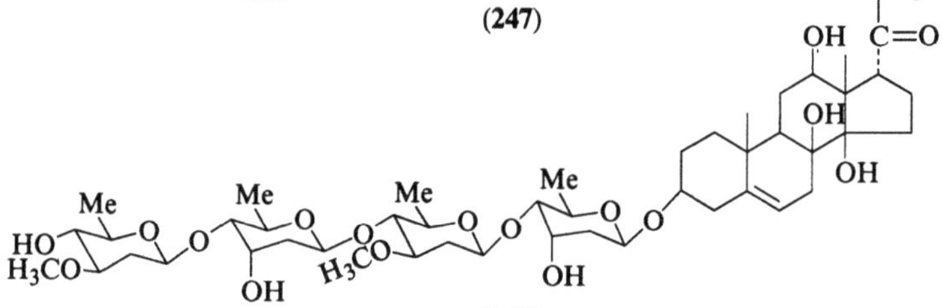

(**248**)

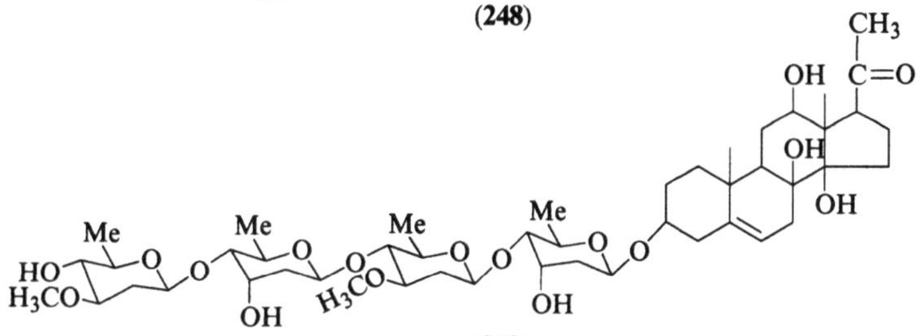

(**249**)

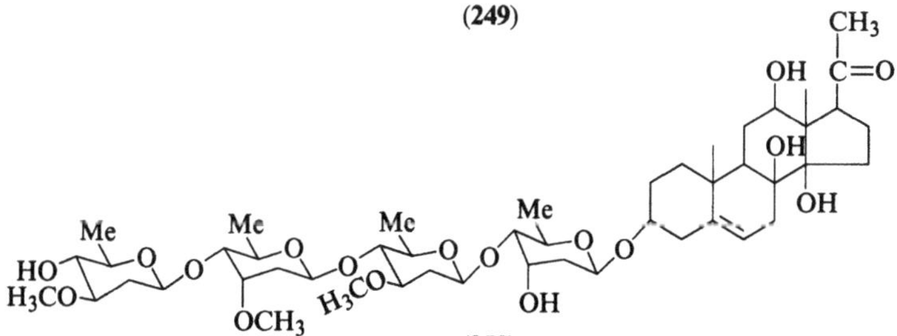

(**250**)

Table 2 (*continued*)

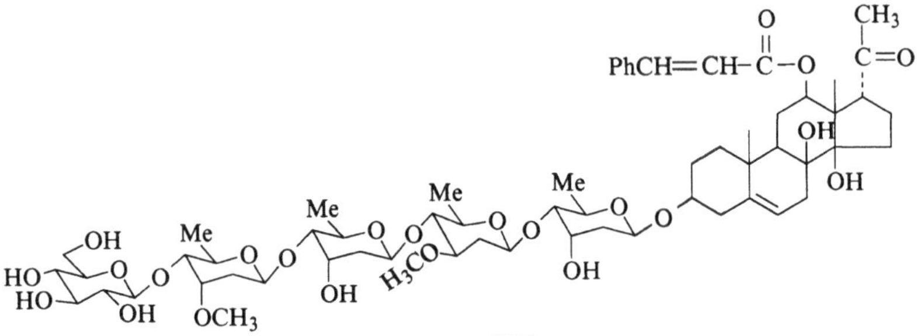

(**251**)

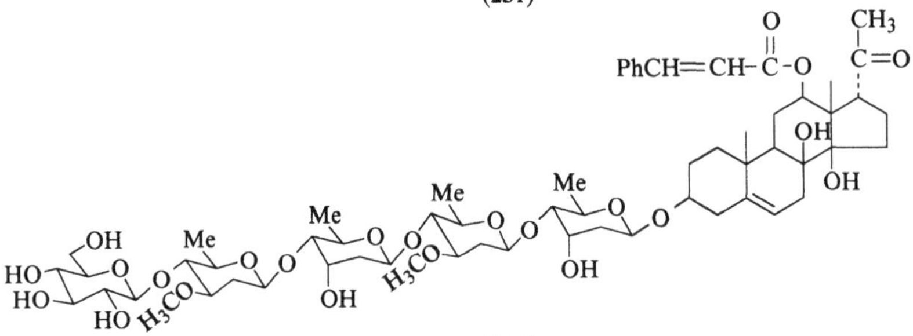

(**252**)

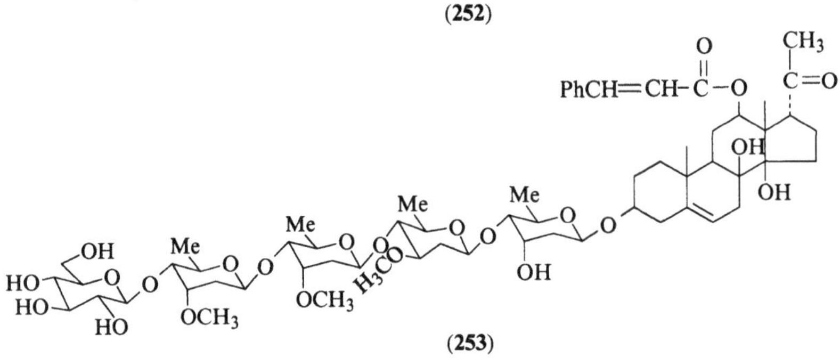

(**253**)

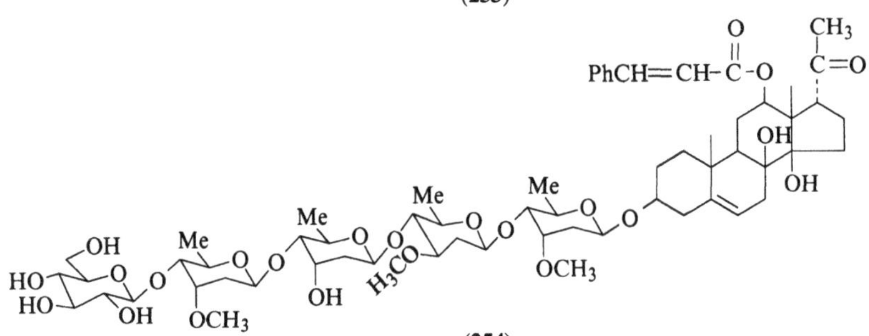

(**254**)

Table 2 (*continued*)

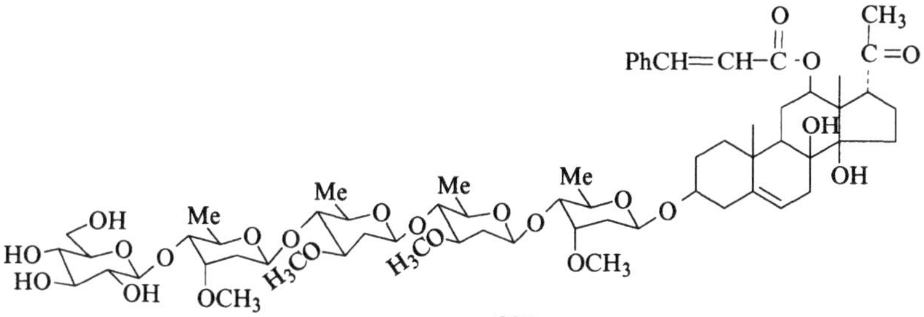

(**255**)

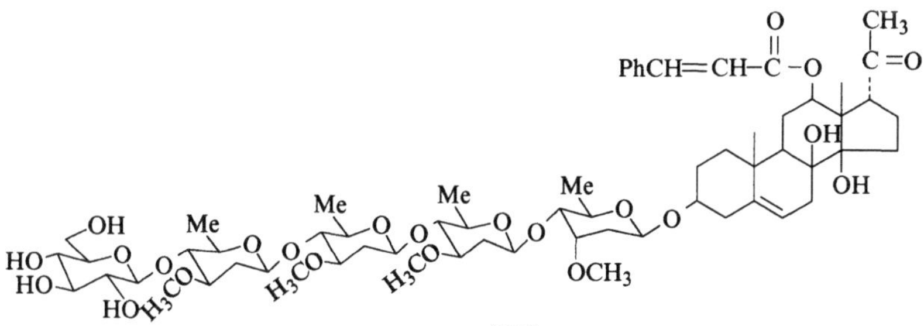

(**256**)

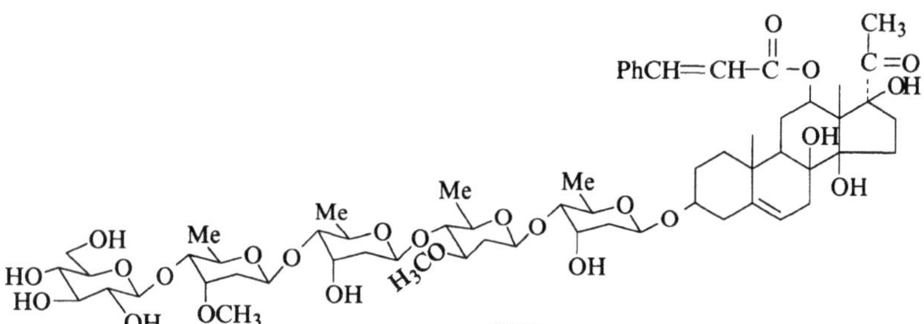

(**257**)

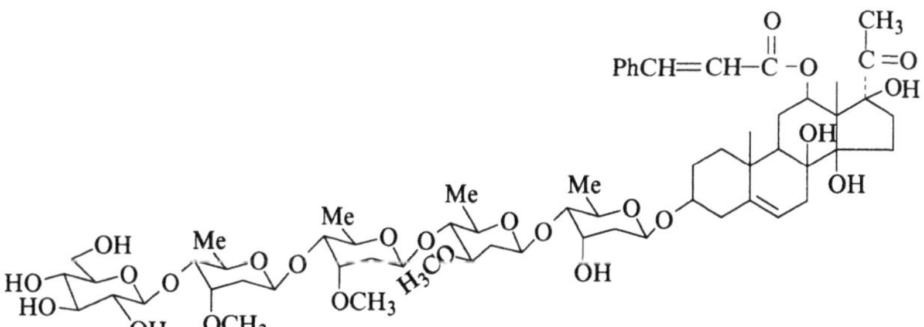

(**258**)

Table 2 (*continued*)

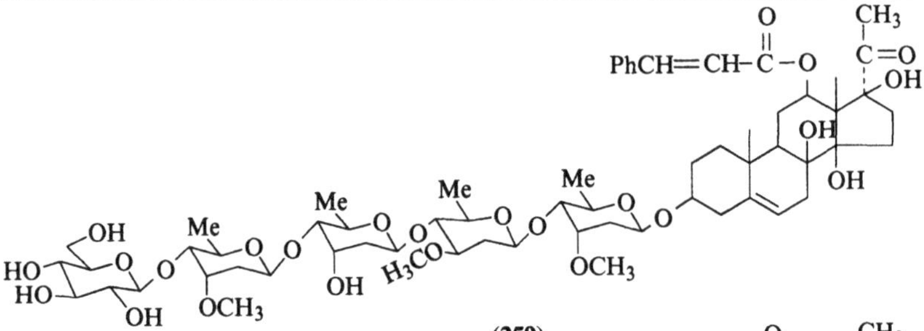

(**259**)

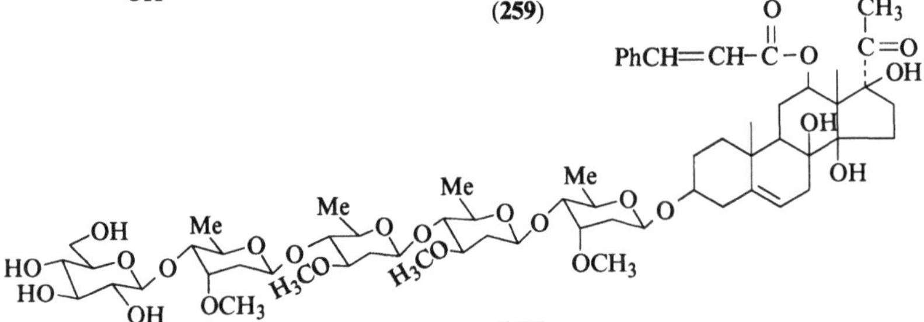

(**260**)

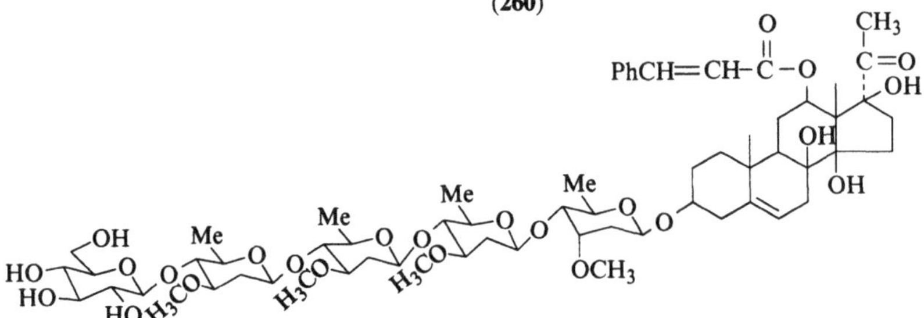

(**261**)

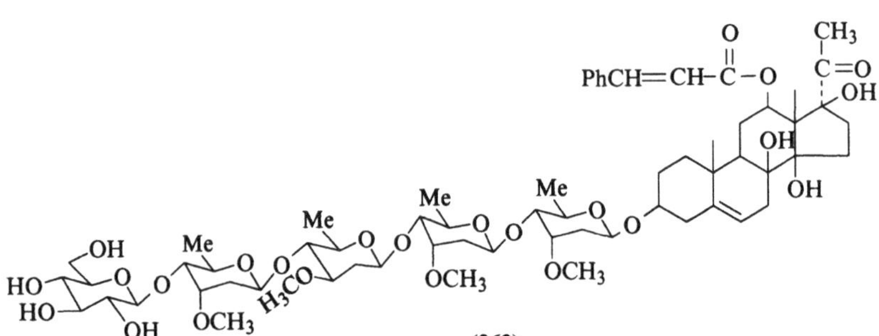

(**262**)

Table 2 (*continued*)

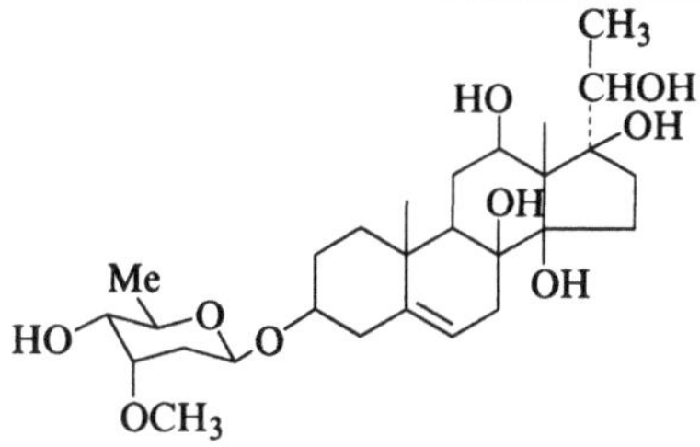

(**263**)

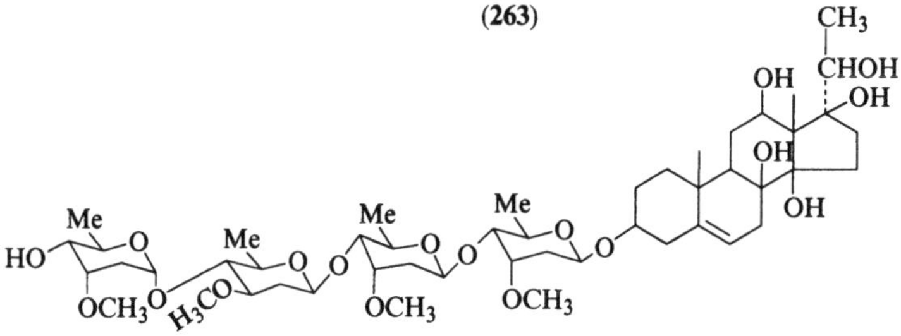

(**264**)

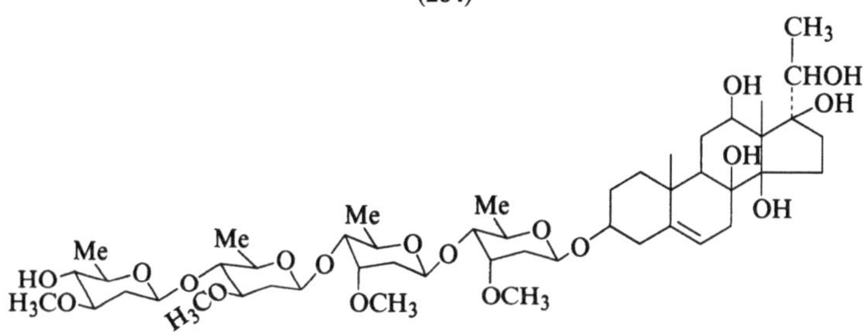

(**265**)

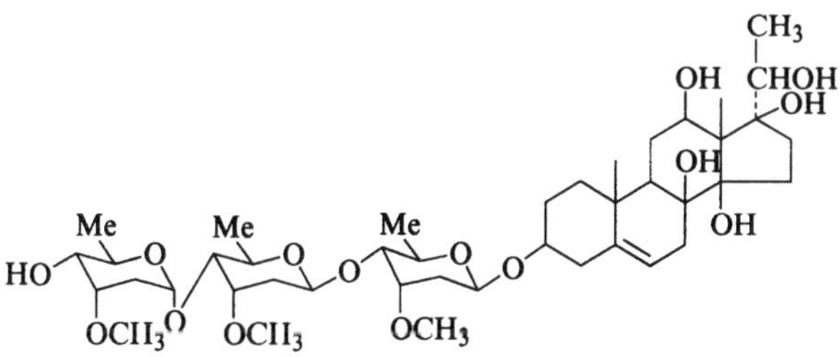

(**266**)

Table 2 (*continued*)

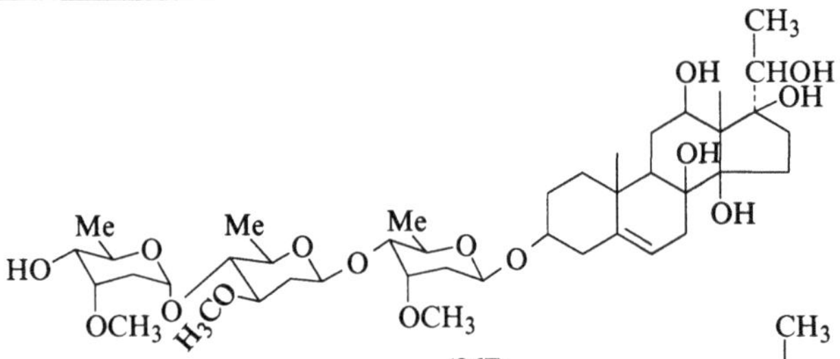

(**267**)

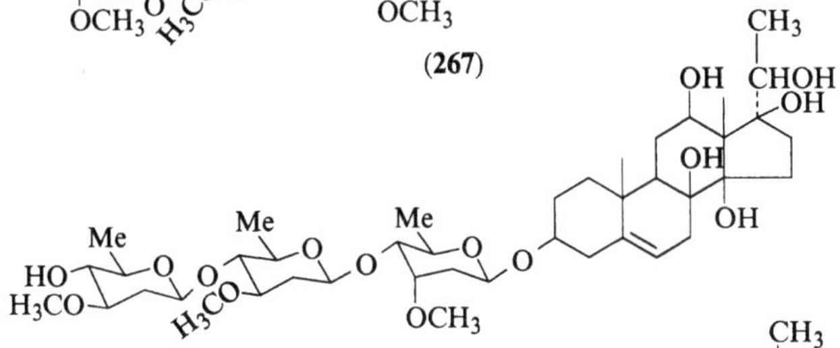

(**268**)

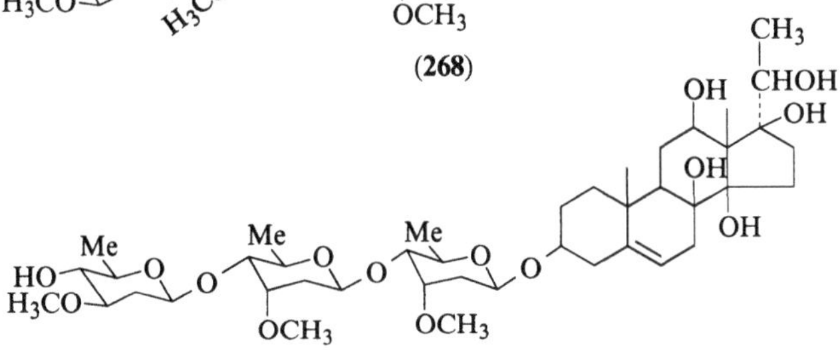

(**269**)

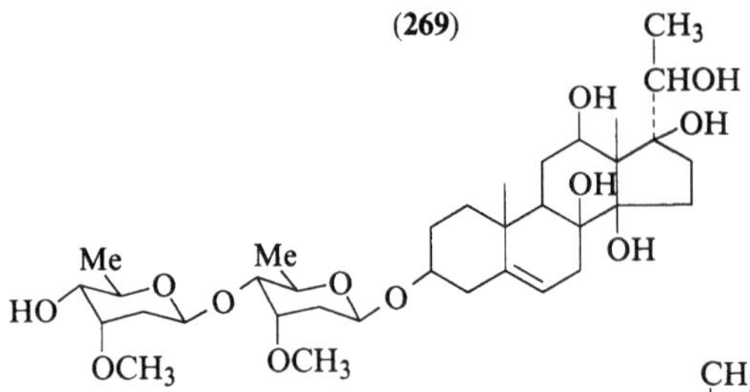

(**270**)

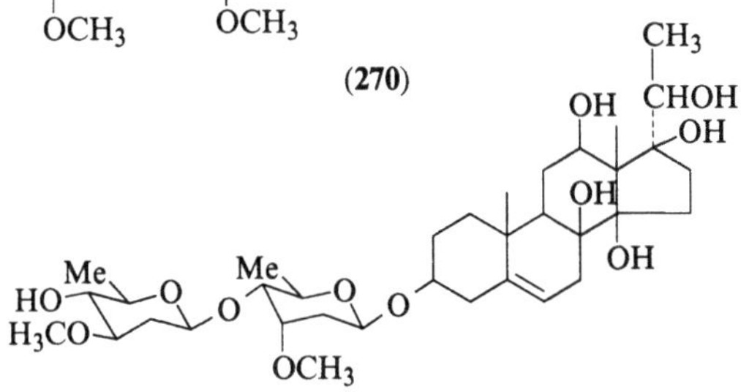

(**271**)

Table 2 (*continued*)

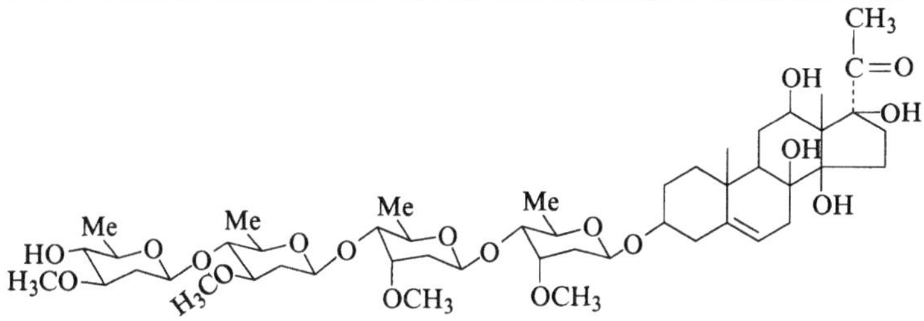

(**272**)

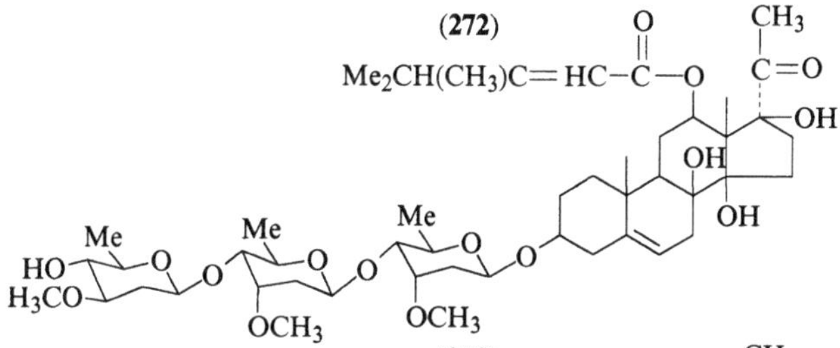

(**273**)

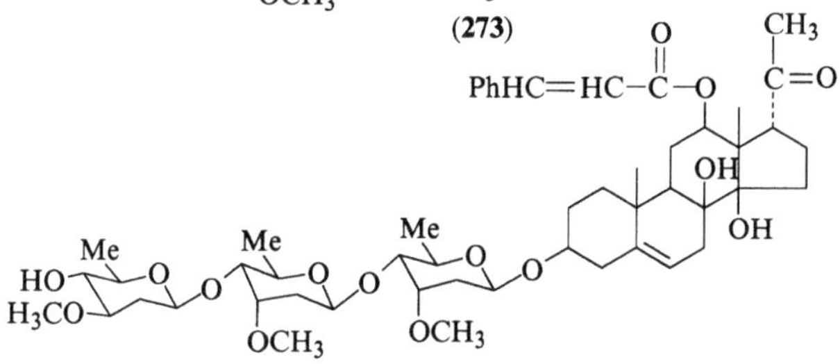

(**274**)

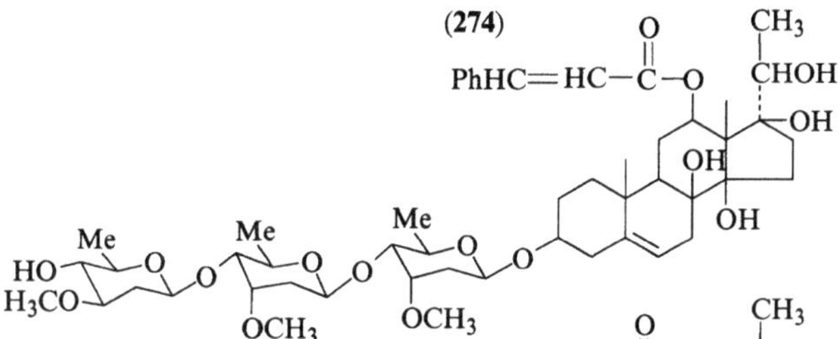

(**275**)

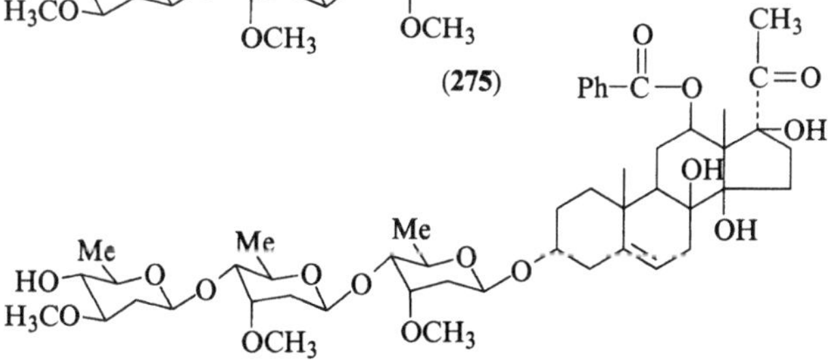

(**276**)

Table 2 (*continued*)

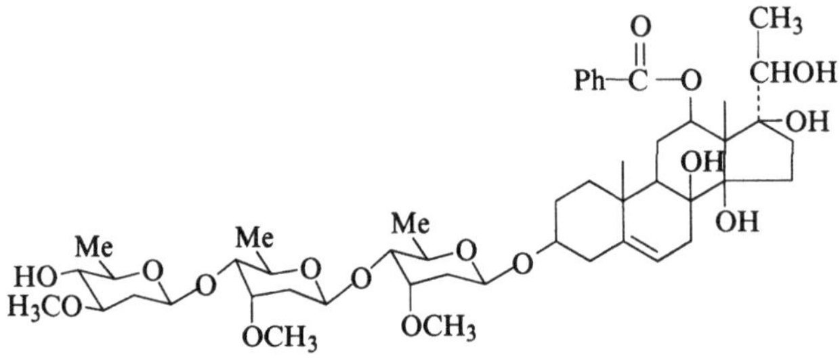

(**277**)

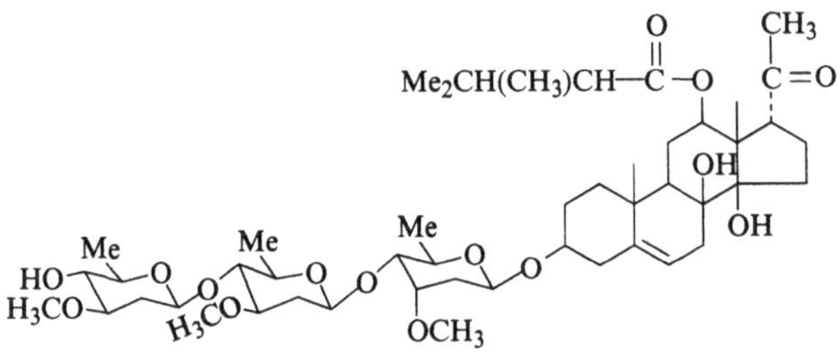

(**278**)

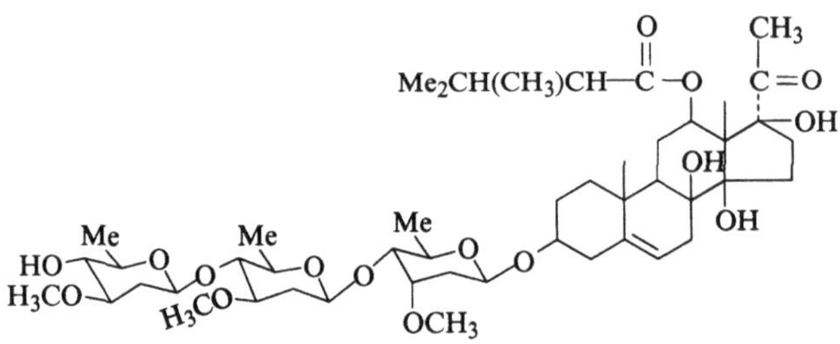

(**279**)

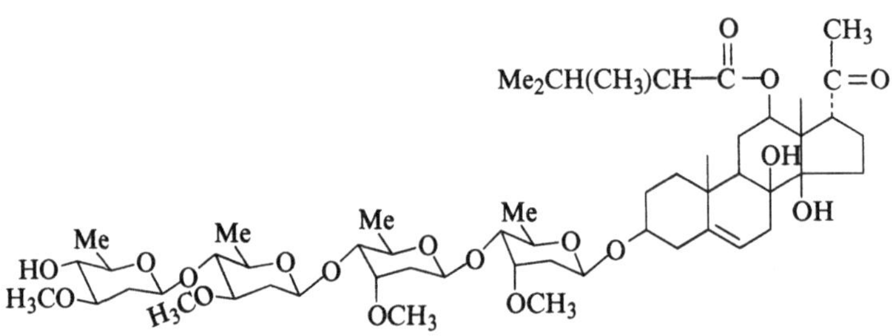

(**280**)

References, pp. 309–325

Table 2 (*continued*)

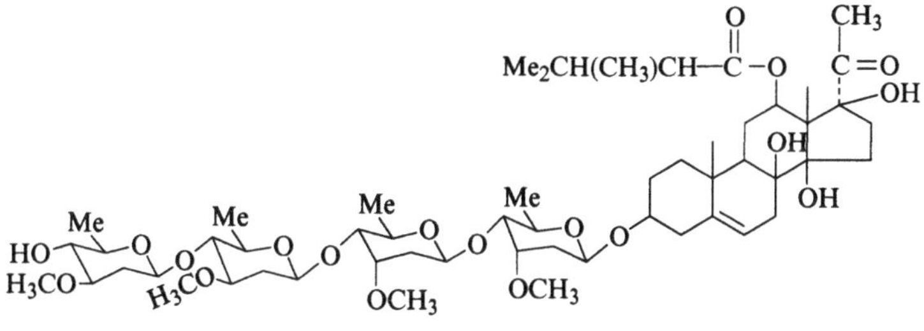

(**281**)

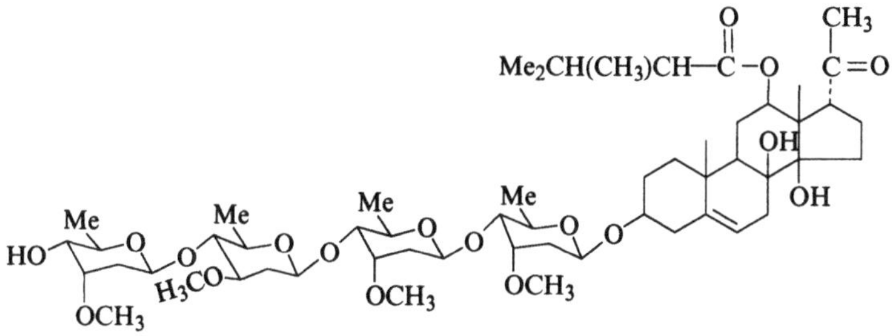

(**282**)

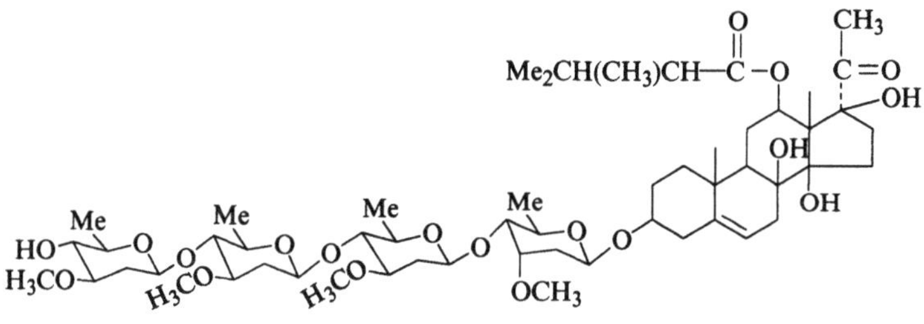

(**283**)

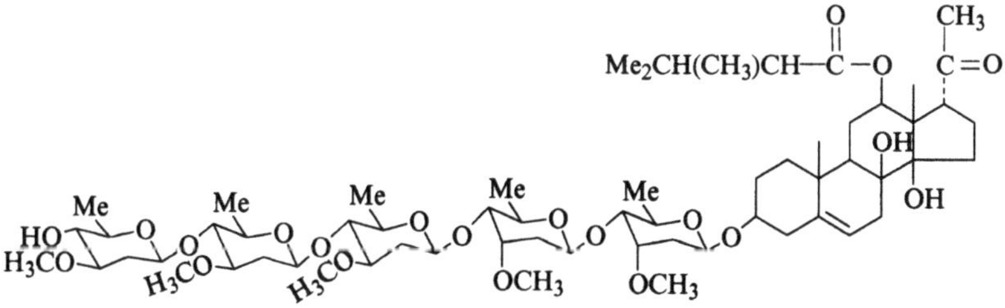

(**284**)

Table 2 (*continued*)

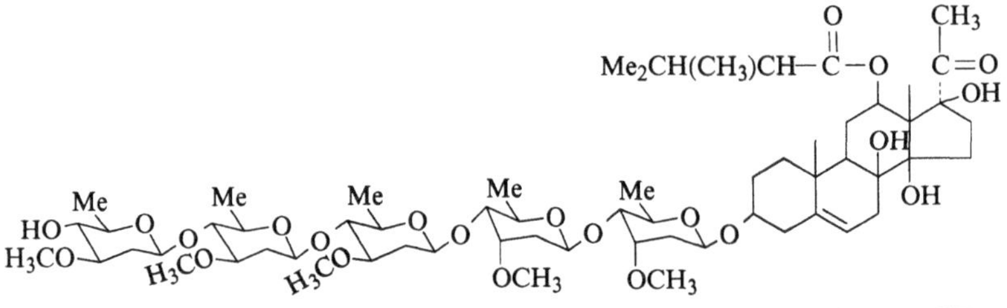

(**285**)

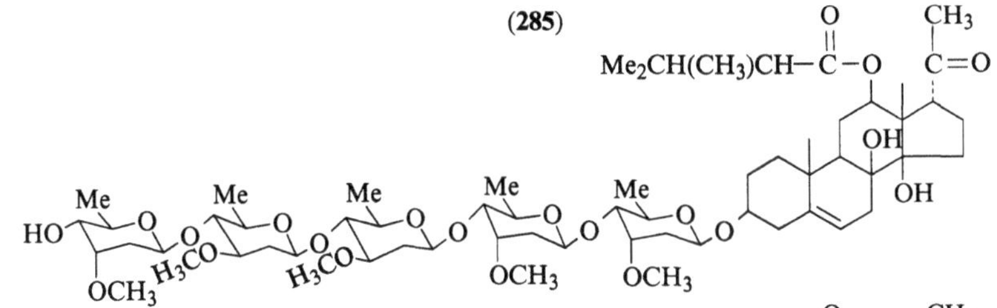

(**286**)

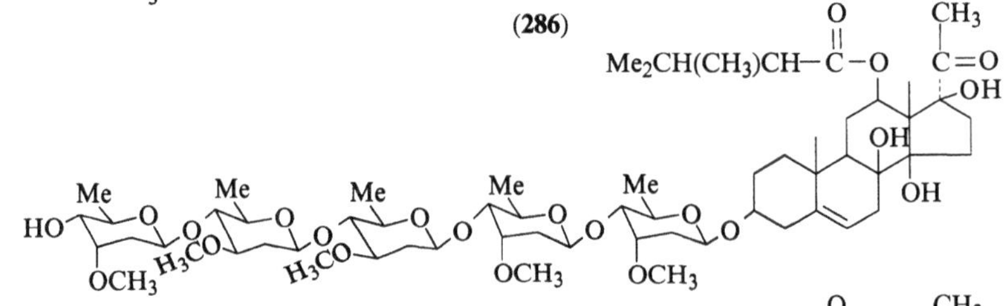

(**287**)

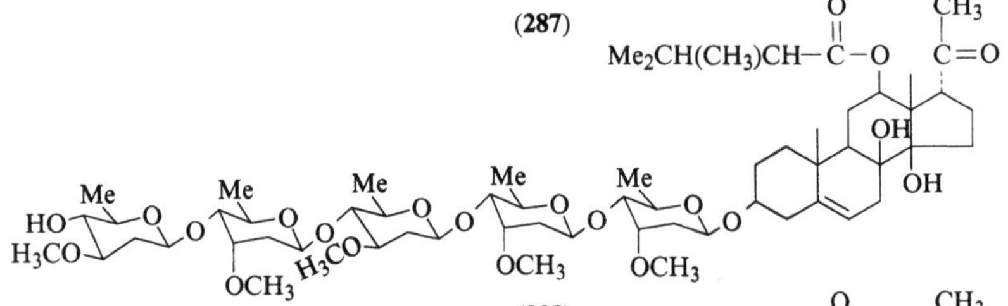

(**288**)

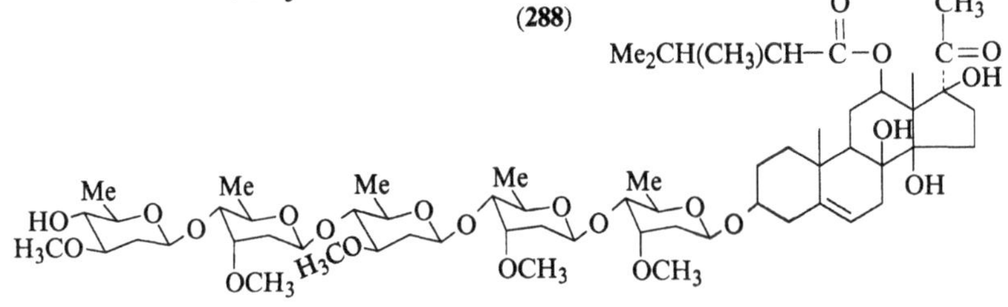

(**289**)

References, pp. 309–325

Table 2 (*continued*)

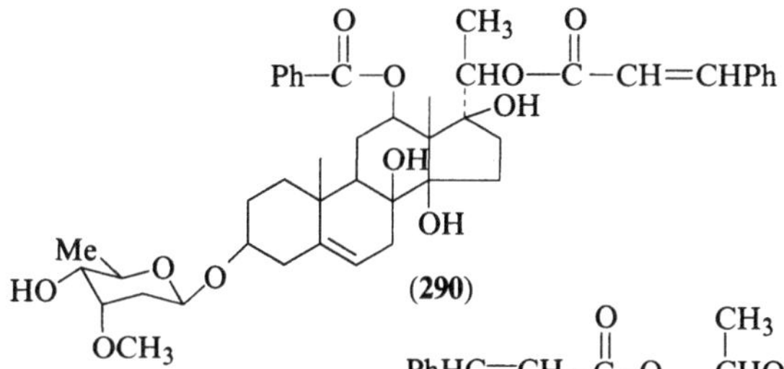

(**290**)

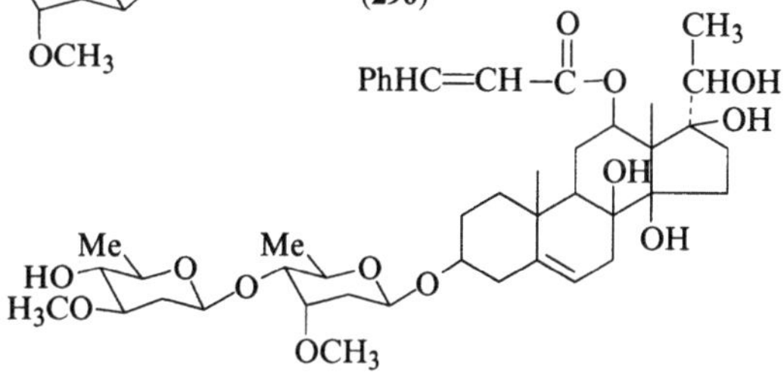

(**291**)

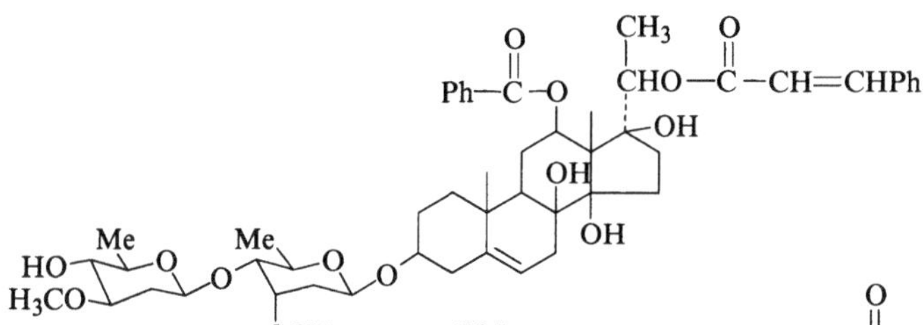

(**292**)

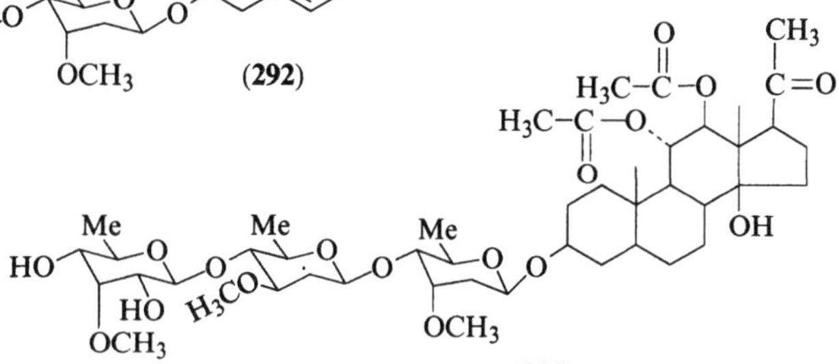

(**293**)

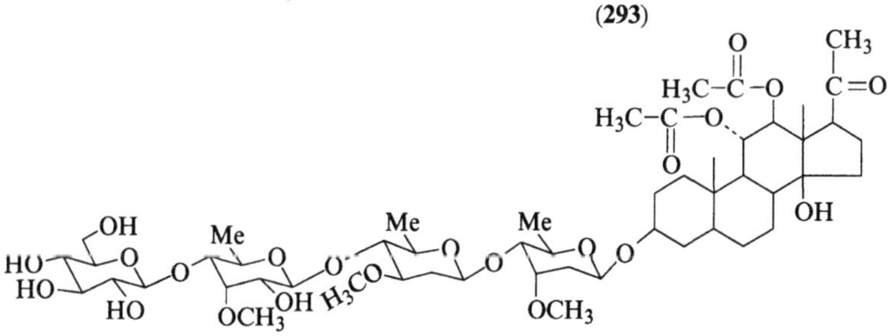

(**294**)

Table 2 (*continued*)

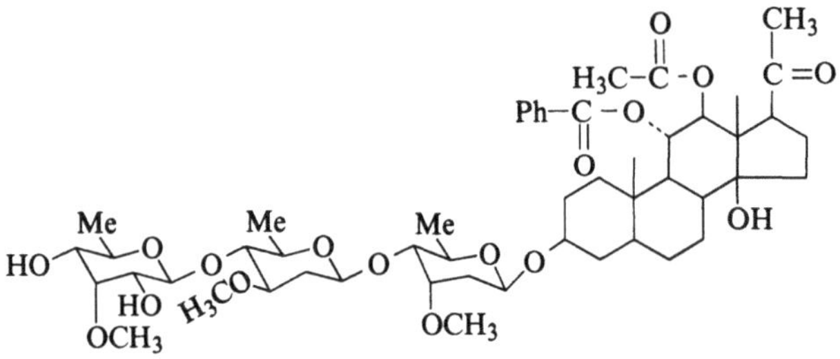

(**295**)

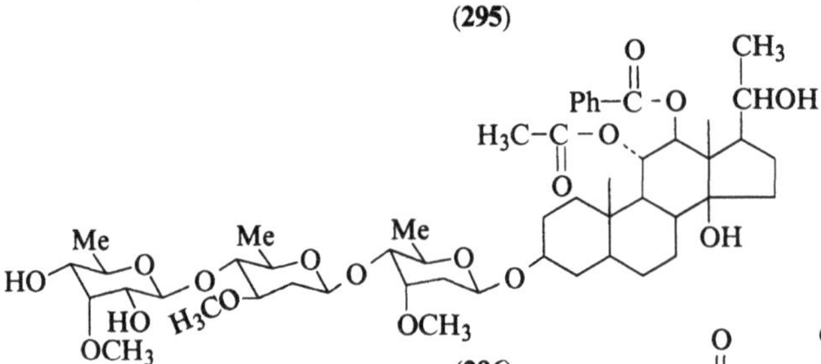

(**296**)

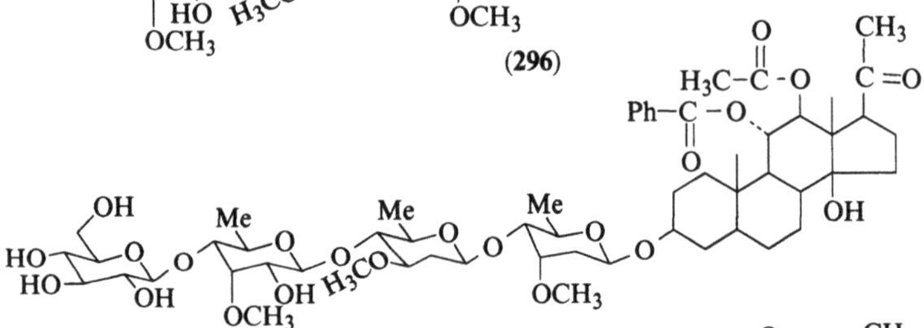

(**297**)

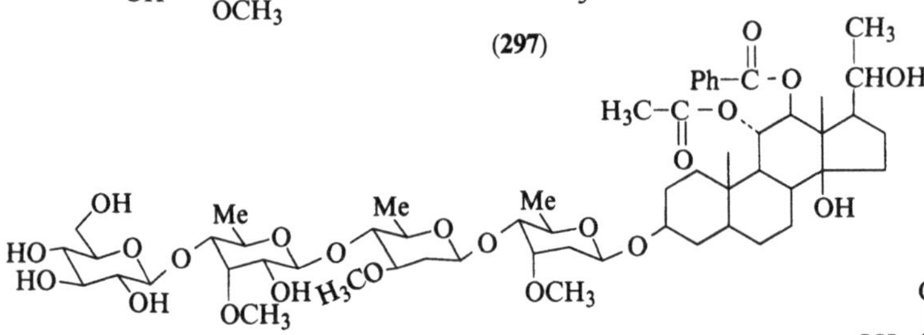

(**298**)

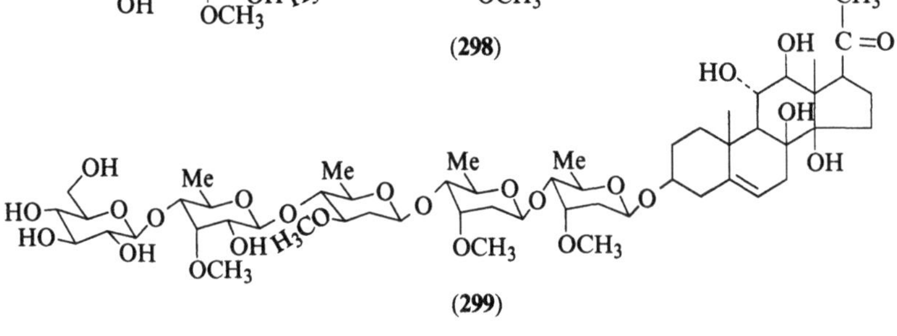

(**299**)

(*215*). A number of pregnane glycosides have been isolated from antitumor active fractions of *Periploca sepium* (*31,69,162,187*). Among these periplocoside A (**217**) showed significant antitumor activity against Sarcoma 180 ascites in mice (*162*). Another pregnane glycoside, periploside A (**211**) from the same source, showed significant anticomplementary activity at a concentration of 1.0 mg/ml (*228*). Recently, pregnane derivatives isolated from *Stizophyllum riparium* (*236*), *Gelsemium sempervirens* (*237*) and *Marsdenia tenacissima* (*65*) showed cytotoxic activity while two pregnane glycosides isolated from *Cynanchum otophyllum* showed antiepilepsy activity (*238*). Marsdekoiside A (**183**) from *Marsdenia koi* has shown good antifertility activity (*17,239*). Verrucoside (**238**), a pregnane glycoside from the gorgonian *Eunicella verrucosa*, possesses cytotoxic activity (*7*) against human lung carcinoma (P-388) and human colon carcinoma (HT-29). Six pregnane glycosides—condurangoglycoside A (**160**), condurangoglycoside C (**161**), condurangoglycoside E_2 (**174**), condurangoside A (**293**), condurangoside B (**295**) and condurangoside C (**296**) obtained from the methanol extract of Condurango cortex (bark of *Marsdenia condurango*), possess differentiation-inducing activity towards mouse myeloid leukemia (M1) cell line (*240*). M1 cells were differentiated into phagocytic cells by these glycosides which were found to be more effective than their aglycons. Kondurangoglycosides A (**160**) and C (**161**) having a cinnamoyl group in their aglycons, were the most potent differentiation inducers and M1 cells became phagocytic cells after 24 hours treatment with these glycosides (*240*).

Acknowledgement

The authors thank Dr. P. K. Agrawal, Central Institute of Medicinal and Aromatic Plants, Lucknow, for helpful discussions and DST and CSIR, New Delhi, for financial assistance.

References

1. REICHSTEIN, T.: Cardenolid- und Pregnanglykoside. Naturwissenschaften **54**, 53 (1967).

2. DEEPAK, D., A. KHARE, and M.P. KHARE: Plant Pregnanes. Phytochemistry **28**, 3255 (1989).

3. CHANDRA, R., D. DEEPAK, and A. KHARE: Pregnane Glycosides from *Hemidesmus indicus*. Phytochemistry **35**, 1545 (1994).

4. PALTER, R., W.F. HADDON, and R.E. LUNDIN: Structure of a Safflower Steroid Cellobioside. Phytochemistry **11**, 2327 (1972).

5. ABE, F., and T. YAMAUCHI: Pregnane and Pregnane Glycosides from *Trachelospermum liukiuense*. Chem. Pharm. Bull. **37**, 33 (1989).
6. NAKANISHI, T., M. KOBAYASHI, H. MURATA, and A. INADA: Phytochemical Studies on Meliaceous Plants. IV. Structure of a New Pregnane Glycoside, Toosendanoside, from Leaves of *Melia toosendan* Sieb. et. Zucc. Chem. Pharm. Bull. **36**, 4148 (1988).
7. KASHMAN, Y., D. GREEN, C. GARCIA, and D.G. AREVALOS: Verrucoside, a New Cytotoxic Pregnane Glycoside from a Gorgonian *Eunicella verrucosa*. J. Nat. Prod. **54**, 1651 (1991).
8. ABE, F., and T. YAMAUCHI: Two Pregnanes from Oleander Leaves. Phytochemistry **31**, 2819 (1992).
9. PRAKASH, K., D. DEEPAK, A. KHARE, and M.P. KHARE: A Pregnane Glycoside from *Streblus asper*. Phytochemistry **31**, 1056 (1992).
10. JIJUN, C., Z. ZHUANGXIN, and Z. JUN: Cynauricuosides A, B and C, Steroid Glycosides from the Root of *Cynanchum auriculatum*. Acta Bot. Yunn. **12**, 197 (1990) [Chem. Abstr. **114**, 98149 (1991)].
11. WETTSTEIN, P.A.: Biosynthèse des hormones stéroides. Experientia **17**, 329 (1961).
12. HEFTMANN, E.: Recent Progress in the Biochemistry of Plant Steroids other than Sterols (Saponins, Glycoalkaloids, Pregnane Derivatives, Cardiac Glycosides, and Sex Hormones). Lipids **9**, 626 (1975) [Chem. Abstr. **82**, 40660 (1975)].
13. YAMAUCHI, T.: Cardenolides, Pregnanes, and Iridoids, Characteristic of Apocynaceae Plants. *Yakugaku Zasshi* **105**, 695 (1985) [Chem. Abstr. **103**, 175350 (1985)].
14. MITSUHASHI, H.: Chemistry and Biological Activity of Polyhydroxypregnane Glucosides. Yaxoue Tangbao **20**, 645 (1985) [Chem. Abstr. **104**, 174414 (1986)].
15. KRISHNA, G., A. KHARE, and M.P. KHARE: Three Pregnane Glycosides from *Dregea lanceolata*. Indian J. Chem. **30B**, 265 (1991).
16. JIN, Q.D., Q.L. ZHOU, and Q.Z. MU: Two New Pregnane Oligoglycosides from *Dregea sinensis* var. *corrugata*. J. Nat. Prod. **52**, 1214 (1989).
17. YUAN, J.-L., Z.-Z. LU, G.-X. CHEN, W.-P. DING, B.-N. ZHOU, C.A.J. ERDELMEIER, M.O. HAMBURGER, H.H.S. FONG, and G.A. CORDELL: The Pregnane Glycoside Marsdekoiside A from *Marsdenia koi*. Phytochemistry **31**, 1058 (1992).
18. HAYASHI, K., I. IIDA, Y. NAKAO, Y. NAKAO, and Y. KANEKO: Four Pregnane Glycosides, Boucerosides AI, AII, BI and BII from *Boucerosia aucheriana*. Phytochemistry **27**, 3919 (1988).
19. TANAKA, T., S. TSUKAMOTO, and K. HAYASHI: Pregnane Glycosides from *Boucerosia aucheriana*. Phytochemistry **29**, 229 (1990).
20. MARSTON, A., and K. HOSTETTMANN: Modern Separation Methods. Nat. Prod. Rep. **8**, 391 (1991).
21. ABE, F., and T. YAMAUCHI: Pregnane Glycosides from *Trachelospermum asiaticum*. Chem. Pharm. Bull. **36**, 621 (1988).
22. CHEN, Z.-S., J.-S. LAI, and Y.-H. KUO: Cynanformosides A and B, Two New Pregnane Glycosides, from the Aerial Part of *Cynanchum formosanum*. Chem. Pharm. Bull. **39**, 3034 (1991).
23. IDAKA, K., Y. HIRAI, and J. SHOJI: Studies on the Constituents of Palmae Plants, IV: The Constituents of the Leaves of *Sabal causiarum* BECC. Chem. Pharm. Bull. **36**, 1783 (1988).
24. STILL, W.C., M. KAHN, and A. MITRA: Rapid Chromatographic Technique for Preparative Separations with Moderate Resolution. J. Org. Chem. **43**, 2923 (1978).
25. CABRERA, G., A.M. SELDES, and E.G. GROS: A Pregnane Glycoside from the Roots of *Mandevilla pentlandiana*. Phytochemistry **32**, 171 (1993).
26. EISENBEISS, F., and H. HENKE: Preparative High Performance Liquid Chromatography

with Reversed-Phase Packed Glass Columns J High Res Chromatogr Chrom Commun **2**, 733 (1979)

27 Simpson, C F Practical High Performance Liquid Chromatography London Heyden 1978

28 Verzele, M Preparative Liquid Chromatography Anal Chem **62**, 265A (1990)

29 Verzele, M, and C Dewaele Preparative High Performance Liquid Chromatography-A Practical Guideline Belgium RSL Europe, Eke 1986

30 Abe, F, T Nagao, Y Mori, T Yamauchi, and Y Saiki Pregnanes and Pregnane Glycosides from the Roots of *Apocynum venetum* var *basikurumon* (Apocynum I) Chem Pharm Bull **35**, 4087 (1987)

31 Itokawa, H, J Xu, and K Takeya Studies on Chemical Constituents of Antitumor Fraction from *Periploca sepium*, V Structures of New Pregnane Glycosides, Periplocosides J, K, F and O Chem Pharm Bull **36**, 4441 (1988)

32 Ahmad, V U, K Usmanghani, and G H Rizwani New Pregnane Glycosides from *Caralluma tuberculata* J Nat Prod **51**, 1092 (1988)

33 Neher, R, and A Wettstein Uber Steroide 107 Mitteilung Farbreaktionen mit Steroiden, insbesondere Corticosteroiden, im Papierchromatogramm Helv Chim Acta **34**, 2278 (1951)

34 Yamauchi, T, M Hara, and K Mihashi Pregnenolone Glucosides of *Nerium odorum* Phytochemistry **11**, 3345 (1972)

35 Voneuw, J V, and T Reichstein Die Glykoside der Samen von *Strophanthus nicholsonii* Holm Glykoside und Aglykone Helv Chim Acta **31**, 883 (1948)

36 Heftmann, E, S T Ko, and R D Benett Response of Steroids to Sulfuric Acid in Thin-Layer Chromatography J Chromatogr **21**, 490 (1966)

37 Abisch, E, and T Reichstein Orientierende Chemische Untersuchung einiger Apocynaceen Helv Chim Acta **43**, 1844 (1960)

38 Mann, F G, and B C Saunders Practical Organic Chemistry 4th edn, p 367 New Delhi Orient Longmann 1990

39 Konda, Y, Y Toda, Y Harigaya, H Lou, X Li, and M Onda Two New Glycosides, Hancoside and Neohancoside A, from *Cynanchum hancockianum* J Nat Prod **55**, 1447 (1992)

40 Tschesche, R, G Grimmer, and F Seehofer Uber pflanzliche Herzgifte, XXIV Mitteilung Die quantitative Trennung und Identifizierung von Herzgiftglykosiden aus *Digitalis purpurea* und *lanata* durch echte Verteilungschromatographie an Papier Ber d Deutsch Chem Ges **86**, 1235 (1953)

41 Webb, J M, and H B Levy A Sensitive Method for the Determination of Deoxyribonucleic Acid in Tissues and Microorganisms J Biol Chem **213**, 107 (1955)

42 Maclenan, A P, H M Randall, and D W Smith Detection and Identification of Deoxysugars on Paper Chromatograms Anal Chem **31**, 2020 (1959)

43 Krishna, G, G V Shinde, M S Shingare, A Khare, and M P Khare A Pregnane Ester Tetraglycoside from *Dregea lanceolata* Phytochemistry **29**, 2961 (1990)

44 Nagata, W, C Tamm, and T Reichstein Die Glykoside von *Erysimum crepidifolium* H G L Reichenbach Glykoside und Aglykone, 169 Mitteilung Helv Chim Acta **40**, 41 (1957)

45 Partridge, S M Aniline Hydrogen Phthalate as a Spraying Reagent for Chromatography of Sugars Nature **164**, 443 (1949)

46 Feigl, F Spot Tests in Organic Analysis, 7th edn, p 337 Amsterdam Elsevier Publications 1975

47 Jaeggi, K A, E Weiss, W Wehrli, and T Reichstein Die Glykoside der Wurzeln von

Gongronema taylorii (SCHLTR and RENDLE) BULLOCK) Glykoside und Aglykone, 292 Mitteilung Helv Chim Acta **50**, 1201 (1967)

48 MARKHAM, K R Techniques of Flavonoid Identification London Academic Press 1982

49 MABRY, T J, K R MARKHAM, and M B THOMAS The Systematic Identification of Flavonoids New York Springer 1970

50 SRIVASTAV, S, D DEEPAK, and A KHARE Structural Studies of Trisaccharide of Leptaculatin J Carbohydr Chem **13**, 75 (1994)

51 SHIBUYA, H, R ZHANG, J D PARK, N I BAEK, Y TAKEDA, M YOSHIKAWA, and I KITAGAWA Indonesian Medicinal Plants, V Chemical Structures of Calotroposides C, D, E, F and G, Five Additional New Oxypregnane Oligoglycosides from the Root of *Calotropis gigantea* (Asclepiadaceae) Chem Pharm Bull **40**, 2647 (1992)

52 AGRAWAL, P K NMR Spectroscopy in the Structural Elucidation of Oligosaccharides and Glycosides Phytochemistry, **31**, 3307 (1992)

53 SRIVASTAV, S, D DEEPAK, and A KHARE Three Novel Pregnane Glycosides from *Leptadenia reticulata* Wight and Arn Tetrahedron **50**, 789 (1994)

54 BERGER, S, P JUNIOR, and L KOPANSKI Structural Revision of Pregnane Ester Glycosides from Condurango Cortex and New Compounds Phytochemistry **27**, 1451 (1988)

55 ABE, F, and T YAMAUCHI Pregnane Glycosides of Teikasides B and C Series, from *Trachelospermum asiaticum* Chem Pharm Bull **36**, 4330 (1988)

56 ALLGEIER, H Struktur der Pachybiose und Asclepobiose Desoxyzucker, 44 Mitteilung Helv Chim Acta **51**, 311 (1968)

57 DEEPAK, D, M P KHARE, and A KHARE A Pregnane Ester Diglycoside from *Periploca calophylla* Phytochemistry **24**, 3015 (1985)

58 ZHANG, Z-X, J ZHOU, K HAYASHI, and K KANEKO Atratosides A, B, C and D, Steroid Glycosides from the Root of *Cynanchum atratum* Phytochemistry **27**, 2935 (1988)

59 SETHI, A, D DEEPAK, M P KHARE, and A KHARE A Novel Pregnane Glycoside from *Periploca calophylla* J Nat Prod **51**, 787 (1988)

60 TRIVEDI, R, A KHARE, and M P KHARE A Pregnane Ester Oligoglycoside from *Oxystelma esculentum* Phytochemistry **28**, 1211 (1989)

61 DEEPAK, D, S SRIVASTAV, and A KHARE Indicusin-A Pregnane Diester Triglycoside from *Hemidesmus indicus* R Br Nat Prod Letts **6**, 81 (1995)

62 TRIVEDI, R, A KHARE, and M P KHARE A Pregnane Triglycoside from *Oxystelma esculentum* Phytochemistry **29**, 3967 (1990)

63 PRAKASH, K, A SETHI, D DEEPAK, A KHARE, and M P KHARE Two Pregnane Glycosides from *Hemidesmus indicus* Phytochemistry **30**, 297 (1991)

64 QIDUAN, J, Z QIANLAN, and M QUANZHANG Structure of Dregeoside A from *Dregea sinensis* var *corrugata* Acta Bot Yunn **10**, 466 (1988) [Chem Abstr **111**, 54152 (1989)]

65 LUO, S-Q, L-Z LIN, G A CORDELL, L XUE, and M E JOHNSON Polyoxypregnanes from *Marsdenia tenacissima* Phytochemistry **34**, 1615 (1993)

66 SUMMONS, R E, J ELLIS, and E GELLERT Steroidal Alkaloids of *Marsdenia rostrata* Phytochemistry **11**, 3335 (1972)

67 SINGHAL, S, M P KHARE, and A KHARE Tenasogenin, A Pregnane Ester from *Marsdenia tenacissima* Phytochemistry **19**, 2431 (1990)

68 BOSE, A K, and P R SRINIVASAN Trichloroacetylisocyanate as an *in situ* Derivatizing Reagent for ^{13}C NMR Spectroscopy of Alcohols, Phenols and Amines Tetrahedron **31**, 3025 (1975)

69 ITOKAWA, H, J XU, and K TAKEYA Studies on Chemical Constituents of Antitumor

Fraction from *Periploca sepium*, IV Structures of New Pregnane Glycosides Periplocosides D, E, L, and M Chem Pharm Bull **36**, 2084 (1988)

70 TSUKAMOTO, S, K HAYASHI, and K KANEKO Studies on the Constituents of Asclepiadaceae Plants, Part 67 Further Studies on Glycosides with a Novel Sugar Chain Containing a Pair of Optically Isomeric Sugars, D- and L- Cymarose, from *Cynanchum africanum* J Chem Soc, Perkin Trans I, 2625 (1988)

71 BOCK, K, C PEDERSEN, and H PEDERSEN Carbon-13 Nuclear Magnetic Resonance Data for Oligosaccharides Adv Carbohydr Chem Biochem **42**, 193 (1984)

72 BRADBURY, J H, and G A JENKINS Determination of the Structures of Trisaccharides by Carbon-13 NMR Spectroscopy Carbohydr Res **126**, 125 (1984)

73 AGRAWAL, P K, D C JAIN, R K GUPTA, and R S THAKUR Carbon-13 NMR Spectroscopy of Steroidal Sapogenins and Steroidal Saponins Phytochemistry **24**, 2479 (1985)

74 AGRAWAL, P K, and R P RASTOGI ^{13}C NMR Spectroscopy of Flavonoids Heterocycles **16**, 2181 (1981)

75 PERLIN, A S, and B CASU In The Polysaccharides (G O ASPINALL, ed), Vol I, p 133 New York Academic Press 1982

76 MITSUHASHI, H, and K HAYASHI Chemistry and Biological Activity of Polyoxypregnane Glycosides Shoyakugaku Zasshi **39**, 1 (1985) [Chem Abstr **103**, 109804 (1985)]

77 KITAGAWA, I, R ZHANG, J D PARK, N I BAEK, Y TAKEDA, M YOSHIKAWA, and H SHIBUYA Indonesian Medicinal Plants, I Chemical Structures of Calotroposides A and B, Two New Oxypregnane-Oligoglycosides from the Root of *Calotropis gigantea* (Asclepiadaceae) Chem Pharm Bull **40**, 2007 (1992)

78 ZURCHER, R F Protonenresonanzspektroskopie und Steroid Struktur II Die Lage der C-18-und C-19-Methylsignale in Abhangigkeit von den Substituenten am Steroidgerust Helv Chim Acta **46**, 2054 (1963)

79 SUGAMA, K, K HAYASHI, H MITSUHASHI, and K KANEKO Studies on the Constituents of Asclepiadaceae Plants, LXVI The Structures of Three New Glycosides, Cynapanosides A, B, and C, from the Chinese Drug "Xu-Change Qing", *Cynanchum paniculatum* KITAGAWA Chem Pharm Bull **34**, 4500 (1986)

80 QIDUAN, J, Z QIANLAN, and M QUANZHANG A Pregnane Triglycoside Ester from *Dregea sinensis* var *corrugata* Phytochemistry **28**, 1273 (1989)

81 TERUI, Y, K TORI, and N TSUJI Esterification Shifts in Carbon-13 NMR Spectra of Alcohols Tetrahedron Letters, 621 (1976)

82 DORMAN, D E, D BAUER, and J D ROBERTS Nuclear Magnetic Resonance Spectroscopy Carbon-13 Chemical and Carbon-13 Proton Couplings in Some Esters and Ethers J Org Chem **40**, 3729 (1975)

83 AGRAWAL, P K, H -J SCHNEIDER, M S MALIK, and S N RASTOGI Stereochemical and Carbon-13 NMR Investigations, 32 Carbon-13 NMR Shifts and Conformations of Substituted Indans Org Magn Reson **21**, 146 (1983)

84 JIJUN, C, Z ZHUANGXIN, and Z JUN The Chemical Constituents of *Cynanchum forrestii* Acta Bot Yunn **11**, 471 (1989) [Chem Abstr **113**, 94734 (1990)]

85 TSUKAMOTO, S, K HAYASHI, and H MITSUHASHI Studies on the Constituents of Asclepiadaceae Plants, LX Further Studies on Glycosides with a Novel Sugar Chain Containing a Pair of Optically Isomeric Sugars, D- and L- Cymarose, from *Cynanchum wilfordi* Chem Pharm Bull **33**, 2294 (1985)

86 LIN, L -J, L -Z LIN, R R GIL, G A CORDELL, M RAMESH, B SRILATHA, B REDDY, and A V N P RAO Pregnane Glycosides from *Caralluma umbellata* Phytochemistry **35**, 1549 (1994)

87 AGRAWAL, P K, and M C BANSAL Studies in Organic Chemistry-Carbon-13 NMR of

Flavonoids (P K AGRAWAL, ed), Vol 39, p 42 Amsterdam Elsevier Publications 1989

88 BAUMANN, H , B ERBING, P E JANSSON, and L KENNE NMR and Conformational Studies of some 3-O-, 4-O-, and 3, 4-Di-O-glycopyranosyl-substituted Methyl-α-D-galactopyranosides J Chem Soc , Perkin Trans I, 2153 (1989)

89 BAUMANN, H , B ERBING, P E JANSSON, and L KENNE Synthesis, NMR and Conformational Studies of some 3, 4-Di-O-glycopyranosyl-substituted Methyl-α-D-galactopyranosides J Chem Soc , Perkin Trans I, 2011 (1989)

90 JANSSON, P E , L KENNE, and H OTTOSSON Synthesis of and Nuclear Magnetic Resonance and Conformational Studies on some 1, 2-Linked Disaccharide Methyl Glycosides Containing D-Mannose and L-Rhamnose J Chem Soc , Perkin Trans I, 2011 (1990)

91 BAUMANN, H , P E JANSSON, and L KENNE Synthesis and Nuclear Magnetic Resonance Studies of Some L- Fucosyl-containing Disaccharides J Chem Soc , Perkin Trans I, 2229 (1991)

92 KOCHETKOV, N K , O S CHIZHOV, and A S SHASHKOV Dependence of Carbon-13 Chemical Shifts on the Spatial Interaction of Protons, and its Application in Structural and Conformational Studies of Oligo- and Polysaccharides Carbohydr Res **133**, 173 (1984)

93 BOCK, K , A BRINGNOLE, and B W SIGUSKJOLD Conformational Dependence of ^{13}C Nuclear Magnetic Resonance Chemical Shifts in Oligosaccharides J Chem Soc , Perkin Trans II, 1711 (1986)

94 SHASHKOV, A S , G M LIPKIND, Y A KNIREL, and N K KOCHETKOV Stereochemical Factors Determining the Effects of Glycosylation on the ^{13}C Chemical Shifts in Carbohydrates Magn Reson Chem **26**, 735 (1988)

95 ROWEN, D D , and R H NEWMAN Noroleanane Saponins from *Celmisia petriei* Phytochemistry **23**, 639 (1984)

96 GUINAUDEAU, H , O SELIGMANN, H WAGNER, and A NESZMELYI Faralatroside and Faratroside, Two Flavonol Triglycosides from *Colubrina faralaotra* Phytochemistry **20**, 1113 (1981)

97 OKADA, Y , S SHIBATA, T IKEKAWA, M J JAVELLANA, and O KAMO Entada Saponin-III, A Saponin isolated from the Bark of *Entada phaseoloides* Phytochemistry **26**, 2789 (1987)

98 MIYAMOTO, T , K TOGAWA, R HIGUCHI, T KOMORI, and T SASAKI Six Newly Identified Biologically Active Triterpenoid Glycoside Sulfates from the Sea Cucumber *Cucumaria echinata* Liebigs Ann Chem 453 (1990)

99 BOCK, K , and C PEDERSEN A Study of ^{13}CH Coupling Constants in Hexopyranoses J Chem Soc , Perkin Trans II, 293(1974)

100 VOLD, R L , J S WAUGH, M P KLEIN, and D E PHELPS Measurements of Spin Relaxation in Complex Systems J Chem Phys **48**, 3831 (1968)

101 ALLERHAND, A , D DODDRELL, V GLUSHKO, D W COCHRAN, E WENKERT, P J LAWSON, and F R N GURD Conformation and Segmental Motion of Native and Denatured Ribonuclease A in Solution Application of Natural Abundance Carbon-13 Partially Relaxed Fourier Transform Nuclear Magnetic Resonance J Am Chem Soc **93**, 544 (1971)

102 DODDRELL, D , and A ALLERHAND Segmental Motion in Liquid 1-Decanol Application of Natural Abundance Carbon-13 Partially Relaxed Fourier Transform Nuclear Magnetic Resonance J Am Chem Soc **93**, 1558 (1971)

103 ALLERHAND, A , and D DODDRELL Strategies in the Application of Partially Relaxed Fourier Transform Nuclear Magnetic Resonance Spectroscopy in Assignments of Carbon-13 Resonances of Complex Molecules Stachyose J Am Chem Soc **93**, 2777 (1971)

104 AGRAWAL, P K, and M C BANSAL Studies in Organic Chemistry-Carbon 13 NMR of Flavonoids (P K AGRAWAL, ed), Vol 39, p 290 Amsterdam Elsevier Publications 1989
105 UZAWA, J, and S TAKEUCHI Application of Selective Carbon-13-Proton Nuclear Overhauser Effects with Low-Power-Proton Irradiation in Carbon-13 NMR Spectroscopy Org Magn Reson **11**, 502 (1978)
106 UZAWA, J, and M URAMOTO Assignment of Indirect Carbon-13-Proton Couplings in the Carbon-13 NMR Spectra of Some Purine and Pyrimidine Nucleosides and their Analogs by Long-range Selective Proton Decoupling Org Magn Reson **12**, 612 (1979)
107 TSUKAMOTO, S, K HAYASHI, K KANEKO, H MITSUHASHI, F O SNYCKERS, and T G FOURIE Studies on the Constituents of Asclepiadaceae Plants, LXIV The Structure Elucidation of Cynafogenin Chem Pharm Bull **34**, 1337 (1986)
108 KOMURA, H, K MIZUKAWA, and H MINAKATA Ceroalbolinic Acid, a Common Body Pigment of Three Ceroplastes Scale Insects in Japan Confirmation of Structure Bull Chem Soc Jpn **55**, 3053 (1982) [Chem Abstr **97**, 213037 (1982)]
109 OHMOTO, T, K YAMAGUCHI, and K IKEDA Constituents of *Hibiscus moscheutos* L I Chem Pharm Bull **36**, 578 (1988)
110 STOTHERS, J B Carbon-13 NMR Spectroscopy, p 38 New York Academic Press 1972
111 RABENSTEIN, D L, and T T NAKASHIMA Spin-Echo Fourier-Transform Nuclear Magnetic Resonance Spectroscopy Analyt Chem **51**, 1465a (1979)
112 PATT, S L, and J N SHOOLERY Attached Proton Test for Carbon-13 NMR J Magn Reson **46**, 535 (1982)
113 WASYLYK, J M, G E MARTIN, A J WEINHEIMER, and M ALAM Isolation and Identification of a New Pregnane Glycoside from the Gorgonian *Pseudoplexaura wagenaari* J Nat Prod **52**, 391 (1989)
114 AQUINO, R, C PIZZA, N DE TOMMASI, and F DE SIMONE New Polyoxypregnane Ester Derivatives from *Leptadenia hastata* J Nat Prod **58**, 672 (1995)
115 DODDRELL, D M, W BROOKS, J FIELD, and R M LYNDENBELL Generation of Heteronuclear Carbon-13/Proton Chemical Shift Correlations Using Soft Pulses J Magn Reson **59**, 384 (1984)
116 DODDRELL, D M, D T PEGG, and M R BENDALL Distortionless Enhancement of NMR Signals By Polarization Transfer J Magn Reson **48**, 323 (1982)
117 BURUM, D P, and R R ERNST Net Polarization Transfer via a *J*-Ordered State for Signal Enhancement of Low-Sensitivity Nuclei J Magn Reson **39**, 163 (1980)
118 DODDRELL, D M, and D T PEGG Assignment of Proton-Decoupled Carbon-13 Spectra of Complex Molecules by Using Polarization Transfer Spectroscopy A Superior Method to Off-Resonance Decoupling J Am Chem Soc **102**, 6388 (1980)
119 MORRIS, G A, and R FREEMAN Enhancement of Nuclear Magnetic Resonance Signals by Polarization Transfer J Am Chem Soc **101**, 760 (1979)
120 BAX, A Structure Determination and Spectral Assignment by Pulsed Polarization Transfer via Long-range Proton-Carbon-13 Couplings J Magn Reson **57**, 314 (1984)
121 CORDELL, G A, and A D KINGHORN One-Dimensional Proton-Carbon Correlations for the Structure Determination of Natural Products Tetrahedron **47**, 3521 (1991)
122 BAX, A, W EGAN and P KOVAC New NMR Techniques for Structure Determination and Resonance Assignments of Complex Carbohydrates J Carbohydr Chem **3**, 593 (1984)
123 CAPEK, P, D UHRIN, J RASIK, A KARDOSOVA, R TOMAN, and V MIHALOV Polysaccharides from the Roots of the Marsh mallow (*Althaea officinalis* var rhobusta) Dianhydrides of Oligosaccharides of the Aldose type Carbohydr Res **182**, 160 (1988)

124. RASOANAIVO, P., N. KANEDA, A.D. KINGHORN, and N.R. FARNSWORTH: Folotsoside A, a New Pregnane Glycoside from *Folotsia sarcostemmoides*. J. Nat. Prod. **54**, 1672 (1991).
125. TSCHESCHE, R., P. WELZEL, and G. SNATZKE: Digitanolglykoside, XII: Die Konstitution von Kondurangogenin A, dem Aglykon eines Esterglykosides der Kondurangorinde. Tetrahedron **21**, 1777 (1965).
126. TSCHESCHE, R., P. WELZEL, and H.W. FEHLHABER: Digitanolglykoside, XIII: Massenspektrometrische Untersuchungen am Kondurangogenin A. Tetrahedron **21**, 1797 (1965).
127. TSCHESCHE, R., H. KOHL, and P. WELZEL: Digitanolglykoside, XVI: Die Struktur der Kondurangogenine A und C. Tetrahedron **23**, 1461 (1967).
128. TSCHESCHE, R., and H. KOHL: Digitanolglykoside, XIX: Die Struktur der Kondurangoglykoside A, A_1 und C, C_1. Tetrahedron **24**, 4359 (1968).
129. HAYASHI, K., K. WADA, H. MITSUHASHI, H. BANDO, M. TAKASE, S. TERADA, Y. KOIDE, T. AIBA, T. NARITA, and D. MIZUNO: Antitumor Active Glycosides from Condurango Cortex. Chem. Pharm. Bull. **28**, 1954 (1980).
130. HAYASHI, K., K. WADA, H. MITSUHASHI, H. BANDO, M. TAKASE, S. TERADA, Y. KOIDE, T. AIBA, T. NARITA, and D. MIZUNO: Further Investigation of Antitumor Condurangoglycosides with C-18 Oxygenated Aglycone. Chem. Pharm. Bull. **29**, 2725 (1981).
131. JEENER, J.: Ampere International Summer School. Basko Polje, Yugoslavia. 1971.
132. AUE, W.P., E. BARTHOLDI, and R.R. ERNST: Two-Dimensional Spectroscopy. Application to Nuclear Magnetic Resonance. J. Chem. Phys. **64**, 2229 (1976).
133. MORRIS, G.A.: Modern NMR Techniques for Structure Elucidation. Magn. Reson. Chem. **24**, 371 (1986).
134. KESSLER, H., M. GEHRKE, and C. GRIESINGER: Two-Dimensional NMR Spectroscopy: Background and Overview of the Experiments. Angew. Chem. Int. Ed. **27**, 490 (1988).
135. ERNST, R.R., G. BODENHAUSEN, and A. WOKAUN: Principle of Nuclear Magnetic Resonance in One and Two Dimensions. London:Oxford University Press (Clarendon). 1987.
136. BAX, A., and D.G. DAVIES: Advanced Magnetic Resonance Techniques in Systems of High Molecular Complexity (N. NICOLAI, and G. VALENSIN, eds.), p. 21. Stuttgart: Birkhäuser. 1986.
137. VLEGGAAR, R., F.R. VANHEERDEN, L.A.P. ANDERSON, and G.L. ERASMUS: Toxic Constituents of the Asclepiadaceae. Structure Elucidation of Sarcovimiside A-C, Pregnane Glycosides of *Sarcostemma viminale*. J. Chem. Soc., Perkin Trans I, 483 (1993).
138. PIANTINI, U., O.W. SORENSEN, and R.R. ERNST: Multiple Quantum Filters for Elucidating NMR Coupling Networks. J. Am. Chem. Soc. **104**, 6800 (1982).
139. RANCE, M., O.W. SORENSEN, G. BODENHAUSEN, G. WAGNER, R.R. ERNST, and K. WUTHRICH: Improved Spectral Resolution in COSY ^1H NMR Spectra of Proteins via Double Quantum Filtering. Biochem. Biophys. Res. Commun. **117**, 479 (1983).
140. MÜLLER, N., R.R. ERNST, and K. WUTHRICH: Multiple Quantum-Filtered Two-Dimensional Correlated NMR Spectroscopy of Proteins. J. Am. Chem. Soc. **108**, 6482 (1986).
141. EDWARDS, M.W., and A. BAX: Complete Proton and Carbon-13 NMR Assignments of the Alkaloid Gephyrotoxin through the Use of Homonuclear Hartmann–Hahn and Two-Dimensional NMR Spectroscopy. J. Am. Chem. Soc. **108**, 918 (1986).
142. SHAKA, A.J., and R. FREEMAN: Simplification of NMR Spectra by Filtration Through Multiple-Quantum Coherence. J. Magn. Reson. **51**, 169 (1983).
143. EICH, G., G. BODENHAUSEN, and R.R. ERNST: Exploring Nuclear Spin Systems by Relayed Magnetization Transfer. J. Am. Chem. Soc. **104**, 3731 (1982).

144 BAX, A , and G DROBNY Optimization of Two-Dimensional Homonuclear Relayed Coherence Transfer NMR Spectroscopy J Magn Reson **61**, 306 (1985)

145 HUGHES, D W Application of Relayed Coherence Transfer Two-Dimensional Nuclear Magnetic Resonance Spectroscopy to the Assignment of ^{1}H Chemical Shifts in Steroids Magn Reson Chem **26**, 214 (1988)

146 BOLTON, P H Assignments and Structural Information via Relayed Coherence Transfer Spectroscopy J Magn Reson **48**, 336 (1982)

147 MACURA, S , and R R ERNST Elucidation of Cross Relaxation in Liquids by Two-dimensional NMR Spectroscopy Mol Phys **41**, 95 (1980)

148 KUMAR, A , R R ERNST, and K WUTHRICH A Two Dimensional Nuclear Overhauser Enhancement (2D NOE) Experiment for the Elucidation of Complete Proton–Proton Cross Relaxation Networks in Biological Macromolecules Biochem Biophys Res Commun **95**, 1 (1980)

149 DAVIS, D , and A BAX Assignment of Complex ^{1}H NMR Spectra via Two-Dimensional Homonuclear Hartmann–Hahn Spectroscopy J Am Chem Soc **107**, 2820 (1985)

150 BRAUNSCHWEILER, L , and R R ERNST Coherence Transfer by Isotropic Mixing Application to Proton Correlation Spectroscopy J Magn Reson **53**, 521 (1983)

151 BAX, A , and D G DAVIS MLEV-17-based Two-Dimensional Homonuclear Magnetization Transfer Spectroscopy J Magn Reson **65**, 355 (1985)

152 INAGAKI, F , I SHIMADA, D KOHADA, A SUZUKI, and A BAX Relayed HOHAHA, a Useful Method for Extracting Subspectra of Individual Components of Sugar Chains J Magn Reson **81**, 186 (1989)

153 AUE, W P , J KARHAN, and R R ERNST Homonuclear Broad Band Decoupling and Two Dimensional J-Resolved NMR Spectroscopy J Chem Phys **65**, 4226 (1976)

154 HALL, L D , S SUKUMAR, and G R SULLIVAN Two-Dimensional J-Spectroscopy Proton NMR Spectra of Mono- and Disaccharides J Chem Soc , Chem Comm , 292 (1979)

155 BERNSTEIN, M A , and L D HALL De Novo Sequencing of Oligosaccharides by Proton NMR Spectroscopy J Am Chem Soc **104**, 5553 (1982)

156 DABROWSKI, J Methods in Stereochemical Analysis Two-Dimensional NMR Spectroscopy for Chemists and Biochemists (R M CARLSON, and W R CAROASMUM, eds) Florida Verlag Chemie (Deerfield Beach) 1987

157 BAX, A , and G A MORRIS An Improved Method for Heteronuclear Chemical Shift Correlation by Two-Dimensional NMR J Magn Reson **42**, 501 (1981)

158 HALL, L D , G A MORRIS, and S SUKUMAR Resolution and Assignment of the 270-MHz Proton Spectrum of Cellobiose by Homo- and Heteronuclear Two-Dimensional NMR J Am Chem Soc **102**, 1745 (1980)

159 KESSLER, H , W BERMEL, C GRIESINGER, and C KOLAR The Elucidation of the Constitution of Glycopeptides by the NMR Spectroscopic COLOC Technique Angew Chem Int Ed **25**, 342 (1986)

160 MORRIS, G A , and L D HALL Experimental Chemical Shift Correlation Maps from Heteronuclear Two-Dimensional NMR Spectroscopy, 1 Carbon-13 and Proton Chemical Shifts of Raffinose and Its Subunits J Am Chem Soc **103**, 4703 (1981)

161 PATT, S L 2-Dimensional NMR in Carbohydrate Structural Analysis J Carbohydr Chem **3**, 493 (1984)

162 ITOKAWA, H , J XU, K TAKEYA, K WATANABE, and J SHOJI Studies on Chemical Constituents of Antitumor Fractions from *Periploca sepium*, II Structures of New Pregnane Glycosides, Periplocosides A, B and C Chem Pharm Bull **36**, 982 (1988)

163 MARTIN, G E , and A S ZEKTZER Long-Range Two-Dimensional Heteronuclear Chemical Shift Correlation Magn Reson Chem **26**, 631 (1988)

164 KESSLER, H , C GRIESINGER, J ZARBOCK, and H R LOOSTI Assignment of Carbonyl Carbons and Sequence Analysis in Peptides by Heteronuclear Shift Correlation via Small Coupling Constants with Broad-band Decoupling in t_1 (COLOC) J Magn Reson **57**, 331 (1984)

165 BAX, A , and R FREEMAN Investigation of Complex Networks of Spin–Spin Coupling by Two-Dimensional NMR J Magn Reson **44**, 542 (1981)

166 SUMMERS, M F , L G MARZILLI, and A BAX Complete ^1H and ^{13}C Assignments of Coenzyme B_{12} through the use of New Two-Dimensional NMR Experiments J Am Chem Soc **108**, 4285 (1986)

167 GRIESINGER, C , and R R ERNST Frequency Offset Effects and their Elimination in NMR Rotating-Frame Cross Relaxation Spectroscopy J Magn Reson **75**, 261 (1987)

168 BAX, A , A ASZALOS, Z DINYA, and K SUDO Structure Elucidation of the Antibiotic Desertomycin through the Use of New Two-Dimensional NMR Techniques J Am Chem Soc **108**, 8056 (1986)

169 DEEPAK, D , S SRIVASTAV, A SETHI, and A KHARE Mass Spectral Studies of Pregnane Glycosides Phytochemical Analysis (Communicated)

170 KHARE, M P , and A KHARE Mass Spectrometry in the Structure Studies of Oligosaccharides J Carbohydr Chem **6**, 523 (1987)

171 HOWE, I , and M JARMANN New Techniques for the Mass Spectrometry of Natural Products Prog Chem Org Nat Prod **47**, 107 (1985)

172 BROWN, P , F BRUSCHWEILER, G R PETTIT, and T REICHSTEIN Field Ionization Mass Spectrometry-III Cardenolides Org Mass Spectrom **5**, 573 (1971)

173 OBERAI, K , M P KHARE, and A KHARE A Pregnane Ester Triglycoside from *Sarcostemma brevistigma* Phytochemistry **24**, 3011 (1985)

174 OBERAI, K , M P KHARE, and A KHARE A Pregnane Ester Diglycoside from *Hemidesmus indicus* Phytochemistry **24**, 2395 (1985)

175 TIWARI, K N , A KHARE, and M P KHARE Structure of Orthenthose Carbohydr Res **123**, 231 (1983)

176 BOSSO, C , F TARAVEL, J ULRICH, and M VIGNON Utilisation du ^{13}C en Spectrometrie de Masse Etude de la Fragmentation de Disaccharides Org Mass Spectrom **13**, 477 (1978)

177 TIWARI, K N , N K KHARE, A KHARE, and M P KHARE Structure of Digoxose Carbohydr Res **129**, 179 (1984)

178 KHARE, D P , S S TIWARI, A KHARE, and M P KHARE Structure of Brevobiose Carbohydr Res **79**, 279 (1980)

179 FUKUOKA, M , K HAYASHI, and H MITSUHASHI Studies on the Constituents of Asclepiadaceae Plants, XXIX Mass Spectra of C/D cis Pregnane Derivatives Chem Pharm Bull **19**, 1469 (1971)

180 BUDZIKIEWICZ, H , C DJERASSI, and H WILLIAMS Structure Elucidation of Natural Products by Mass Spectrometry, Vol II Steroids, Terpenoids, Sugars, and Miscellaneous Classes London Holden-Day, Inc 1964

181 KHARE, N K , R KUMAR, M P KHARE, and A KHARE Sarcogenin, A Pregnane Derivative from *Pergularia pallida* and *Sarcostemma brevistigma* Phytochemistry **25**, 491 (1986)

182 KHARE, N K , R KUMAR, M P KHARE, and A KHARE A Novel Pregnane Derivative from *Sarcostemma brevistigma* J Nat Prod **50**, 600 (1987)

183 MU, Q , and Q ZHOU Studies on Constituents of *Cynanchum otophyllum* Schneid Roots Acta Bot Yunn **5**, 99 (1983) [Chem Abstr **99**, 10735 (1983)]

184 KAUR, K J , M P KHARE, and A KHARE A Novel Polyhydroxy Pregnane Ester from *Orthenthera viminea* Phytochemistry **27**, 1809 (1988)

185 Tiwari, K N , A Khare, and M P Khare Orgogenin, a Pregnane Derivative from *Orthenthera viminea* Phytochemistry **24**, 2391 (1985)

186 Wood, G W Some Recent Applications of Field Ionization/Field Desorption Mass Spectrometry to Organic Chemistry Tetrahedron **38**, 1125 (1982)

187 Itokawa, H , J Xu, and K Takeya Pregnane Glycosides from an Antitumor Fraction of *Periploca sepium* Phytochemistry **27**, 1173 (1988)

188 Nagai, U , and H Iga Optical Rotatory Dispersion of Nitrobenzene Derivatives-I *o*-Nitrobenzoates of Secondary Alcohols Tetrahedron **26**, 725 (1970)

189 Hayashi, K , and H Mitsuhashi Studies on the Constituents of Asclepiadaceae Plants XXXIV Chemistry of Sarcostin Chem Pharm Bull **23**, 1845 (1975)

190 Mitsuhashi, H , T Nomura, and M Fukuoka Studies on the Constituents of Asclepiadaceae Plants, XIII Epimerization at C-17 and Optical Rotatory Dispersion Study of C/D *cis* Pregnane 20-One Derivatives Steroids **4**, 483 (1964)

191 Rangaswami, S , and T Reichstein Konstitution von Odorosid A und Odorosid B Die Glykoside von *Nerium odorum* Sol , 2 Mitteilung Glykoside und Aglykone, 45 Mitteilung Helv Chim Acta **32**, 939 (1949)

192 Deepak, D , M P Khare, and A Khare A New Pregnane Glycoside from *Periploca calophylla* Indian J Chem **25B**, 44 (1986)

193 Srivastava, S , M P Khare, and A Khare Cardenolide Diglycosides from *Oxystelma esculentum* Phytochemistry **32**, 1019 (1993)

194 Kiliani, H Uber Digitalinum Verum Ber Deutsch Chem Ges **63**, 2866 (1930)

195 Krishna, G , G V Shinde, M S Shingare, A Khare, and M P Khare Two Pregnane Ester Triglycosides from *Dregea lanceolata* J Nat Prod **53**, 1399 (1990)

196 Kaur, K J , M P Khare, and A Khare A Pregnane Ester and its Glycoside from *Orthenthera viminea* Phytochemistry **24**, 3007 (1985)

197 Trivedi, R , A Khare, and M P Khare A Pregnane Ester Tetraglycoside from *Oxystelma esculentum* Phytochemistry **27**, 2297 (1988)

198 Oliver, J E , W R Lusby, R M Waters, and M J Thompson Structures of the Pregnenediol Tri- and Di-Glucosides from Eggs of the Tobacco Hornworm, *Manduca sexta* J Nat Prod **51**, 103 (1988)

199 Tsukamoto, S , K Hayashi, K Kaneko, and H Mitsuhashi Studies on the Constituents of Asclepiadaceae Plants, LXV The Optical Resolution of D- and L-Cymaroses Chem Pharm Bull **34**, 3130 (1986)

200 Mannich, C , and G Siewert Uber g-Strophanthin (Ouabain) und g-Strophanthidin Ber Deutsch Chem Ges **75**, 737 (1942)

201 Hakomori, S A Rapid Permethylation of Glycolipid and Polysaccharide Catalysed by Methyl Sulfinyl Carbanion in Dimethyl Sulfoxide J Biochem **55**, 205 (1964)

202 Tursunova, R N , V A Maslennikova, and N K Abubakirov Pregnane Glycosides of *Cynanchum sibiricum*, III Structure of Sibricosides D and E Khim Prirod Soedin **11**, 171 (1975) [Chem Abstr **83**, 114803 (1975)]

203 Fujimoto, H , K Suzuki, H Hagiwara, and M Yamazaki New Toxic Metabolites from a Mushroom, *Hebeloma vinosophyllum*, I Structures of Hebevinosides I, II, III, IV and V Chem Pharm Bull **34**, 88 (1986)

204 Kennard, O , S K Fawcett, D G Watson, K A Kerr, K Stockel, W Stocklin, and T Reichstein Hirundigenin and Anhydrohirundigenin, Two Natural 15-Oxasteroids of Plant Origin Chemical and X-Ray Investigation Tetrahedron Letters, 3799 (1968)

205 Eppenberger, U , H Kaufmann, W Stocklin, and T Reichstein Die Glykoside der Samen von *Stapelia gigantea* N E Br Glykoside und Aglykone, 275 Mitteilung Helv Chim Acta **49**, 1492 (1966)

206 EPPENBERGER, U , W VETTER, and T REICHSTEIN Stapelogenın, Vermutlıche Struktur Glykosıde und Aglykone, 276 Mıtteılung Helv Chım Acta **49**, 1505 (1966)
207 ZHANG, Z -X , J ZHOU, K HAYASHI, and H MITSUHASHI Studıes on the Constıtuents of Asclepıadaceae Plants, LVIII The Structures of Fıve Glycosıdes, Cynatratosıde-A, –B, –C, –D, and –E, from the Chınese Drug "Paı-Weı", *Cynanchum atratum* Bunge Chem Pharm Bull **33**, 1507 (1985)
208 ZHANG, Z -H , J ZHOU, K HAYASHI, and H MITSUHASHI Studıes on the Constıtuents of Asclepıadaceae Plants, LXI The Structure of Cynatratosıde-F from the Chınese Drug "Paı-Weı", Drıed Root of *Cynanchum atratum* Bunge Chem Pharm Bull **33**, 4188 (1985)
209 NAKAGAWA, T , K HAYASHI, K WADA, and H MITSUHASHI Studıes on the Constıtuents of Asclepıadaceae Plants, LII The Structures of Fıve Glycosıdes, Glaucosıde-A, –B, –C, –D, and –E, from the Chınese Drug "Paı-Ch' ıen" *Cynanchum glaucescens* Hand-Mazz Tetrahedron **39**, 607 (1983)
210 NAKAGAWA, T , K HAYASHI, and H MITSUHASHI Studıes on the Constıtuents of Asclepıadaceae Plants, LIV The Structures of Glaucosıdes-F and –G from the Chınese Drug "Paı-Ch' ıen", *Cynanchum glaucescens* Hand-Mazz Chem Pharm Bull **31**, 879 (1983)
211 NAKAGAWA, T , K HAYASHI, and H MITSUHASHI Studıes on the Constıtuents of Asclepıadaceae Plants, LV The Structures of Three New Glycosıdes, Glaucosıde-H, –I, and –J from the Chınese Drug "Paı-Ch' ıen", *Cynanchum glaucescens* Hand-Mazz Chem Pharm Bull **31**, 2244 (1983)
212 NAKAGAWA, T , K HAYASHI, and H MITSUHASHI Studıes on the Constıtuents of Asclepıadaceae Plants, LIII The Structures of Glaucogenın-A, -B, and -C Mono-D-thevetosıde from the Chınese Drug "Paı-Ch' ıen", *Cynanchum glaucescens* Hand-Mazz Chem Pharm Bull **31**, 870 (1983)
213 LEE, M D , T S DUNNE, M M SIEGEL, C C CHANG, G O MORTON, and D B BORDERS Calıchemıcıns, a Novel Famıly of Antıtumor Antıbıotıcs, 1 Chemıstry and Partıal Structure of Calıchemıcın J Am Chem Soc **109**, 3464 (1987)
214 BUNDLE, D R Topıcs ın Current Chemıstry, Vol 154 p 1 Berlın Sprınger-Verlag 1990
215 YOSHIMURA, S -I , H NARITA, K HAYASHI, and H MITSUHASHI Studıes on the Constıtuents of Asclepıadaceae Plants, LVI Isolatıon of New Antıtumor Actıve Glycosıdes from *Dregea volubılıs* (L) Benth Chem Pharm Bull **31**, 3971 (1983)
216 NOHARA, T , H YABUTA, M SUENOBU, R HIDA, K MIYAHARA, and T KAWASAKI Steroıd Glycosıdes ın *Parıs polyphylla* Sm Chem Pharm Bull **21**, 1240 (1973)
217 KAUR, K J , M P KHARE, and A KHARE A Novel Pregnane Ester Tetraglycosıde from *Orthenthera vımıneae* J Nat Prod **48**, 928 (1985)
218 JANOT, M M , Q KHUONG-HUU, C MONNERET, I KABORE, J HILDESHEIM, S D GERO, and R GOUTAREL Alcaloıdes Steroıdıques-C[1] Les Holantosınes A et B, Nouveaux Amıno-Glycosteroıdes ısoles des Feuılles de L' *Holarrhena antıdysenterıca* (Roxb) Wall (Apocynacees) Tetrahedron **26**, 1695 (1969)
219 SAUER, H H , E WEISS, and T REICHSTEIN Dıe Struktur der Drevogenıne, 2 Mıtteılung Struktur von Drevogenın P Glykosıde und Aglykone, 279 Mıtteılung Helv Chım Acta **49**, 1632 (1966)
220 SAUER, H H , E WEISS, and T REICHSTEIN Dıe Struktur der Drevogenıne, 3 Mıtteılung Struktur von Drevogenın A, B und D Glykosıde und Aglykone, 279 Mıtteılung Helv Chım Acta **49**, 1655 (1966)
221 ABISCH, E , C TAMM, and T REICHSTEIN Dıe Glykosıde der Wurzeln von *Pachycarpus*

lineolatus (Decne) Bullock (oder P Schweinfurthii (N E Br) Bullock Glykoside und Aglykone, 201 Mitteilung Helv Chim Acta **42**, 1014 (1959)
222 SAWLEWICZ, L, E WEISS, and T REICHSTEIN Die Cardenolide und Pregnan-Glykoside der Wurzeln von *Asclepias lilacina* Weimarck, I Isolierungen Glykoside und Aglykone, 290 Mitteilung Helv Chim Acta **50**, 504 (1967)
223 SAWLEWICZ, L, E WEISS, and T REICHSTEIN Die Pregnanderivative der Wurzeln von *Asclepias lilacina* Weimarck, II Strukturbestimmungen Glykoside und Aglykone, 291 Mitteilung Helv Chim Acta **50**, 530 (1967)
224 ALLGEIER, H Struktur der Drebyssobiose, Lilacinobiose und Viminose Desoxyzucker, 45 Mitteilung Helv Chim Acta **51**, 668 (1968)
225 NAKAGAWA, T, K HAYASHI, K WADA, and H MITSUHASHI A New Disaccharide, Glaucobiose from Chinese Drug "Pai-Ch' ien" A Comparison of ^{13}C NMR with its Diastereomeric Isomer, Strophanthobiose Tetrahedron Letters **23**, 5431 (1982)
226 KAWANISHI, S, S SAKUMA, H OKINO, and J SHOJI Constituents of Chinese Crude Drug "Wujiapi", IV On the Structure of a New Acetylbiose from Steroidal Glycosides of "Bei-Wujiapi" Chem Pharm Bull **20**, 93 (1972)
227 ZHANG, Z, and J ZHOU Structure of Wallicoside Acta Chim Sinica **41**, 1058 (1983) [Chem Abstr **100**, 99896 (1984)]
228 OSHIMA, Y, T HIROTA, and H HIKINO Periplosides A, B and C, Steroidal Glycosides of *Periploca sepium* Root-Barks Heterocycles **26**, 2093 (1987)
229 SINGH, B, and R P RASTOGI Cardenolides-Glycosides and Genins Phytochemistry **9**, 315 (1970)
230 DEEPAK, D, S SRIVASTAVA, N K KHARE, and A KHARE Cardiac Glycosides Prog Chem Org Nat Prod
231 ABBOTT, B J, J LEITER, J L HARTWELL, M E CALDWELL, and S A SCHEPARTZ Cancer Chemotherapy Screening Data, XIV Screening Data from the Cancer Chemotherapy National Service Center Screening Laboratories XXXIII Plant Extracts Cancer Research **26**, 587 (1966)
232 MITSUHASHI, H, D MIZUNO, K HAYASHI, S ABE, M TAKASE, and T NARITA Condurangoglycosides, their Use as Antitumor Agents and Composition Containing them Chem Abstr **98**, 149582 (1983), Condurangoglycoside E_{01} Chem Abstr **98**, 132306 (1983)
233 AHSAN, A M, D M PIATAK, and P D SORENSEN Isolation and Structure of Amplexoside A A New Glycoside from *Asclepias amplexicaulis* Experientia **29**, 788 (1973)
234 HAYASHI, K, A NAKAO, and H MITSUHASHI Studies on the Constituents of Asclepiadaceae Plants, XXVI Isolation of a New Glycoside from *Dregea volubilis* (L) Benth Chem Pharm Bull **17**, 2629 (1969)
235 YOSHIMURA, S-I, H NARITA, K HAYASHI, and H MITSUHASHI Studies on the Constituents of Asclepiadaceae Plants, LIX The Structures of Five New Glycosides from *Dregea volubilis* (L) Benth Chem Pharm Bull **33**, 2287 (1985)
236 DUH, C Y, J M PEZZUTO, A D KINGHORN, S L LEUNG, and N R FARNSWORTH Plant Anticancer Agents, XLIV Cytotoxic Constituents from *Stizophyllum riparium* J Nat Prod **50**, 63 (1987)
237 SCHUN, Y, and G A CORDELL Cytotoxic Steroids of *Gelsemium sempervirens* J Nat Prod **50**, 195 (1987)
238 MU, Q, J LU, and Q ZHOU Two New Antiepilepsy Compounds-Otophyllosides A and B Sci Sin Ser B **29**, 295 (1986) [Chem Abstr **106**, 81545 (1987)]
239 YUAN, J-L, W-P DING, J-P SHI, Z-Z LU, B-N ZHOU, C A J ERDELMEIER, G A CORDELL, H H S FONG, and N R FARNSWORTH Studies on the Antifertility Compo-

nents from *Marsdenia koi* J Tong-Ji Med Univ **11**, 165 (1991) [Chem Abstr **117**, 83593 (1992)]

240 Umehara, K, M Endoh, T Miyase, M Kuroyanagi, and A Ueno Studies on Differentiation Inducers, IV Pregnane Derivatives from Condurango Cortex Chem Pharm Bull **42**, 611 (1994)

241 Cannon, J R, E L Ghisalberti, and V Lojanapiwatna The Alkaloids of *Holarrhena curtisii* King and Gamble J Sci Soc Thailand **6**, 81 (1980) [Chem Abstr **93**, 217922 (1980)]

242 Samikov, K, R Shakirov, D U Abdullaeva, and S Y Yunusov Alkaloids of *Korolkowia sewertzovii* Structure of Sevkorine Khim Prirod Soedin 269 (1976) [Chem Abstr **85**, 108928 (1976)]

243 Medina, J D, and R Bracho Constituents of the Bark of *Malouetia glandulifera* Planta Med **29**, 367 (1976)

244 Abe, F, and T Yamauchi Teikaside A, a Pregnane Glycoside of *Trachelospermum asiaticum* Chem Pharm Bull **29**, 416 (1981)

245 Warashina, T, and T Noro Steroidal Glycosides from *Asclepias fruticosa* L Chem Pharm Bull **42**, 322 (1994)

246 Warashina, T, and T Noro Steroidal Glycosides and Cardenolide Glycosides from *Asclepias fruticosa* Phytochemistry **37**, 217 (1994)

247 Abe, F, Y Mori, H Okabe, and T Yamauchi Steroidal Constituents from the Roots and Stems of *Asclepias fruticosa* Chem Pharm Bull **42**, 1777 (1994)

248 Tsukamoto, S, K Hayashi, H Mitsuhashi, F O Snyckers, and T G Fourie Studies on the Constituents of Asclepiadaceae Plants, LXII The Structures of Two Glycosides, Cynafoside-A and B, with a Novel Sugar Chain Containing a Pair of Optically Isomeric Sugars, D- and L-Cymaroses, from *Cynanchum africanum* R Br Chem Pharm Bull **33**, 4807 (1985)

249 Wada, K, K Hayashi, H Mitsuhashi, and H Bando Studies on the Constituents of Asclepiadaceae Plants, L Two New Oligoglycosides, Cynanchoside C_2 and Cynanchoside C_1, from *Cynanchum caudatum* Max Chem Pharm Bull **30**, 3500 (1982)

250 Wada, K, K Hayashi, H Mitsuhashi, and H Bando Cynanchoside C_2, a New Steroidal Oligoglycoside from *Cynanchum caudatum* Max Application of ^{13}C-NMR Spectroscopy to the Structural Elucidation of Plant Glycosides Chem Pharm Bull **27**, 2252 (1979)

251 Warashina, T, and T Noro Steroidal Glycosides from *Cynanchum caudatum* Phytochemistry, **39**, 199 (1995)

252 Warashina, T, and T Noro Steroidal Glycosides from the Root of *Cynanchum caudatum* M Chem Pharm Bull **43**, 977 (1995)

253 Nakagawa, T, K Hayashi, and H Mitsuhashi The Structures of Glaucogenin-A, Glaucogenin-B, and Glaucogenin-C Mono D- Thevetoside from Chinese Drug "Pai-Ch' ien" *Cynanchum glaucescens* Hand-Mazz Tetrahedron Letters **23**, 757 (1982)

254 Tursunova, R N, V A Maslennikova, and N K Abubakirov Pregnane Glycosides of *Cynanchum maximoviczii* Khim Prirod Soedin **11**, 522 (1975) [Chem Abstr **83**, 203757 (1975)]

255 Tsukamoto, S, K Hayashi, and H Mitsuhashi Studies on the Constituents of Asclepiadaceae Plant, LVII The Structures of Six Glycosides, Wilfoside C1N, C2N, C3N, C1G, C2G, and C3G, with Novel Sugar Chain Containing a Pair of Optically Isomeric Sugars Tetrahedron **41**, 927 (1985)

256 Bhatnagar, A S, W Stocklin, and T Reichstein Die Pregnanderivate der Samen von *Dregea abyssinica* (Hochst) K Schum (Asclepiadaceae), II Strukturbestimmungen Helv Chim Acta **51**, 133 (1968)

257 ZHANG, Z , J CHEN, and J ZHOU Gymnemaroside A and B from *Gymnema yunnanense* Acta Bot Yunn **13**, 75 (1991) [Chem Abstr **115**, 68474 (1991)]

258 DEEPAK, D , S SRIVASTAVA, and A KHARE Three Novel Pregnane Glycosides from *Hemidesmus indicus* R Br Phytochemistry (in Press)

259 KAPUR, B M , H ALLGEIER, and T REICHSTEIN Die Glykoside der Wurzeln von *Kanalia laniflora* (Forssk) R Br , 2 Mitteilung Struktur von Kalanosid-H und Kalanosid-K Helv Chim Acta **50**, 2171 (1967)

260 ITO, K , J LAI, and K USUDA Studies on the Constituents of *Marsdenia formosana* Musamune, III Isolation and Structural Elucidation of Some New Steroidal Glycosides Chem Pharm Bull **26**, 3189 (1978)

261 LAI, J Studies on the Components of Steroidal Glycoside of *Marsdenia formosana* Musamune Proc Natl Sci Counc Repub China Part B **6**, 51 (1981) [Chem Abstr **96**, 139660 (1982)]

262 CHEN, J , Z ZHANG, J ZHOU, D WANG, L ZHOU, and G TAO A Novel C_{21} Steroidal Glycoside from *Marsdenia incisa* Acta Bot Yunn **13**, 231 (1991) [Chem Abstr **115**, 275751 (1991)]

263 RUAN, J, X -Q XU, G CHEN, W -P DING, and B -N ZHOU Chemical Constituents of Kocondorvine (*Marsdenia koi*) Zhougcaoyao **23**, 6 (1992) [Chem Abstr **117**, 147165 (1992)]

264 RUAN, J , J WANG, and W -P DING Structure Determination of Marsdekoiside E Zhongguo Yaoxue Zazhi **28**, 213 (1993) [Chem Abstr **119**, 91260 (1993)]

265 ZHANG, Y , J YUAN, and W -P DING Structural Elucidation of Marsdeoreophiside A Zhongcaoyao **24**, 171 (1993) [Chem Abstr **119**, 199529 (1993)]

266 MIYAKAWA, S , K YAMAURA, K HAYASHI, K KANEKO, and H MITSUHASHI Five Glycosides from the Chinese Drug "Tong–Guang–San" The Stems of *Marsdenia tenacissima* Phytochemistry **25**, 2861 (1986)

267 SHEN, Y , Q ZHOU, Q MU, Y HU, and X SHEN Chemical Constituents of *Hemsley metaplexis* (*Metaplexis hemsleyana*) I Zhongcaoyao **23**, 622 (1992) [Chem Abstr **118**, 260783 (1993)]

268 KAUR, K J , M P KHARE, and A KHARE A Pregnane Ester Glycoside from *Orthenthera viminea* Indian J Chem **24B**, 1053 (1985)

269 TRIVEDI, R , A KHARE, and M P KHARE A Pregnane Ester Tetraglycoside from *Oxystelma esculentum* Phytochemistry **27**, 2297 (1988)

270 KHARE, N K , M P KHARE, and A KHARE Two Pregnane Ester Glycosides from *Pergularia pallida* Phytochemistry **23**, 2931 (1984)

271 SRIVASTAVA, O P , A KHARE, and M P KHARE Structure of Calocin J Nat Prod **45**, 211 (1982)

272 DEEPAK, D , M P KHARE, and A KHARE A Pregnane Ester Glycoside from *Periploca calophylla* Phytochemistry **24**, 1037 (1985)

273 SAKUMA, S , H ISHIZONE, R KASAI, S KAWANISHI, and J SHOJI On the Structure of Glycoside G and K of Bei-Wujiapi Chem Pharm Bull **17**, 2183 (1969)

274 SAKUMA, S , H ISHIZONE, R KASAI, S KAWANISHI, and J SHOJI Constituents of Chinese Crude Drug "Wujiapi", III On the Structure of Glycoside G and K of Bei-Wujiapi Chem Pharm Bull **19**, 52 (1971)

275 KAWANISHI, S , S SAKUMA, and J SHOJI Constituents of Chinese Crude Drug "Wujiapi", V On the Structure of Glycoside H_1 of Bei-Wujiapi Chem Pharm Bull **20**, 469 (1972)

276 ISHIZONE, H , S SAKUMA, S KAWANISHI, and J SHOJI Constituents of Chinese Crude Drug "Wujiapi", VII On the Structure of Glycoside E of Bei-Wujiapi Chem Pharm Bull **20**, 2402 (1972)

277 SAKUMA, S , S KAWANISHI, and J SHOJI Constituents of the Chinese Crude Drug

"Wujiapi", IX Structure of Glycoside H_2, a Potentiator of NGF-mediated Nerve Fibre Outgrowth Chem Pharm Bull **28**, 163 (1980)

278 Oberai, K, M P Khare, and A Khare A Pregnane Ester Diglycoside from *Sarcostemma brevistigma* Phytochemistry **24**, 1341 (1985)

279 Inada, A, K Mari, and T Nakanishi Phytochemical Studies on Meliaceous Plants, III Structures of Two New Pregnane Steroids, Toosendansterols A and B, from Leaves of *Melia toosendan* Sieb et Zucc Chem Pharm Bull **36**, 609 (1988)

280 Adam, G, H T Huong, and N H Khoi Isolation of 3β-Hydroxy-5-pregnane-16-one from *Solanum hainanense* Phytochemistry **17**, 1802 (1978)

281 Chiplunkar, Y G, B A Nagasampagi, S S Tavale, and V G Puranik Villosterol, 3β, 5β-Dihydro-20-Pregnane-6-one, Steroid from *Turraea villosa* Phytochemistry **33**, 901 (1993)

282 Xu, J, K Takeya, and H Itokawa Pregnanes and Cardenolides from *Periploca sepium* Phytochemistry **29**, 344 (1990)

283 Jin, Q D, and Q Z Mu C_{21} Steroids from *Dregea sinensis* var *corrugata* Yaoxue Xuebao **24**, 587 (1989) [Chem Abstr **112**, 115727 (1990)]

284 Seto, H, K Hayashi, and H Mitsuhashi Constituents of Asclepiadaceae Plants, XXXV Component of *Marsdenia tomentosa* Decne Structure of Tomentin and Dehydrotomentin Chem Pharm Bull **23**, 2397 (1975)

285 Seto, H, K Hayashi, and H Mitsuhashi Studies on the Constituents of Asclepiadaceae Plants, XLIII Component of *Marsdenia tomentosa* Decne Structure of Tomentomin Chem Pharm Bull **25**, 876 (1977)

286 Seto, H, K Hayashi, and H Mitsuhashi Studies on the Constituents of Asclepiadaceae Plants, XXXVI Component of *Marsdenia tomentosa* Decne Structure of Tomentomin, Tomentodin and Dehydrotomentosin and Difference in the Reactivity between Utendin and Tomentogenin Diesters on Mild Alkaline Hydrolysis Chem Pharm Bull **24**, 443 (1976)

287 Seto, H, K Hayashi, and H Mitsuhashi Constituents of Asclepiadaceae Plants, XXXIII Component of *Marsdenia tomentosa* Decne Structure of Tomentosin Chem Pharm Bull **23**, 1552 (1975)

288 Seto, H, K Hayashi, and H Mitsuhashi Studies on the Constituents of Asclepiadaceae Plants, XLII Component of *Marsdenia tomentosa* Decne Structure of 12β-O-Acetyltomentogenin and Hypothetical Biogenetic Pathway of Polyoxypregnanes in *M tomentosa* Chem Pharm Bull **25**, 611 (1977)

289 Schaub, F, H Kaufmann, W Stocklin, and T Reichstein Die Pregnanglykoside der oberirdischen Teile von *Sarcostemma viminale* (L) R Br Helv Chim Acta **51**, 738 (1968)

290 Hayashi, K, K Sugama, Z Zhang, S Tsukamoto, H Nakaya, K Sasaki, T Nakagawa, and H Mitsuhashi On the Pregnane Glycosides from the Plants belonging to the Genus *Cynanchum* (Asclepiadaceae) Tennen Yuki Kogobutsu Toronkai Koen Yoshishu **28**, 216 (1986) [Chem Abstr **106**, 135258 (1987)]

291 Jin, Q D, and Q Mu The Constituent of C_{21} Steroids from *Dregea sinensis* var *corrugata* Acta Bot Yunn **9**, 227 (1987) [Chem Abstr **111**, 171179 (1989)]

292 Jin, Q D, and Q Mu Structure of Dresgenin from *Dregea sinensis* var *corrugata* Zhiwu Xuebao **31**, 874 (1989) [Chem Abstr **113**, 94756 (1990)]

293 Chen, J, S Qin, Z Zhang, and J Zhou The Chemical Constituents of *Gymnema yunnanense* Acta Bot Yunn **11**, 203 (1989) [Chem Abstr **112**, 52246 (1990)]

294 Yamagishi, T, K Hayashi, and H Mitsuhashi The Structure and Internal Acyl Migration of Gagaminin Chem Pharm Bull **20**, 2289 (1972)

295 Sasaki, T, K Hayashi, and H Mitsuhashi On the Structure of Kidjolanin and the Position of the Ester linkage of Penupogenin Chem Pharm Bull **20**, 628 (1972)

296 MITSUHASHI, H, and Y SHIMIZU The Isolation and Structure of Penupogenin Chem Pharm Bull **8**, 565 (1960)

297 SETO, H, T SASAKI, K HAYASHI, and H MITSUHASHI Studies on the Constituents of Asclepiadaceae Plants, XXXIX Component of *Marsdenia tomentosa* Decne Structure of Deacetyldehydrotomentodin, 20-O-Acetylpenupogenin, Deacetylkidjoladinin and Kidjoladinin Chem Pharm Bull **24**, 2185 (1976)

298 SHIMIZU, Y, Y SATO, and H MITSUHASHI Isolation and Structures of New Pregnane Derivatives from *Adonis amurensis* et Radd Chem Pharm Bull **17**, 2391 (1969)

299 MITSUHASHI, H, and Y SHIMIZU Structure of Cynanchogenin and Sarcostin Steroids **2**, 373 (1963)

300 HAYASHI, K, and H MITSUHASHI Studies on the Constituents of Asclepiadaceae Plants, XXXII Aglycones from *Cynanchum wilfordi* Hemsley Chem Pharm Bull **23**, 139 (1975)

301 MASLENNIKOVA, V A, R N TURSUNOVA, and N K ABUBAKIROV Structure of Sibirigenin Khim Prirod Soedin **6**, 322 (1970) [Chem Abstr **73**, 77477 (1970)]

302 YAMAGISHI, T, and H MITSUHASHI Structure of Ikemagenin and Isoikemagenin Chem Pharm Bull **20**, 2070 (1972)

303 BANDO, H, and H MITSUHASHI Studies on the Constituents of Asclepiadaceae Plants XLIV Components of *Cynanchum caudatum* Max Structure of 20-O-Cinnamoylsarcostin, 12-O-Cinnamoylikemagenol and 20-O-Cinnamoylikemagenol Chem Pharm Bull **26**, 2128 (1978)

304 YAMAGISHI, T, and H MITSUHASHI The Structure of Caudatin Chem Pharm Bull **20**, 625 (1972)

305 BANDO, H, T AMIYA, E SATO, and H MITSUHASHI Studies on the Constituents of Asclepiadaceae Plants XLVIII 5α, 6α-Epoxycaudatin, a New Polyoxypregnane Derivative from *Cynanchum caudatum* Max Chem Pharm Bull **28**, 2258 (1980)

306 SINGHAL, S, G MITTAL, M P KHARE, and A KHARE Constituents of *Marsdenia tenacissima* Structure of a New Genin Drevogenin Q Indian J Chem Sect B **19B**, 178 (1980)

307 YAMAGISHI, T, K HAYASHI, H MITSUHASHI, M IMANARI, and K MATSUSHITA Carbon-13 Nuclear Magnetic Resonance Spectroscopy of C/D *cis*-Polyoxypregnanes, III Structure of 12β-O-Cinnamoyl-20-O-Acetylsarcostin Tetrahedron Letters, 4735 (1973)

308 STEYN, P S, F R VANHEERDEN, R VLEGGAAR, G L ERASMUS, and L A P ANDERSON Toxic Constituents of the Asclepiadaceae Structure Elucidation of the Cynafosides, Toxic Pregnane Glycosides of *Cynanchum africanum* R Br S Afr J Chem **42**, 29 (1989) [Chem Abstr **111**, 130746 (1989)]

309 YANG, R, T YANG, and J ZHOU Structures of Tenacigenin A, B and C Memory of Professor TSAI XITAU Yun Nan Zhi Wu Yan Jiu **3**, 271 (1981) [Chem Abstr **96**, 65666 (1982)]

310 SINGHAL, S, M P KHARE, and A KHARE Cissogenin, a Pregnane Genin from *Marsdenia tenacissima* Phytochemistry **19**, 2427 (1980)

311 LUO, S-Q, L-Z LIN, G A CORDELL, L XUE, and M E JOHNSON Assignment of the ^{1}H and ^{13}C NMR Spectra of the C_{21} Steroids 12-β-O-Acetyltenacigenin A and Tenacigenin A by Two-Dimensional NMR Techniques and Computer Modeling Magn Reson Chem **31**, 215 (1993)

(Received March 6, 1996)

Author Index

Page numbers printed in *italics* refer to References

Subject Index

SpringerChemistry

Fortschritte der Chemie organischer Naturstoffe

Progress in the Chemistry of Organic Natural Products

Founded by L. Zechmeister
Edited by W. Herz, G. W. Kirby, R. E. Moore, W. Steglich, and Ch. Tamm

Volume 70

1997. 86 partly coloured figures. VII, 307 pages.
Cloth DM 290,–, öS 2030,–
Subscription price:
Cloth DM 261,–, öS 1827,–
ISBN 3-211-82825-7

Contents:
G.R. Pettit: The Dolastatins.
A. Cavé, B. Figadère, A. Laurens, and D. Cortes: Acetogenins from Annonaceae.

Volume 69

1996. 17 figures. IX, 268 pages.
Cloth DM 250,–, öS 1750,–
Subscription price:
Cloth DM 225,–, öS 1575,–
ISBN 3-211-82824-9

Contents:
J.F. Grove: Non-Macrocyclic Trichothecenes, Part 2.
D. Deepak, S. Srivastava, N.K. Khare, A. Khare: Cardiac Glycosides.
E. Haslam: Aspects of the Enzymology of the Shikimate Pathway.

SpringerWienNewYork

Sachsenplatz 4-6, P O Box 89, A-1201 Wien, Fax +43-1-330 24 26,
e-mail springer@springer co at, Internet http //www springer co at
New York, NY 10010, 175 Fifth Avenue • Heidelberger Platz 3, D-14197 Berlin
Tokyo 113, 3 13, Hongo 3 chome, Bunkyo ku

Springer-Verlag and the Environment

MIX
Papier aus verantwortungsvollen Quellen
Paper from responsible sources
FSC® C105338

If you have any concerns about our products,
you can contact us on
ProductSafety@springernature.com

In case Publisher is established outside the EU,
the EU authorized representative is:
Springer Nature Customer Service Center GmbH
Europaplatz 3, 69115 Heidelberg, Germany

Printed by Libri Plureos GmbH
in Hamburg, Germany